# Numerical Solution of
# Ordinary Differential Equations

# Numerical Solution of Ordinary Differential Equations

L. FOX AND D.F. MAYERS

*Oxford University*

LONDON    NEW YORK

CHAPMAN AND HALL

First published in 1987 by Chapman and Hall Ltd
11 New Fetter Lane, London EC4P 4EE

Published in the USA by Chapman and Hall
29 West 35th Street, New York NY 10001

Printed in Great Britain
by J.W. Arrowsmith Ltd., Bristol

ISBN  0  412  22650  2

British Library Cataloguing in Publication Data

Fox, L.
  Numerical solution of ordinary differential
  equations.
  1. Differential equations—Numerical
  solutions—Data processing
  I. Title     II. Mayers, D.F.
  515.3′5      QA370

  ISBN 0-412-22650-2

Library of Congress Cataloging in Publication Data

Fox, L. (Leslie)
  Numerical solution of ordinary differential equations.

  Bibliography: p.
  Includes index.
  1. Differential equations—Numerical solutions.
I. Mayers, D.F. (David Francis), 1931–     . II. Title.
QA372.F69  1987     515.3′5     87-5191
ISBN  0-412-22650-2

# Contents

# Preface

Nearly 20 years ago we produced a treatise (of about the same length as this book) entitled *Computing methods for scientists and engineers*. It was stated that most computation is performed by workers whose mathematical training stopped somewhere short of the 'professional' level, and that some books are therefore needed which use quite simple mathematics but which nevertheless communicate the essence of the 'numerical sense' which is exhibited by the real computing experts and which is surely needed, at least to some extent, by all who use modern computers and modern numerical software. In that book we treated, at no great length, a variety of computational problems in which the material on ordinary differential equations occupied about 50 pages. At that time it was quite common to find books on numerical analysis, with a little on each topic of that field, whereas today we are more likely to see similarly-sized books on each major topic: for example on numerical linear algebra, numerical approximation, numerical solution of ordinary differential equations, numerical solution of partial differential equations, and so on. These are needed because our numerical education and software have improved and because our relevant problems exhibit more variety and more difficulty. Ordinary differential equations are obvious candidates for such treatment, and the current book is written in this sense.

Some mathematics in this sort of writing is of course essential, and we were somewhat surprised 20 years ago to hear of professors of engineering being excited by the title of our book but alarmed by the standard of mathematics needed for its understanding. There, and here, little more is needed than the ability to differentiate ordinarily and partially, to know a little about the Taylor series, and a little elementary material on matrix algebra in the solution of simultaneous linear algebraic equations and the determination of eigenvalues and eigenvectors of matrices. Such knowledge is surely needed by anybody with differential equations to solve, and surely no apology is needed for the inclusion of some mathematics at this level. If occasionally the mathematics is a little more difficult the reader should take it as read, and proceed confidently to the next section.

Though there are two main types of ordinary differential equation, those of initial-value type and those of boundary-value type, most books until quite recently have concentrated on the former, and again until recently the boundary-value problem has had little literature. Very few books, even now, treat both problems between the same two covers, and this book, which does this in quite an elementary way, must be almost unique in these respects.

After a simple introduction there is quite a substantial chapter on the sensitivity or the degree of 'condition' of both types of problem, mainly with respect to the corresponding difference equations but also, analogously and more briefly, with respect to differential equations. We believe that this feature is very important but largely neglected, and that it is desirable to know (and to practise that knowledge) how to determine the effects of small changes in the relevant data on the answers, and how, if necessary, some reformulation of the original problem can reduce the sensitivity. Next there are two chapters on the initial-value problem solved by one-step and multistep finite difference methods, the Taylor series method and Runge–Kutta methods, with much concentration on the possibility of induced instability (instability of the method) which can produce very poor results even for well-conditioned problems.

Two other chapters spend some 75 pages on the discussion and analysis of two main methods for boundary-value problems. The first uses techniques which produce results from a combination of initial-value problem solutions, but which can be guilty of significant induced instability. The second uses global methods, including the solution of sets of algebraic equations, which give good results unless the problem is very ill-conditioned. This chapter also describes two methods for improving the accuracy of first approximate solutions which, though mentioned briefly in the first initial-value chapter, are far more useful in the boundary-value case. There is also some discussion of techniques for producing a good first approximation, usually very necessary for the successful iterative solution of nonlinear boundary-value problems. A short chapter then discusses methods of polynomial approximation for both initial-value and boundary-value problems, here with virtually the same numerical techniques in both cases.

Almost every technique in every chapter is illustrated with numerical examples, with special reference to inherent instability (problem sensitivity) and to induced instability (a function of the chosen computational method). Exercises for the reader are presented at the end of each chapter, and the final chapter gives solutions to a selected number of these exercises, particularly those which involve some non-trivial computation. We recommend the reader to try both the worked examples and these exercises on a computer, using any relevant software (programs) available, or in some cases with routines written by the reader. The computer should not be expected to give exactly our quoted results, since these depend on the machine's word length, rounding rules, and so on. If, however, the general shape of the solution does not agree with ours it is likely that the reader has made a mistake.

In the eighth chapter we give a short account of the nature of the routines which are likely to be available currently in the best international program libraries. For the initial-value problem we also indicate some of the methods which are used to produce solutions with guaranteed accuracy at each mesh point, the production of economic relevant methods being one of the outstanding problems in this field. At present there is far more numerical software for initial-value problems compared with boundary-value problems, and in our treatment of the latter we mention some points not yet treated in library routines, including for example the difficulties and some possible techniques for dealing with problems with singularities or similar difficult behaviour at boundary points, perhaps particularly with eigenvalue problems.

The ninth chapter mentions other problems which were treated rather briefly in the main text, or not at all, but its main content is a bibliography of some sixty books and papers, with commentaries on each entry relevant to its content and its mathematical level. Most readers are unlikely to read all of these but the specific problems of some scientific disciplines may be eased by a study of some of the relevant literature. The final piece of advice we would give to a scientific or engineering reader is to use his mathematics or consult a professional mathematician (and even a numerical analyst!) before starting any serious computation. Some qualitative knowledge about the nature of the solution can make a significant improvement in the choice of the best and most economic relevant computing method and program, on both the time taken and the accuracy of the results.

Due to various causes and illnesses we have taken quite a long time in the writing of this book. We are very grateful for the close attention of a reader, selected by the publishers, who reported rather many weaknesses in the first draft on its style, ambiguities, topics perhaps wrongly included or omitted, and even a few errors. In our eyes the final version is a significant improvement. We also thank Chapman and Hall for the patience they have showed as the writing time got longer and longer, and for the quick application of the publishing sequence when they finally received the typescript.

*L. Fox*
*D.F. Mayers*
May 1987

# 1
# Introduction

## 1.1 DIFFERENTIAL EQUATIONS AND ASSOCIATED CONDITIONS

A simple example of an ordinary differential equation is

$$\frac{dy}{dx} = f(x, y), \tag{1.1}$$

where $f(x, y)$ is a specified function of $x$ and $y$, such as

$$f(x, y) = x^2 + e^x \tan y. \tag{1.2}$$

From the differential equation, and any other information associated with it, we want to find the solution

$$y = g(x), \tag{1.3}$$

in which the *dependent* variable $y$ is expressed as a function of the *independent* variable $x$. There may be more than one dependent variable, for example in the set of two coupled ordinary differential equations

$$\left. \begin{array}{l} \dfrac{dy_1}{dx} = a_1(x)y_1 + b_1(x)y_2 + c_1(x) \\[2mm] \dfrac{dy_2}{dx} = a_2(x)y_1 + b_2(x)y_2 + c_2(x) \end{array} \right\}, \tag{1.4}$$

but there is still only one independent variable with respect to which all derivatives are taken. The adjective *ordinary* springs from this fact.

Equations (1.1) and (1.4) are of *first order* because this is the highest order of any derivative involved. A corresponding single second-order equation would have the form

$$\frac{d^2 y}{dx^2} = f\left( x, y, \frac{dy}{dx} \right). \tag{1.5}$$

It is worth noting that in the literature there are various notations for derivatives. Commonly we write

$$\frac{dy}{dx} = y', \frac{d^2 y}{dx^2} = y'', \ldots, \frac{d^4 y}{dx^4} = y^{iv}, \ldots, \frac{d^{12} y}{dx^{12}} = y^{xii}, \tag{1.6}$$

with the order of derivative denoted by a roman superscript, or sometimes

$$\frac{d^n y}{dx^n} = y^{(n)} \tag{1.7}$$

for a more general order lacking a general roman equivalent. Equation (1.5) would then be written in the space-saving form

$$y'' = f(x, y, y'). \tag{1.8}$$

It may be asked why we start with (1.1) in which the derivative is written *explicitly* in terms of $x$ and $y$, instead of the obviously more general form

$$\phi(x, y, y') = 0, \tag{1.9}$$

which we might call the *implicit* form. The reason is that the implicit form involves a number of mathematical complications which makes its solution difficult, and fortunately it turns out that in the majority of scientific and engineering problems which can be modelled in terms of differential equations these equations are in explicit form.

We used the phrase 'any other information associated with' a differential equation, and such information is always needed to provide the required solution. A differential equation of order $n$ will require $n$ associated conditions to guarantee this solution. We shall not prove this, but it is easy to see why the result would be expected. Consider, for example, the simple fourth-order differential equation

$$y^{iv} = 1, \tag{1.10}$$

and integrate four times. The result is

$$\left.\begin{aligned}
y''' &= x + A \\
y'' &= \tfrac{1}{2}x^2 + Ax + B \\
y' &= \tfrac{1}{6}x^3 + \tfrac{1}{2}Ax^2 + Bx + C \\
y &= \tfrac{1}{24}x^4 + \tfrac{1}{6}Ax^3 + \tfrac{1}{2}Bx^2 + Cx + D
\end{aligned}\right\}, \tag{1.11}$$

and the last of these gives the *general solution* of the equation (1.10). It contains four arbitrary constants. We therefore need four associated conditions for the determination of these constants, which will then define a unique solution.

In this example the simplest conditions would specify the values of $y''', y'', y'$ and $y$ at some particular point $x = a$. For then $A, B, C$ and $D$ can be calculated directly from the successive equations (1.11). Such a problem, in which all the necessary conditions are given at the same point, is called an *initial-value* problem, and the conditions are called *initial conditions*. An important feature of the initial-value problem is that if all derivatives of $y$ are finite at $x = a$ we can immediately compute the value of $y$ at a nearby point $x = a + h$ from the

Taylor series

$$y(a + h) = y(a) + hy'(a) + \tfrac{1}{2}h^2 y''(a) + \cdots. \tag{1.12}$$

For the second-order equation (1.8), for example, the specification of $y$ and $y'$ at $x = a$ permits the immediate computation of $y''(a)$, and successive derivatives are calculated from successive differentiations of (1.8), the first of which gives

$$y''' = \frac{\partial f}{\partial x} + \frac{\partial f}{\partial y}y' + \frac{\partial f}{\partial y'}y'', \tag{1.13}$$

where everything on the right-hand side has been specified or already computed at $x = a$.

When all the associated conditions are not specified at the same point the problem is called a *boundary-value* problem, and the conditions are *boundary conditions*. For a second-order equation of boundary-value type we must have two boundary points $a$ and $b$. The simplest linear conditions would be

$$y(a) = \alpha, \quad y(b) = \beta, \tag{1.14}$$

where $\alpha$ and $\beta$ are given numbers, but at either end we could have a more general form of the type

$$a_0 y(a) + a_1 y'(a) = c, \tag{1.15}$$

or even more generally

$$a_0 y(a) + a_1 y'(a) + b_0 y(b) + b_1 y'(b) = d. \tag{1.16}$$

Conditions like (1.14) or (1.15), each of which involves one boundary point, are called *separated*, while those of type (1.16) are called *non-separated* or *unseparated*.

For the simultaneous equations (1.4) we would again have two boundary points, and it is easy to see that separated conditions would have the form

$$a_0 y_1(a) + a_1 y_2(a) = c, \tag{1.17}$$

and non-separated conditions

$$a_0 y_1(a) + a_1 y_2(a) + b_0 y_1(b) + b_1 y_2(b) = d. \tag{1.18}$$

Notice that the specification of $y_1$ at both $x = a$ and $x = b$ is a perfectly good set of associated conditions, with $y_2$ not specified at either boundary.

For third-order equations we could have one condition at each of three points, or more usually two conditions at one point and one condition at another point, and for higher-order equations the number of possibilities obviously increases quite rapidly. A single first-order equation, on the other hand, is essentially of initial-value type.

An important feature of boundary-value problems is that we cannot immediately compute the Taylor series at any boundary point, or any other point, since we do not have sufficient information. If, for example, we have the second-order equation (1.8), with the values of $y(a)$ and $y(b)$ specified, we cannot construct the Taylor series at the point $x = a$ until the value of $y'(a)$ has been computed. As we shall see later, we therefore have to use different numerical methods to solve the two different types of problem.

## 1.2 LINEAR AND NON-LINEAR DIFFERENTIAL EQUATIONS

In mathematical analysis there is a great difference between linear and non-linear differential equations. In a linear equation $y$ and its derivatives appear only linearly. The most general linear second-order equation, for example, is

$$y'' + f(x)y' + g(x)y = k(x), \tag{1.19}$$

whereas the first-order equation

$$y' = \tan y + x \tag{1.20}$$

and the third-order equation

$$y''' + yy'' + y'^2 = 1 \tag{1.21}$$

are both non-linear. Equation (1.1) with $f(x, y)$ defined by (1.2) is non-linear, and the simultaneous equations (1.4) are linear.

The solution of a linear equation like (1.19) can quite easily be written in a rather general form. We consider at the same time the *homogeneous* equation

$$y'' + f(x)y' + g(x)y = 0. \tag{1.22}$$

Now equation (1.22) will have two independent solutions $y^{(1)}$ and $y^{(2)}$ (the danger of ambiguity with the notation in (1.7) being very small). Since the equation is homogeneous any constant multiple of $y^{(1)}$ or $y^{(2)}$ is also a solution, so that the most general solution of (1.22) is

$$y = Ay^{(1)} + By^{(2)}, \tag{1.23}$$

where $A$ and $B$ are arbitrary constants; this solution is called the *complementary function* of equation (1.19). If we now choose *any* solution of (1.19), which we call a *particular* solution $y^{(p)}$, then the general solution of (1.19) will be given by

$$y = y^{(p)} + Ay^{(1)} + By^{(2)}. \tag{1.24}$$

The values of the constants $A$ and $B$ will be determined by the two associated conditions.

For equation (1.10), which is of fourth order, we can use the same type of analysis. Four independent solutions of the corresponding homogeneous equation

$$y^{iv} = 0 \tag{1.25}$$

are

$$y^{(1)} = 1, \quad y^{(2)} = x, \quad y^{(3)} = x^2, \quad y^{(4)} = x^3, \tag{1.26}$$

and a particular solution of (1.10) is

$$y^{(p)} = \tfrac{1}{24}x^4.$$
(1.27)

Equations (1.26) and (1.27) give the general solution agreeing with equation (1.11).

For linear problems the arbitrary constants always appear as multiples of solutions of the corresponding homogeneous equation, but in non-linear problems they may appear in quite different ways. In fact, in a non-linear problem there is nothing corresponding to the complementary function, or to a corresponding homogeneous equation.

For example, the general solution of the first-order equation

$$y' = x^3 y^2$$
(1.28)

is

$$y = \frac{-4}{x^4 + A}$$
(1.29)

where $A$ is the expected arbitrary constant. The solution of

$$y' = \frac{y - x}{y + x}$$
(1.30)

cannot be expressed in a simple form at all, but is given by

$$\log(x^2 + y^2) + 2\tan^{-1}(y/x) + A = 0.$$
(1.31)

For a second-order equation we expect two arbitrary constants. For example, the general solution of

$$y'' + 3yy' + y^3 + y = 0$$
(1.32)

is

$$y = (\cos x - A\sin x)/(\sin x + A\cos x + B).$$
(1.33)

Non-linear equations are therefore much more complicated than linear equations; moreover, their solutions are more likely to have some form of singularity. For a linear equation such as

$$y'' + f(x)y' + g(x)y = 0$$
(1.34)

the solution $y$, or its derivatives, can be infinite only at points at which $f(x)$ or $g(x)$ is infinite, and these points are known in advance. But for a non-linear equation the position of the singularities will be unknown until the solution has been found, and will often depend on the initial conditions. For example, for equation (1.32) with the initial conditions $y = 1$ and $y' = -2$ at $x = 0$, the solution is

$$y = \frac{\cos x - \sin x}{\sin x + \cos x},$$
(1.35)

and this becomes infinite at $x = \frac{3}{4}\pi$. Similarly, the first-order equation

$$y' = 1 + y^2 \tag{1.36}$$

has the general solution $y = \tan(x + A)$, where $A$ is the arbitrary constant. This solution becomes infinite whenever $x + A$ is an odd multiple of $\pi/2$; for example, the initial condition $y(0) = 1$ gives $A = \pi/4$, so that the solution becomes infinite at $x = \pi/4$.

## 1.3 UNIQUENESS OF SOLUTIONS

The question of whether or not a differential equation with the correct number of associated conditions has or has not a unique solution is really quite complicated. Generally speaking, both linear and non-linear initial-value problems, such as

$$y^{(n)} = f(x, y, y', \ldots, y^{(n-1)}), \tag{1.37}$$

with $y$ and its first $n - 1$ derivatives specified at a particular point $x = a$, will have a unique solution if the function $f$ is sufficiently 'smooth' in each of its arguments. This covers nearly all of the initial-value problems which we need to solve.

Boundary-value problems are rather different. A linear equation with the correct number of boundary conditions may have no solution at all, a unique solution, or many different solutions. For example, the equation

$$y'' + y = 1 \tag{1.38}$$

has the general solution

$$y = 1 + A \sin x + B \cos x. \tag{1.39}$$

The conditions

$$y(0) = 0, \quad y(\pi/2) = 0 \tag{1.40}$$

give the unique solution with $A = -1$, $B = -1$, but the conditions

$$y(0) = 0, \quad y(\pi) = 0 \tag{1.41}$$

lead to the inconsistent equations

$$B = -1, \quad B = 1 \tag{1.42}$$

for $B$, and no equations for $A$. Finally, the boundary conditions

$$y(0) = 0, \quad y(\pi) = 2 \tag{1.43}$$

give the solution

$$y = A \sin x - \cos x + 1, \tag{1.44}$$

which satisfies the differential equation, and the boundary conditions, for any value of the arbitrary constant $A$.

Linear differential equations with non-linear boundary conditions may give several solutions of quite different form. For example, the same equation (1.38) with the boundary conditions

$$y'(0) = \{y(0)\}^2, \qquad y\left(\frac{3\pi}{2}\right) = 0 \qquad (1.45)$$

has the two solutions

$$y = 1 + \sin x, \qquad y = 1 + \sin x - 2\cos x. \qquad (1.46)$$

In such a case we should need more information, perhaps derived from the scientific problem modelled by the equation, to decide which of the two solutions to choose. And of course non-linear equations, whether the boundary conditions are linear or not, may also have several solutions.

## 1.4 MATHEMATICAL AND NUMERICAL METHODS OF SOLUTION

We have remarked that for a differential equation and its associated conditions we wish to determine the dependent variable $y$ in terms of the independent variable $x$, for a range of $x$ determined by any particular requirements in an initial-value problem, and most likely in the range given by the extreme boundary points in a boundary-value problem. The way in which the solution is to be presented will vary with the particular problem, but in most practical cases we shall wish to present the result in such a way that a numerical value of $y$ is easily obtainable for any given value of $x$. Possibilities therefore include:

(a) the formation of a numerical table of values of $y$ for selected values of $x$, from which any intermediate value can be found by easy interpolation;
(b) a graph of $y$ against $x$, from which any value can be obtained with the required accuracy;
(c) direct computation of a simple solution $y = g(x)$, if such a simple solution exists;
(d) the computation of the coefficients $c_0, c_1, \ldots, c_k$ in for example a polynomial approximation $c_0 + c_1 x + \cdots + c_k x^k$ to the solution, valid to the required accuracy over some range of values of $x$.

In previous sections we have given what are called 'exact' solutions, obtained by mathematical analysis, for a few simple examples. These exact solutions have two important properties. Firstly, they are general solutions of the differential equations, and secondly, in a mathematical sense the hardest part of the work is completed. For numerical purposes we must then use the associated conditions to determine the arbitrary constants in the general solutions, and then compute whatever values of $y(x)$ are required.

Unfortunately this programme can rarely be carried out. In the first place there are very few differential equations which arise in scientific and engineering problems for which the exact solution can be obtained. In the second place the use of the exact solution in the computation of $y(x)$ may involve a large amount of computation, as for example in the solution (1.31) of equation (1.30). Consider also the solution of

$$y' = \frac{2y}{(1-x^4)},$$
(1.47)

which is

$$y = A\left(\frac{1+x}{1-x}\right)^{\frac{1}{4}} e^{\tan^{-1}x}.$$
(1.48)

Given as associated condition the value of $y$ for some value of $x$ in $-1 < x < 1$ we can reasonably easily compute the value of $A$, and then the value of $y$ corresponding to any other value of $x$ assuming the availability of some means of finding square roots, the exponential function, etc. But now consider the slightly different equation

$$y' = \frac{2y}{(1-x^4)} + x.$$
(1.49)

This has the much more fearsome-looking exact solution

$$y = \left(\frac{1+x}{1-x}\right)^{\frac{1}{4}} e^{\tan^{-1}x}\left\{B + \int x\left(\frac{1-x}{1+x}\right)^{\frac{1}{4}} e^{-\tan^{-1}x}\,dx\right\}.$$
(1.50)

Here we are in a much more difficult situation, since we cannot express the integral in (1.50) in terms of 'simple' functions. Mathematically (1.50) might be 'exact', but for numerical purposes the mathematics is incomplete.

For these reasons we generally use a so-called 'approximate' numerical method. In most cases we produce a numerical table of values of $y$ without using any sort of exact solution. More rarely, in some particular cases, we find a function $y = \phi(x)$ which, over the required range of $x$, is a sufficiently accurate approximation to the solution, and which is very easy to compute. In contrast to the analytic methods, our numerical methods obtain particular solutions separately for each given set of associated conditions; these associated conditions are essentially introduced right from the start. Moreover, the method which will be used for (1.47), for example, is virtually identical to that used for the slight variation (1.49), and indeed will not be very different in principle from that used for any first-order equation of type (1.1).

The terms 'exact' and 'approximate' need some clarification. Whereas a solution like (1.48) is mathematically exact, we shall be able to compute the arbitrary constant $A$ correct only to a certain number of figures, depending on the accuracy of the computed square roots, inverse tangents, etc. The computation of the value of $y$ for a given value of $x$ will again be affected by the inexact values of similar quantities, and by the error in the value of $A$. In this sense, therefore, most mathematically exact solutions will only be numerically approximate. The errors in our 'approximate' numerical methods arise for quite different reasons, but with sufficient work the approximate methods can be made as accurate as exact methods, and in any case accurate enough for the particular purposes required.

We shall therefore concentrate on numerical methods, and confine our discussion of exact solutions to one very special case, in the next section. This does not indicate any lack of importance of all kinds of mathematical analysis. Some types of analysis are very important for the insight they give into the general behaviour of the solution, and especially the behaviour of solutions near singular points. This will not only help in the choice of numerical methods, but also, in collaboration with numerical methods, give accurate treatment near singular points, which may not be possible with the use of either mathematical or numerical methods alone.

## 1.5 DIFFERENCE EQUATIONS

There is a close analogy between differential equations and difference equations, and indeed many of our numerical methods will deliberately solve the latter to produce approximations to the solutions of the former. A first-order linear *difference equation*, or *recurrence relation*, has the form

$$y_{r+1} + a_r y_r = b_r, \tag{1.51}$$

where $a_r$ and $b_r$ are functions of the integer variable $r$. Clearly one initial condition, say a given $y_0$, is needed to produce a definite solution. The corresponding second-order equation is

$$y_{r+1} + a_r y_r + b_r y_{r-1} = c_r, \tag{1.52}$$

and here we would expect to need two associated conditions. These may be 'initial values', say given values of $y_0$ and $y_1$, or 'boundary values', such as given values of $y_0$ and $y_{20}$. The general solution of a linear difference equation is analogous to that of a linear differential equation. For (1.52), for example, the general solution is

$$y_r = y_r^{(p)} + A y_r^{(1)} + B y_r^{(2)}, \tag{1.53}$$

where $y_r^{(p)}$ is any particular solution of (1.52), $y_r^{(1)}$ and $y_r^{(2)}$ are two independent solutions of the homogeneous form of (1.52), with $c_r$ replaced by zero, and $A$ and $B$ are arbitrary constants.

For the linear difference equation with constant coefficients the complementary solutions are easily obtained, again by analogy with the differential equation. The independent solutions of the corresponding differential equation

$$y'' + ay' + by = 0, \qquad (1.54)$$

where $a$ and $b$ are constants, are

$$y^{(1)} = e^{m_1 x}, \qquad y^{(2)} = e^{m_2 x}, \qquad (1.55)$$

where $m_1$ and $m_2$ are the roots of the quadratic equation

$$m^2 + am + b = 0. \qquad (1.56)$$

If the roots $m_1$ and $m_2$ are both equal to $m$ the two solutions in (1.55) are not independent, and in this case an independent second solution is given by

$$y^{(2)} = xe^{mx}. \qquad (1.57)$$

Similarly, for the difference equation

$$y_{r+1} + ay_r + by_{r-1} = 0 \qquad (1.58)$$

the solutions are

$$y_r^{(1)} = m_1^r, \quad y_r^{(2)} = m_2^r, \qquad (1.59)$$

where $m_1$ and $m_2$ are again the roots of the quadratic equation (1.56). In the case of equal roots an independent second solution is

$$y_r^{(2)} = rm^r. \qquad (1.60)$$

The need for associated conditions, and the number of conditions required, is really quite obvious for both linear and non-linear difference equations. For the linear equation (1.52), for example, it is clear that a knowledge of $y_0$ and $y_1$ permits the successive computation of $y_2, y_3$, and so on. For a corresponding boundary-value problem, the system of equations (1.52), for $r = 1, 2, \ldots, N - 1$, gives $N - 1$ simultaneous linear algebraic equations for the determination of the $N + 1$ unknowns $y_0, y_1, \ldots, y_N$, so that two independent associated conditions are needed to produce a solution. In a non-linear boundary-value problem we shall still have $N + 1$ algebraic equations in $N + 1$ unknowns, but the equations will be non-linear, with the possibility of the existence of more than one solution.

Again we shall not consider any exact methods of solving difference equations. Indeed it is probably correct to say that few problems are modelled in terms of recurrence relations; they commonly arise as an easy way of solving problems modelled in other ways. For example, the computation of

$$I_r = \int_0^1 x^r e^{x-1} \, dx, \qquad r = 0, 1, \ldots, \qquad (1.61)$$

can be simplified very considerably by the observation that $I_r$ satisfies the recurrence relation

$$I_{r+1} + (r+1)I_r = 1. \tag{1.62}$$

It is also well known that the Bessel function $J_r(x)$, for successive values of the order $r$, and fixed argument $x$, satisfies the recurrence relation

$$J_{r+1} - \frac{2r}{x} J_r + J_{r-1} = 0. \tag{1.63}$$

The uneducated use of (1.62) and (1.63) can, in fact, give very poor results, but with some care, as discussed in the next chapter, they provide very accurate and economic methods for computing respectively $I_r$ and $J_r(x)$.

## 1.6 ADDITIONAL NOTES

*Section 1.2* It is interesting to examine the performance of reputable computer routines for solving some of these nonlinear problems. For the first-order equation (1.28), for example, a typical routine works satisfactorily with the initial condition $y(0) = -4$, but with $y(0) = +4$ it will produce very poor results near $x = 1$, at which point the correct solution has an infinite value. Commonly the routines will come to a grinding halt somewhere before reaching $x = 1$, with a suitable warning comment. Similarly, for the second-order equation (1.32), routines are happy with the initial conditions $y(0) = \frac{1}{4}$, $y'(0) = -\frac{5}{16}$, for which the solution is (1.33) with $A = 1$, $B = 3$, and the solution has no singularities. But with the initial conditions $y(0) = 1$, $y'(0) = -2$, giving the solution (1.35) with a singularity at $x = \frac{3}{4}\pi$, all routines break down in the neighbourhood of this point.

It is also interesting to consider the initial conditions for a non-linear differential equation like (1.30). In Section 1.1 we said that a differential equation of order $n$ will require $n$ associated conditions to provide a solution, but for some problems we may need a little more than one condition. For example, the equation $xy' = y(y-1)$ has the solution $y = (1 + Ax)^{-1}$, but the condition $y(0) = 1$ does not determine $y'$ uniquely and the condition $y(1) = 0$ does not determine $A$.

*Section 1.3* Numerical treatment of the boundary-value problems in this section reveals the difficulty of producing robust software for these problems, the word 'robust' indicating that the routines virtually never fail. It also reveals the importance of using any associated mathematics we may have at our disposal or may be able to produce with suitable mathematical analysis.

For example, problem (1.38) with conditions (1.40) gives no difficulties, but with conditions (1.41) a standard routine tries to find a solution, puts the difficulties down to the use of too large a finite-difference interval (see Chapter 6), and stops when this interval gets too small for the machine store. For conditions (1.43), which give solution (1.44) for any $A$, the same standard routine quickly produces a solution with $A = 0$ and gives no hint of any other solution.

Perhaps even more interesting is problem (1.38) with the non-linear conditions (1.45), for which there are two definite solutions. There are in fact various different methods for non-linear problems, and they may produce one or other of the solutions but again without a warning about a second solution.

---

### EXERCISES

1. Show that the $n$th-order system

$$y^{(n)} = f(x, y, y', \ldots, y^{(n-1)}), \quad y(a) = \alpha_1, \quad y'(a) = \alpha_2, \ldots, \quad y^{(n-1)}(a) = \alpha_n$$

   can be replaced by a system of $n$ simultaneous first-order equations for which the unknown functions have specified initial values.

2. The differential equation (1.10) has associated initial conditions

$$y(1) = 1, \quad y'(1) = -1, \quad y''(1) = 0, \quad y'''(1) = 1.$$

   Write down the Taylor series for the value of $y(1 + h)$, and verify that this is the same as the last of equations (1.11) for $x = 1 + h$.
   (But note that the general solution in (1.11) is a polynomial of degree four, and the computation is simplified if, for this purpose, it is expressed as $y = \frac{1}{24}(x - 1)^4 + A'(x - 1)^3 + B'(x - 1)^2 + C'(x - 1) + D'$.)

3. Find a few terms of the Taylor series at $x = 0$ for the solution of the problem

$$y' = 1 + y^2, \quad y(0) = 0.$$

   (Note that this is the Taylor series for $y = \tan x$, which converges for $|x| < \pi/2$.)

4. For the boundary-value problem with equation (1.19) and first boundary-condition (1.15) show that the Taylor series at $x = a$ can be expressed in terms of the single unknown value $y(a)$.

5. Find the general solution of the equation

$$y''' = x + 1$$

   and also the particular solution for the case $y(0) = y'(0) = y''(0) = 1$.

6. Find the solution of the pair of first-order equations

$$\left. \begin{array}{l} y' = z \\ z' = -y \end{array} \right\}$$

   with $y(0) = 0$, $y(\pi/2) = 1$.

Comment on the solution of this problem with boundary conditions $y(0) = 0$, $y(\pi) = 1$.

7. For equation (1.28) show that the initial condition $y(0) = \frac{1}{4}$ produces a singularity (an infinite value of $y$) at the point $x = 2$. Comment on what happens with the condition $y(0) = 0$ and the condition $y(0) = -1$.

8. Find the general solution of the difference equation

$$y_{r+3} - 3y_{r+2} + 3y_{r+1} - y_r = 0.$$

9. Members of the sequence $y_0, y_1, y_2, \ldots, y_r, \ldots$, satisfy the recurrence relation $y_{r+2} = y_{r+1} + y_r$. If $y_0 = 0$, show that for any non-zero value of $y_1$ the ratio $y_{r+1}/y_r$ tends to the limit $\frac{1}{2}(1 + \sqrt{5})$ as $r \to \infty$.

10. Consider the behaviour of the solution of Exercise 9 if $y_0$ is not zero. In particular, find out what happens if $y_0 = 1$ and $y_1 = \frac{1}{2}(1 - \sqrt{5})$. (But see Exercise 13 in Chapter 2.)

# 2

# Sensitivity analysis: inherent instability

## 2.1 INTRODUCTION

In most scientific problems not all the data are known exactly. They will contain 'physical constants' like gravity, density, latent heat, or the speed of light or sound, whose values are generally obtained by experiment and with a precision which depends on our experimental ability. They may also contain the results of measurement, say with ruler or protractor, and again the resulting precision depends on the fine tuning of the relevant measuring equipment.

The data therefore have uncertainties, hopefully reasonably small, and as a result the required answers will also have uncertainties which may or may not be small but which we clearly ought to investigate. In particular, if a solution $x$ depends on a particular piece of data $c$ through the equation

$$x = f(c),\tag{2.1}$$

then an uncertainty of size $\delta c$ in $c$ will give to first order (that is with neglect of the square and higher powers of $\delta c$) an uncertainty in $x$ of size

$$\delta x = \frac{\mathrm{d}f}{\mathrm{d}c}\delta c.\tag{2.2}$$

Commonly $\delta c$ could have either sign and so could $\delta x$, and we therefore write (2.2) in the form

$$|\delta x| = \left|\frac{\mathrm{d}f}{\mathrm{d}c}\delta c\right|.\tag{2.3}$$

If $x$ depends on several pieces of data, each of which has a small uncertainty, then (2.3) is replaced by the sum of each individual contribution. For example, if

$$x = f(a, b, c),\tag{2.4}$$

then to first order

$$\delta x = \frac{\partial f}{\partial a}\delta a + \frac{\partial f}{\partial b}\delta b + \frac{\partial f}{\partial c}\delta c \Bigg\}$$

$$|\delta x| \geqslant \left|\frac{\partial f}{\partial a}\delta a\right| + \left|\frac{\partial f}{\partial b}\delta b\right| + \left|\frac{\partial f}{\partial c}\delta c\right| \Bigg\} \quad , \tag{2.5}$$

where the derivatives on the right are now partial derivatives, each representing the differentiation of $f(a, b, c)$ with respect to one of $a, b, c$, with the others held temporarily constant.

The important quantities in (2.3) and (2.5), of course, are the derivatives, the amounts by which small uncertainties in the data are multiplied to give the uncertainty in the answer. Their determination, for obvious reasons, is sometimes called *sensitivity analysis*, and this analysis has important consequences. For example, if $x$ computed from (2.1) is 20.243 and $|\delta x|$ from (2.3) is 0.042 then we can make the confident assertion that the true $x$ is somewhere between 20.2 and 20.3. A result correct to nearly three significant figures is all that is meaningful, and there is no point in trying to determine $x$ to more figures than this.

A small *absolute uncertainty*, as here, may not be satisfactory if we are interested in the *relative uncertainty* in the answer. The relative uncertainty is given by $|\delta x/x|$, and if, for example, $|\delta x| = 0.042$ but $|x| = 0.243$ then the relative uncertainty is about 17%, and hardly one significant figure is guaranteed in our answer.

If the absolute or relative uncertainty is small we say that the problem is *inherently* absolutely or relatively *stable*, or is *well-conditioned* in the relevant sense. If the relevant uncertainty is large we say that the problem is *inherently unstable*, or *ill-conditioned*. When several pieces of data are involved, as in (2.4), we may have inherent instability with respect to some pieces and inherent stability with respect to others, this depending on the sizes of the derivatives in (2.5).

So far we have discussed what we might call *physical* problems, where the uncertainties in the data are essentially unavoidable and there are definite limits to the number of figures worth quoting in the answers. The adjective *inherent* in this type of instability stresses this point.

There are other problems, which we call *mathematical*, in which all the data are known exactly so that we are justified in quoting as many figures in our answers as anybody might require for any particular purpose. An obvious question is whether or not we can obtain economically the accuracy we might require for a mathematical problem when we use a digital computer for its solution. For the mathematical problem has its own version of inherent instability, in the fact that such a modern computing machine with a finite word length cannot store exactly all mathematically exact data. Numbers like $e$ or $\pi$ cannot be stored exactly however big our word length, and neither can the exact decimal number 0.9 in a machine which uses binary arithmetic. We may therefore have *mathematical uncertainty* in some of the data stored in the machine, and if the problem is ill-conditioned with respect to any of these data there may be a significant uncertainty, now more correctly described as 'inaccuracy', in the answers we produce. Notice, however, that the mathematical uncertainty in the data is likely to be very much smaller than the physical uncertainty of our previous discussion, being only a rounding error in the machine's storage register. Moreover, we now have every justification in using double-length or even multi-length arithmetic if this is needed to obtain the required accuracy.

We have tacitly assumed that we essentially solve exactly the problem, either physical or mathematical, which is originally stored in the machine, and that the only defects in the solution spring from the fact that some of our initial data are to some extent uncertain. In the solution process, however, there are other errors such as those which may be caused simply by the inability of the computer to perform exact arithmetic, or through the too early truncation of a Taylor series, or the inadequate representation of a differential equation by a difference equation, and so on, and all these will depend on our chosen method of solution. Dangers arise when any of these errors are large, or when a sequence of small individual errors combine to produce large errors in the final answer. These errors are *induced* by our method, and if they are absolutely or relatively large we say that our method of solution exhibits *induced instability*. Here we have hopes of using methods which are satisfactory in this respect, and in subsequent chapters we discuss this in some detail in relation to the numerical methods we introduce. We shall see that there are unstable methods which we discard because, even for very well-conditioned problems, they fail to give a single accurate figure in the computed solution!

In this chapter we concentrate on sensitivity analysis or inherent instability, first with an idealized game of darts and then, more systematically, with linear recurrence relations and differential equations of first and second order, the latter with both initial-value and boundary-value problems. We also note the possibility of sometimes reducing the inherent error by some small reformulation of the original problem. There is no detailed discussion of induced instability but the additional notes in Section 2.7 include several important relevant comments.

## 2.2 A SIMPLE EXAMPLE OF SENSITIVITY ANALYSIS

Consider a dart, which we throw in a vertical plane from a point at height $H$ from the ground, with velocity $V$ at angle $\alpha$ from the horizontal, and let it strike the dart board at a distance $D$ away in the horizontal direction. An interesting question is the uncertainty of its vertical position on the board with respect to small uncertainties in the parameters $g$ (gravity), $V$ and $\alpha$. We assume that the dart is a very simple and very small body which satisfies the simple 'projectile' equations

$$
\left.\begin{array}{l}
\dfrac{d^2x}{dt^2} = 0, \quad x = 0, \quad \dfrac{dx}{dt} = V\cos\alpha \quad \text{at } t = 0 \\[2mm]
\dfrac{d^2y}{dt^2} = -g, \quad y = H, \quad \dfrac{dy}{dt} = V\sin\alpha \quad \text{at } t = 0
\end{array}\right\} \tag{2.6}
$$

where $x$ and $y$ are the horizontal and vertical cordinates, respectively. Since we are interested in what happens at a particular value of $x$ it is more convenient to have the dependent variable $y$ expressed in terms of the independent variable $x$. From the first of (2.6) we see that $dx/dt = V\cos\alpha$ for all $t$, so that

$$
\frac{dy}{dt} = \frac{dx}{dt}\frac{dy}{dx} = V\cos\alpha\,\frac{dy}{dx}, \quad \frac{d^2y}{dt^2} = (V\cos\alpha)^2\frac{d^2y}{dx^2}, \tag{2.7}
$$

and we can replace all of (2.6) by the single differential equation and initial conditions given by

$$
\frac{d^2y}{dx^2} = -\frac{g}{(V\cos\alpha)^2}, \quad y = H, \quad \frac{dy}{dx} = \tan\alpha \quad \text{at } x = 0. \tag{2.8}
$$

For this problem we can do all the required work analytically, and we integrate (2.8) twice and apply the initial conditions to produce the required solution

$$
y = -\frac{gx^2}{2(V\cos\alpha)^2} + x\tan\alpha + H. \tag{2.9}
$$

Simple partial differentiation with respect to $g$, $V$ and $\alpha$ then gives for the first-order uncertainty in $y$ at $x = D$ the expression

$$
\delta y = \frac{-D^2}{2V^2\cos^2\alpha}\,\delta g + \frac{gD^2}{V^3\cos^2\alpha}\,\delta V + \frac{D}{\cos^2\alpha}\left(1 - \frac{gD}{V^2}\tan\alpha\right)\delta\alpha. \tag{2.10}
$$

Typical figures, in units of feet and seconds, could be

$$
g = 32, \quad D = 8, \quad V = 20, \quad H = 6, \quad \alpha = \pi/12 \text{ radians}, \tag{2.11}
$$

and (2.10) then gives approximately

$$
\delta y = -0.09\delta g + 0.27\delta V + 7.10\delta\alpha. \tag{2.12}
$$

The darts player, of course, would be more interested in knowing how to improve a previous throw by changing the initial speed and direction of the missile, confident in his ability accurately to make such changes! Here, of course, we are not concerned with changes in gravity, since the board is presumably in the same place for all throws, and we need consider only the last two terms in (2.10). Equation (2.12) shows that with conditions (2.11) a change in $V$ of 0.2, that is 1% of the original velocity, would change the vertical position of the dart by about 0.65 inches, and the same percentage change in the original angle of throw would change its vertical position by about 0.22 inches.

The value of $y$ from (2.9) is about 5.40 feet $\sim 64.8$ inches, so that the relative uncertainties in $y$ are approximately and respectively 1% and 0.33%. But perhaps it is here more important to express the uncertainty in $y$ relative to the size of the dart board, and if this is say 20 inches the uncertainties in $y$ relative to this are approximately 3.2% and 1.1% respectively. The problem is not relatively badly conditioned, and indeed if it were nobody would want to play darts!

A rather more interesting set of data, though one hardly likely to be used by the experts, is

$$g = 32, \quad D = 8, \quad V = 16, \quad \alpha = \pi/4, \tag{2.13}$$

for which, again with neglect of changes in gravity, (2.10) gives the very simple estimate

$$\delta y = \delta V. \tag{2.14}$$

Here a slight change in the angle achieves nothing, and a 1% change in velocity changes the height of the embedded dart by nearly $2\frac{1}{2}$ inches. With these starting conditions the problem is very well conditioned with respect to the angle but rather ill conditioned with respect to the velocity.

Notice also that with the $g$, $D$ and $V$ of (2.13) but with $\alpha > \pi/4$ a small positive increase in $\alpha$ will reduce the height achieved by the dart, and this is competitive with a positive increase in the velocity which has an opposite effect.

We were able to analyse this problem quite easily because we could solve all the relevant equations analytically, which as we stated in the first chapter is quite unusual. In the next section we discuss methods of sensitivity analysis when we have to use numerical methods for solving the differential equations.

## 2.3 VARIATIONAL EQUATIONS

We assume that we can solve numerically all the differential equations which we meet, and the crucial trick in sensitivity analysis is to deduce from the original differential equations some other differential equations for the partial derivatives involved in this analysis and to solve them numerically also. This will become clear with an examination of our darts problem, in which $y$ is given in terms of $x$ by equation (2.8) and we want to find the uncertainty in $y$ when $x = D$ in respect of uncertainties in $g$, $V$ and $\alpha$. For this purpose we want the relevant partial derivatives evaluated at $x = D$, and we find the differential equations satisfied by these partial derivatives, and their initial conditions, by differentiating partially the whole of (2.8) separately with respect to the various parameters.

For example, denoting $\partial y / \partial g$ by $y_g$, the problem for $y_g$ is clearly given by

$$\frac{d^2 y_g}{dx^2} = -\frac{1}{(V \cos \alpha)}, \quad y_g = 0, \quad \frac{dy_g}{dx} = 0 \quad \text{at } x = 0. \tag{2.15}$$

With a similar notation, we find the other equations

$$\frac{d^2 y_V}{dx^2} = \frac{2g}{V^3 \cos^2 \alpha}, \quad y_V = 0, \quad \frac{dy_V}{dx} = 0 \quad \text{at } x = 0, \tag{2.16}$$

and

$$\frac{d^2 y_\alpha}{dx^2} = -\frac{2g \tan \alpha}{V^2 \cos^2 \alpha}, \quad y_\alpha = 0, \quad \frac{dy_\alpha}{dx} = \sec^2 \alpha \quad \text{at } x = 0. \tag{2.17}$$

Equations (2.15), (2.16) and (2.17) are called the *variational equations*, and just as we can integrate (2.8) numerically so we can integrate the variational equations numerically to find the required quantities at $x = D$. Here, of course, we can also integrate them analytically, finding results which reproduce (2.10) exactly.

Notice that the original function $y$ does not here appear in the variational equations. This is unusual, and $y$ itself must appear if the original differential equation is non-linear in $y$ or contains any multiple of $y$ and some function of one or more parameters. For example, the differential equation and condition

$$\frac{dy}{dx} - y^2 = 1, \quad y(0) = \lambda \tag{2.18}$$

has with respect to $\lambda$ the variational equation and condition

$$\frac{dy_\lambda}{dx} - 2yy_\lambda = 0, \quad y_\lambda(0) = 1, \tag{2.19}$$

and the variational equation of

$$\frac{dy}{dx} - y \sin \lambda = 1, \quad y(0) = \lambda \tag{2.20}$$

is

$$\frac{dy_\lambda}{dx} - y_\lambda \sin \lambda - y \cos \lambda = 0, \quad y_\lambda(0) = 1. \tag{2.21}$$

We note further that the variational equation is always linear. Moreover, if the original equation is linear the variational equation is of precisely the same form in the terms involving the dependent variable. This is illustrated in (2.21) with respect to (2.20), the term $y \cos \lambda$ in (2.21) being a known function of $x$ in this context. It is not true in the non-linear case, as exemplified by (2.19) with respect to (2.18), and this remark may have some effect on our computational technique. We refer to this again in Section 2.7.

In large problems with many parameters, sensitivity analysis can be difficult and time consuming. We consider, however, that it is well worth while when the number of parameters is quite small, as in our 'darts' example. The variational equations, moreover, have other important uses which we shall note in subsequent chapters.

## 2.4 INHERENT INSTABILITY OF LINEAR RECURRENCE RELATIONS: INITIAL-VALUE PROBLEMS

Some special interest attaches to the inherent instability of linear initial-value problems with respect to uncertainties in the initial conditions. Here we consider recurrence relations, and in the next section make the necessary brief amendments for the corresponding treatment of ordinary differential equations.

### (i) First-order problems

We analyse first the recurrence relation and initial condition given by

$$y_{r+1} + a_r y_r = b_r, \ y_0 = \alpha. \tag{2.22}$$

The uncertainty $\delta y_r$ in $y_r$, with respect to the uncertainty $\delta\alpha$ in $\alpha$, we now know to be given by

$$\delta y_r = \frac{dy_r}{d\alpha}\delta\alpha, \tag{2.23}$$

and the coefficient $dy_r/d\alpha$ in (2.23) can be computed from the variational equation obtained from (2.22), which is

$$z_{r+1} + a_r z_r = 0, \quad z_0 = 1, \quad \text{where } z_r = dy_r/d\alpha. \tag{2.24}$$

From what we learnt in Chapter 1 we recognize $z_r$ as the *complementary solution* of the original equation (2.22), suitably scaled to have the value unity at $r = 0$, and hence the uncertainty in $y_r$ is the corresponding value of this complementary solution multiplied by the uncertainty in the initial value $y_0$. From (2.24) we see that we can write this complementary solution as

$$z_r = (-1)^r a_0 a_1 \cdots a_{r-1}, \tag{2.25}$$

and if this quantity is large in size for any $r$ the corresponding $y_r$ has a large absolute uncertainty. Whether or not it is also relatively ill-conditioned depends on the size of $y_r$ itself.

Consider, for example, the recurrence

$$y_{r+1} - 10y_r = -\tfrac{3}{2}r - \tfrac{4}{3}, \tag{2.26}$$

for which it is easy to verify that the general solution is

$$y_r = \tfrac{1}{6}(r + 1) + A \cdot 10^r, \tag{2.27}$$

the constant $A$ being determined from the initial condition $y_0 = \alpha$. An uncertainty

$\delta\alpha$ in $\alpha$ produces the uncertainty in $y_r$ of amount

$$\delta y_r = 10^r \delta\alpha, \tag{2.28}$$

and even for small $\delta\alpha$ this is large and increases rapidly as $r$ increases. But for most values of $\alpha$ the solution (2.27) also has a rapidly increasing term $A \cdot 10^r$, where clearly

$$A = \alpha - \tfrac{1}{6}. \tag{2.29}$$

For most values of $\alpha$ the constant $A$ is not very small, and the ratio $\delta\alpha/A$, which clearly measures the relative uncertainty quite closely for large $r$, is of no great size. The worst and quite serious case arises when $\alpha$ should be exactly $\tfrac{1}{6}$ so that the true result is the slowly increasing $y_r = \tfrac{1}{6}(r+1)$, but we either 'measure' the $\tfrac{1}{6}$ with a small uncertainty in a physical problem or store it in the computer with a small error in a mathematical problem. There is then the quite definite uncertainty (2.28), and even with a mathematical $\delta\alpha$ both the absolute and relative errors are very large for large $r$. In fact the 'shapes' of the exact and computed solutions may be completely different for large $r$. Table 2.1 indicates this for the mathematical problem 'solved' with three different computers, with respective word lengths equivalent to about 7, 15 and 31 significant decimals. The values $y_r$ are the correct values rounded to three decimal places, and the $\bar{y}_r$ are the machine values obtained starting with the corresponding $\bar{y}_0 = \bar{\alpha}$, the stored 'values' of $\tfrac{1}{6}$ in the three machines. With the poorest machine the results diverge from the truth quite rapidly beyond $r = 6$, not a single figure being correct even for $r = 8$. With the more accurate computers serious errors appear somewhat later, but in due course they are inevitable.

Notice, however, that if the initial $y_0 = \tfrac{1}{6}$ is measured with some physical apparatus restricted to three figures, so that $|\delta\alpha|$ is about $\tfrac{1}{3} \times 10^{-3} \ldots$, then the uncertainty in $y_r$ given by (2.28) is something like $\tfrac{1}{3} \times 10^{r-3}$, and even as early as $r = 3$ the relative uncertainty in the answer is 50%, with not even the first digit being reliable. And of course in this respect the weaker machine is effectively as accurate as its more powerful neighbours.

**Table 2.1**

| $r$ | 0 | 5 | 10 | 15 | 20 | 35 |
|---|---|---|---|---|---|---|
| $y_r$ | $\alpha = \tfrac{1}{6}$ | 1.000 | 1.833 | 2.667 | 3.500 | 6.000 |
| | | 0.996 | − 439 | | | |
| $\bar{y}_r$ | $\bar{\alpha}$ | 1.000 | 1.833 | 2.660 | − 654 | |
| | | 1.000 | 1.833 | 2.667 | 3.500 | − 3.130 |

## (ii) Homogeneous second-order problems

Another interesting case is the homogeneous second-order problem given by

$$y_{r+1} + a_r y_r + b_r y_{r-1} = 0, \quad y_0 = \alpha, \quad y_1 = \beta. \tag{2.30}$$

Here the general solution is just the complementary solution

$$y_r = A_1 y_r^{(1)} + A_2 y_r^{(2)}, \tag{2.31}$$

where $y_r^{(1)}$ and $y_r^{(2)}$ are independent solutions of the first of (2.30). Independence is guaranteed with the respective choice of initial conditions

$$y_0^{(1)} = 1, \quad y_1^{(1)} = 0; \quad y_0^{(2)} = 0, \quad y_1^{(2)} = 1. \tag{2.32}$$

We then find that $A_1 = \alpha$, $A_2 = \beta$, and differentiation of (2.30) shows that $y_r^{(1)}$ and $y_r^{(2)}$ are the respective solutions of the variational equations with respect to $\alpha$ and $\beta$. The uncertainty in $y_r$, caused by respective uncertainties $\delta\alpha$ and $\delta\beta$ in $\alpha$ and $\beta$, is then

$$\delta y_r = \delta\alpha y_r^{(1)} + \delta\beta y_r^{(2)}. \tag{2.33}$$

Now if either contributor to the complementary solution in (2.33) increases rapidly with $r$, the absolute uncertainty $\delta y_r$ also increases. Whether it increases seriously in a relative sense depends on the corresponding behaviour of the solution (2.31) for $y_r$ with $A_1 = \alpha$, $A_2 = \beta$, and the situation, as in our first-order problem, may become quite serious for particular values of $\alpha$ and $\beta$.

Consider, for example, the problem

$$y_{r+1} - 10.1 y_r + y_{r-1} = 0, \quad y_0 = \alpha, \quad y_1 = \beta. \tag{2.34}$$

The analysis of Chapter 1 shows that the exact solution is

$$y_r = \frac{\alpha}{99}(-10^r + 100 \cdot 10^{-r}) + \frac{\beta}{99}(10 \cdot 10^r - 10 \cdot 10^{-r})$$

$$= \frac{(-\alpha + 10\beta)}{99} 10^r + \frac{(100\alpha - 10\beta)}{99} 10^{-r}. \tag{2.35}$$

Here independent uncertainties $\delta\alpha$ and $\delta\beta$ will cause large uncertainties in $y_r$ for large $r$, but the relative uncertainty is tolerable unless $-\alpha + 10\beta$ is small, and if $-\alpha + 10\beta = 0$ the situation is catastrophic. In other words, we are now looking for the decreasing solution of (2.34), but slight physical uncertainties or mathematical errors in $\alpha$ and $\beta$ will introduce a 'spurious' and certainly unwanted rapidly increasing solution.

For example, with $\alpha = 1$ and $\beta = \frac{1}{10}$ our computer will fail to store $\beta$ exactly, and the true solution $y_r = 10^{-r}$ is rapidly 'swamped' by the rapidly increasing error so caused. Table 2.2, with the notation of Table 2.1, gives some illustrative

**Table 2.2**

| $r$ | 0 | 1 | 2 | 5 | 10 | 15 |
|---|---|---|---|---|---|---|
| $y_r$ | $\alpha = 1$ | $\beta = 0.1$ | 0.01 | $10^{-5}$ | $10^{-10}$ | $10^{-15}$ |
| $\bar{y}_r$ | $\bar{\alpha} = 1$ | $\bar{\beta}$ | 0.01 | $-0.00035$ | $-36.35$ | $-3.6 \times 10^6$ |

results. Here, and in the sequel unless some specific remark is made, we use our small seven-decimal computer.

Some problems, of course, are essentially inherently stable. For example, the general solution of

$$y_{r+1} - y_r + y_{r-1} = 0 \tag{2.36}$$

is

$$y_r = A_1 \cos r\frac{\pi}{3} + A_2 \sin r\frac{\pi}{3}, \tag{2.37}$$

which does not get large for increasing $r$ with any initial conditions. The absolute uncertainty also remains small and the idea of relative uncertainty is hardly meaningful.

### (iii) Non-homogeneous second-order problems •

The solution of the inhomogeneous problem

$$y_{r+1} + a_r y_r + b_r y_{r-1} = c_r, \quad y_0 = \alpha, \quad y_1 = \beta \tag{2.38}$$

is

$$y_r = y_r^{(p)} + A_1 y_r^{(1)} + A_2 y_r^{(2)}, \tag{2.39}$$

where $y_r^{(p)}$ is a particular solution and $y_r^{(1)}$ and $y_r^{(2)}$ are two independent contributors to the complementary solution. In fact if we choose for $y_r^{(p)}$ the initial conditions $y_0^{(p)} = 0$, $y_1^{(p)} = 0$ we can take $y_r^{(1)}$ and $y_r^{(2)}$ to have the definitions of the previous subsection and again $A_1 = \alpha$, $A_2 = \beta$, and the uncertainty in the solution due to those in $\alpha$ and $\beta$ is identical with that of the homogeneous problem.

For the problem

$$y_{r+1} - 10.1 y_r + y_{r-1} = -1.35r, \quad y_0 = \alpha, \quad y_1 = \beta \tag{2.40}$$

the true solution is easily found to be

$$y_r = \tfrac{1}{6}\{r + \tfrac{10}{99}(-10^r + 10^{-r})\} + \frac{\alpha}{99}(-10^r + 100 \cdot 10^{-r}) + \frac{\beta}{99}(10 \cdot 10^r - 10 \cdot 10^{-r})$$

$$= \tfrac{1}{6}r + \frac{10^r}{99}\{-\alpha + 10(\beta - \tfrac{1}{6})\} + \frac{10^{-r}}{99}\{100\alpha - 10(\beta - \tfrac{1}{6})\}. \tag{2.41}$$

We deduce, with respect to uncertainties in $\alpha$ and $\beta$, absolute ill-conditioning in all cases and catastrophic relative ill-conditioning, for large $r$, when

$$\alpha = 10(\beta - \tfrac{1}{6}), \tag{2.42}$$

that is when the true solution has no exponential-type increase with $r$.

For example, with $\alpha = 10$ and $\beta = \tfrac{7}{6}$ the true solution is $\tfrac{1}{6}r + 10 \cdot 10^{-r}$. Some results for this solution and our computer's attempt to achieve it are illustrated

**Table 2.3**

| $r$ | 0 | 1 | 2 | 5 | 10 | 15 |
|---|---|---|---|---|---|---|
| $y$ | $\alpha = 10$ | $\beta = \frac{7}{6}$ | 0.433 | 0.833 | 1.667 | 2.500 |
| $\bar{y}_r$ | $\bar{\alpha} = 10$ | $\bar{\beta}$ | 0.433 | 0.827 | $-642$ | $-644 \times 10^5$ |

in Table 2.3. For $r \geqslant 5$ the term $10 \cdot 10^{-r}$ makes no contribution to the rounded solution $y_r$ tabulated there.

### (iv) Problem reformulation

In all cases our problems have been absolutely and relatively ill-conditioned, at least for large $r$, when the specified initial conditions were expected to suppress a rapidly increasing term in the complementary solution. The question arises as to whether we have a better chance of obtaining good approximations to these required solutions with some simple reformulation of the problems. Quite commonly this is true, and the decisive reformulation is the solution of the recurrence relation with at least one different associated condition applied at a different point.

For the first-order equation (2.26), for example, the specification not of $y_0$ but of $y_N$, for some large number $N$, gives a problem with little inherent instability. If we specify $y_N = \beta$ the exact solution of (2.26) is

$$y_r = \tfrac{1}{6}(r + 1) + 10^{r-N}\{\beta - \tfrac{1}{6}(N + 1)\}, \tag{2.43}$$

and an uncertainty $\delta\beta$ in $\beta$ gives in $y_r$ the uncertainty $\delta y_r = 10^{r-N}\delta\beta$. For large $N$ and somewhat smaller $r$ this can be very small even for quite a large $|\delta\beta|$, and it gets smaller as $r$ decreases. We may not be able to find a very accurate value of $\beta$ either by physical measurement or mathematical analysis, but we do know that the suppression of the increasing exponential solution means that $\beta$ cannot be large for large $N$. Even the choice $\beta = 0$ is quite satisfactory in this respect, with $|\delta\beta|$ equal to the true $|y_N|$.

The arithmetic is performed by recurring backwards, with

$$y_r = \tfrac{1}{10}(y_{r+1} + \tfrac{3}{2}r + \tfrac{4}{3}), \quad y_N = \beta, \tag{2.44}$$

and Table 2.4 gives two illustrative computer results, all numbers being rounded to five decimals, with $y_N = 0$ at $N = 14$ and again at $N = 10$. In both cases the results are quite accurate, increasingly so as $r$ decreases, and their degree of consistency gives a measure of their accuracy for successive values of $r$. Notice that they produce almost as a by-product a very accurate approximation to the $y_0$ which was specified in the original problem. This is hardly surprising, since this is the only $y_0$ which suppresses the unwanted exponentially increasing solution.

**Table 2.4**

| $r$ | 14 | 12 | 10 | 8 | 6 | 4 | 0 |
|---|---|---|---|---|---|---|---|
| $\bar{y}_r$ | 0 | 2.14167 | 1.83308 | 1.50000 | 1.16667 | 0.83333 | 0.16667 |
| $\tilde{y}_r$ | | 0 | 1.48167 | 1.16648 | 0.83333 | 0.16667 | |
| $\frac{1}{6}(r+1)$ | 2.5 | 2.16667 | 1.83333 | 1.50000 | 1.16667 | 0.83333 | 0.16667 |

For the homogeneous second-order problem (2.34), which is very ill-conditioned for a combination of $\alpha$ and $\beta$ which suppresses the rapidly increasing contributor to the complementary solution, we might by analogy with our first-order reformulation suppose that the specification $y_N = 10^{-N}$, $y_{N+1} = 10^{-(N+1)}$ for large $N$ would give a well-conditioned problem for the determination of the solution $y_r = 10^{-r}$. If the stored $y_N$ and $y_{N+1}$ have respective errors $\delta\beta$ and $\delta\alpha$ it is not difficult to see that the corresponding error in $y_r$ is

$$\delta y_r = \{(10^{r-N} - 10^{N-r})\delta\alpha + (10^{N+1-r} - 10^{r-N-1})\delta\beta\}/(10^1 - 10^{-1}). \quad (2.45)$$

At first sight this would appear to be quite large for large $N$ and small $r$, but of course in our specification we have very small $\delta\alpha = k_1 10^{-(N+1)}$, $\delta\beta = k_2 10^{-N}$, where $k_1$ and $k_2$ are themselves small, and (2.45) then simplifies approximately to

$$\delta y_r \sim 10^{-r}(k_2 - 10^{-2}k_1), \quad (2.46)$$

with the relative error or uncertainty also remaining satisfactorily small. In the numerical work, of course, we use the recurrence relation in the reverse direction $r = N+1, N, N-1, \ldots, 0$.

A somewhat better reformulation, with less uncertainty about what to do for large $N$, uses the specified $y_0$ and a specified $y_N = \beta$, with large $N$ and small $|\beta|$ which essentially suppresses the increasing solution. This reformulated problem is of boundary-value type, but here we can easily find its mathematical solution with $y_0 = \alpha$ and $y_N = \beta$, and if these have respective uncertainties $\delta\alpha$ and $\delta\beta$ then that in $y_r$ is

$$\delta y_r = \{(10^{N-r} - 10^{r-N})\delta\alpha + (10^r - 10^{-r})\delta\beta\}/(10^N - 10^{-N}). \quad (2.47)$$

With $\alpha = 1$ and say $\beta = 0$ so that $\delta\alpha = 0$ and $|\delta\beta| = y_N = 10^{-N}$, equation (2.47) simplifies to an approximate value of $\delta y_r \sim 10^{r-2N}$, giving with respect to $y_r = 10^{-r}$ the relative error $\delta y_r/y_r \sim 10^{2(r-N)}$ which is small for $r < N$ and decreases rapidly as $r$ decreases. Our reformulated problem is satisfactorily well conditioned in both senses, and it is interesting to note the size of the denominator in (2.47) compared with that in (2.45).

For the numerical work we can still use the recurrence method, taking $\bar{y}_N = 0$, $\bar{y}_{N-1} = 1$ and recurring backwards to $r = 0$. The results obtained are then multiplied by a constant $k$ so that $k\bar{y}_0 = 1$, the specified value at $r = 0$, and the required solution is just $k\bar{y}_r$. This is valid, of course, because of the linearity and homogeneity of the recurrence relation. The method is so good for the current example that numbers are hardly worth quoting, but results for a more interesting problem are given in Table 2.8 (Section 2.7).

Finally, it is interesting to consider the determination of the solution $y_r = \frac{1}{6}r$ of the non-homogeneous problem (2.40) with $\alpha = 0$, $\beta = \frac{1}{6}$, conditions which suppress both terms $10^r$ and $10^{-r}$ of the complementary solution. As we have seen, this particular formulation fails, since even if $\delta\alpha = 0$ there is a non-zero $\delta\beta$, we have introduced a small multiple of the dangerous $10^r$, and for large $r$ we see from (2.41) that the uncertainty in $y$ is close to $10^{r-1}\delta\beta$. We have an exactly similar problem with the same recurrence with conditions $y_N = \frac{1}{6}N + \delta\beta$, $y_{N+1} = \frac{1}{6}(N + 1) + \delta\alpha$. The uncertainties $\delta\alpha$ and $\delta\beta$ produce in $y_r$ the uncertainty

$$\delta y_r = \frac{10^r}{9.9}\{10^{-N}\delta\alpha - 10^{-(N+1)}\delta\beta\} + \frac{10^{-r}}{9.9}\{10^{N+1}\delta\beta - 10^N\delta\alpha\}. \quad (2.48)$$

Here $\delta\alpha$ and $\delta\beta$ have not introduced a significant multiple of the dangerous $10^r$, but they have introduced a dangerous multiple of the humble $10^{-r}$, which now increases as $r$ decreases from $N + 1$ to zero. The two formulations manifest a fairly obvious symmetry. If we allow a $\delta\alpha$ in the first formulation the uncertainties in the two cases are respectively very close to

$$\delta y_r = \frac{10^r}{99}(-\delta\alpha + 10\delta\beta), \quad \delta y_r = 10^{N-r-1}(-\delta\alpha + 10\delta\beta). \quad (2.49)$$

Again the most useful reformulation specifies both $y_0$ and $y_N$ for some large $N$. With $y_0 = \alpha$ and $y_N = \beta$ the uncertainty or error is again given by (2.47), with respective maxima $|\delta\alpha|$ near $r = 0$ and $|\delta\beta|$ near $r = N$. In fact if $\delta\alpha = 0$ and we take $\beta = 0$ so that $|\delta\beta| = y_N = \frac{1}{6}N$, the error is very close to $\frac{1}{6}N10^{r-N}$, which is quite small for $r$ significantly smaller than $N$. The corresponding relative error is $(N/r)10^{r-N}$, and the problem is well conditioned in both senses.

There is no easy and satisfactory way of using the recurrence relation in stable computation, for reasons which are given in Chapter 5, but a very stable method solves the simultaneous linear algebraic equations obtained from the recurrence relation and boundary conditions, and given here by

$$\left.\begin{array}{rl} -10.1y_1 + \quad y_2 & = -1.35 - \alpha \\ y_1 - 10.1y_2 + y_3 & = -2.70 \\ \vdots & \\ y_{N-3} - 10.1y_{N-2} + y_{N-1} & = -1.35(N-2) \\ y_{N-2} - 10.1y_{N-1} & = -1.35(N-1) - \beta \end{array}\right\} \quad (2.50)$$

Some results for various values of $\alpha$ and $\beta$ are given in Table 2.5. They confirm our analysis and the inherent stability of the reformulated problem.

**Table 2.5**

| $r$ | $y_r = \frac{1}{6}r$ | $\bar{y}_r$ | $\bar{y}_r$ | $\bar{y}_r$ |
|---|---|---|---|---|
| 0 | 0.000000 | 0.000000 | 0.100000 | 0.000000 |
| 1 | 0.166667 | 0.166667 | 0.176667 | 0.166667 |
| 2 | 0.333333 | 0.333333 | 0.334333 | 0.333333 |
| 3 | 0.500000 | 0.500000 | 0.500100 | 0.500000 |
| 4 | 0.666667 | 0.666667 | 0.666677 | 0.666667 |
| 5 | 0.833333 | 0.833333 | 0.833334 | 0.833333 |
| 6 | 1.000000 | 1.000000 | 1.000000 | 1.000000 |
| 7 | 1.166667 | 1.166667 | 1.166667 | 1.166667 |
| 8 | 1.333333 | 1.333333 | 1.333333 | 1.333333 |
| 9 | 1.500000 | 1.500000 | 1.500000 | 1.500000 |
| 10 | 1.666667 | 1.666667 | 1.666667 | 1.666664 |
| 11 | 1.833333 | 1.833334 | 1.833334 | 1.833307 |
| 12 | 2.000000 | 2.000010 | 2.000010 | 1.999733 |
| 13 | 2.166667 | 2.166767 | 2.166767 | 2.164000 |
| 14 | 2.333333 | 2.334333 | 2.334333 | 2.306667 |
| 15 | 2.500000 | 2.510000 | 2.510000 | 2.233333 |
| 16 | 2.666667 | 2.766667 | 2.766667 | 0.000000 |

## 2.5 INHERENT INSTABILITY OF LINEAR DIFFERENTIAL EQUATIONS. INITIAL-VALUE PROBLEMS

Both the theory and the practice of the material of the last section carry over in the most obvious way to the corresponding treatment of differential equations with initial conditions. Here we give some examples, leaving the reader to fill in the details.

The first-order problem

$$y' - 10y = -(31 + 10x)(3 + x)^{-2} \tag{2.51}$$

has the general solution

$$y = (3 + x)^{-1} + Ae^{10x}. \tag{2.52}$$

It is absolutely ill conditioned with respect to uncertainty in any starting condition $y(0) = \alpha$, increasingly so for increasing $x$, and it is relatively unstable for a condition which produces a very small $A$. The catastrophic case is when $y(0) = \frac{1}{3}$ so that $A = 0$ and the true solution is $1/(3 + x)$. We can, however, obtain this solution quite accurately with the reformulation which specifies $y(X)$ accurately for some $X$, and even $y(X) = 0$, for some large $X$, gives good and increasingly better results as $x$ decreases from $X$ to 0.

The homogeneous second-order problem

$$y'' - \pi^2 y = 0, \quad y(0) = \alpha, \quad y'(0) = \beta \tag{2.53}$$

has the solution

$$y = \frac{1}{2\pi}\{(\pi\alpha + \beta)e^{\pi x} + (\pi\alpha - \beta)e^{-\pi x}\}. \qquad (2.54)$$

It is absolutely ill conditioned with respect to uncertainties in $\alpha$ and $\beta$ for any values of $\alpha$ and $\beta$, and catastrophically relatively ill conditioned if $\pi\alpha + \beta = 0$. For $\alpha = 1, \beta = -\pi$, the reformulation

$$y(X) = e^{-\pi X}, \quad y'(X) = -\pi e^{-\pi X} \qquad (2.55)$$

produces a very accurate $e^{-\pi x}$ for $0 < x < X$, and the reformulation

$$y(X) = 0, \quad y'(X) = k, \quad X \text{ large}, \qquad (2.56)$$

with $k$ chosen so that $y(0) = 1$, will produce a very good approximation to $y(x) = e^{-\pi x}$ if $x$ is not too close to $X$. As $x$ gets smaller, the better this approximation becomes.

The non-homogeneous second-order problem

$$y'' - \pi^2 y = -\tfrac{1}{6}\pi^2(x - 1), \quad y(0) = \alpha, \quad y'(0) = \beta \qquad (2.57)$$

has the solution

$$y = \tfrac{1}{6}(x - 1) + \frac{1}{2\pi}\{(\tfrac{1}{6}\pi + \alpha\pi + \beta - \tfrac{1}{6})e^{\pi x} + (\tfrac{1}{6}\pi + \alpha\pi + \tfrac{1}{6} - \beta)e^{-\pi x}\}. \qquad (2.58)$$

There is absolute ill conditioning for large $x$ for any $\alpha$ and $\beta$, and there is a catastrophic amount of relative ill conditioning when

$$\tfrac{1}{6}\pi + \alpha\pi + \beta - \tfrac{1}{6} = 0, \qquad (2.59)$$

that is when the true solution has no exponentially increasing component. This happens particularly when $\alpha = -\tfrac{1}{6}, \beta = \tfrac{1}{6}$, for which the true solution is just $\tfrac{1}{6}(x - 1)$. We cannot get this solution accurately with the reformulated conditions

$$y(X) = \tfrac{1}{6}(X - 1) + \delta\alpha, \quad y'(X) = \tfrac{1}{6} + \delta\beta, \qquad (2.60)$$

since the physical or mathematical uncertainties $\delta\alpha$ and $\delta\beta$ produce in $y(x) = \tfrac{1}{6}(x - 1)$ the uncertainty

$$\delta y(x) = \frac{1}{2\pi}(\pi\delta\alpha + \delta\beta)e^{\pi(x - X)} + \frac{1}{2\pi}(\pi\delta\alpha - \delta\beta)e^{-\pi(x - X)}, \qquad (2.61)$$

of which the second term will be large for large $X$ and small $x$.

Success is, however, achieved with the reformulation as the boundary-value problem

$$y(0) = -\tfrac{1}{6} + \delta\alpha, \quad y(X) = \tfrac{1}{6}(X - 1) + \delta\beta, \quad X \text{ large}, \qquad (2.62)$$

since the uncertainties $\delta\alpha$ and $\delta\beta$ produce in $y(x) = \tfrac{1}{6}(x - 1)$ the uncertainty

$$\delta y(x) = \frac{1}{e^{\pi X} - e^{-\pi X}}\{(-\delta\alpha e^{-\pi X} + \delta\beta)e^{\pi x} + (\delta\alpha e^{\pi X} - \delta\beta)e^{-\pi x}\}, \qquad (2.63)$$

which is approximately $\delta\alpha e^{-\pi x} + \delta\beta e^{\pi(x-X)}$ for large $X$, verifying the essential stability of this formulation.

## 2.6 INHERENT INSTABILITY
### BOUNDARY-VALUE PROBLEMS

Turning now to the inherent instability of boundary-value problems, we consider linear problems of the second order, first for the recurrence relation and boundary conditions given by

$$y_{r+1} + a_r y_r + b_r y_{r-1} = c_r, \quad y_0 = \alpha, \quad y_N = \beta, \tag{2.64}$$

where here $N$ represents the second boundary point and is not necessarily large. From equation (2.38) and its associated definitions we know that the general solution of this recurrence is in fact (2.39) where $y_r^{(p)}$ satisfies the full recurrence with $y_0^{(p)} = y_1^{(p)} = 0$, $y_r^{(1)}$ satisfies the homogeneous recurrence $(c_r = 0)$ with $y_0^{(1)} = 1, y_1^{(1)} = 0$, and $y_r^{(2)}$ satisfies the homogeneous recurrence with $y_0^{(2)} = 0$, $y_1^{(2)} = 1$.

It is then easy to see that (2.39) satisfies the full recurrence relation and it satisfies the first boundary condition $y_0 = \alpha$ if we take $A_1 = \alpha$. Moreover, the contribution to the solution (2.39) of the part

$$y_r^{(q)} = y_r^{(p)} + \alpha y_r^{(1)}, \quad \text{with} \quad y_0^{(q)} = \alpha, \quad y_1^{(q)} = 0, \tag{2.65}$$

can be computed by direct recurrence from the full recurrence relation, up to the evaluation of $y_N^{(q)}$. The contribution $y_r^{(2)}$ can obviously be computed similarly, up to $y_N^{(2)}$, from the homogeneous recurrence with $y_0^{(2)} = 0$, $y_1^{(2)} = 1$, and the only unknown quantity is the constant $A_2$. The only thing not yet satisfied is the second boundary condition at $r = N$, and we therefore choose $A_2$ so that

$$y_N^{(q)} + A_2 y_N^{(2)} = \beta, \tag{2.66}$$

and this gives the required solution

$$y_r = y_r^{(q)} + A_2 y_r^{(2)}, \quad A_2 = (\beta - y_N^{(q)})/y_N^{(2)}. \tag{2.67}$$

In (2.67) the terms $y_r^{(q)}$ and $y_N^{(q)}$ depend only on $\alpha$, and $\beta$ appears only in an explicit form in $A_2$. We also observe from (2.65) that $\partial y_r^{(q)}/\partial\alpha = y_r^{(1)}$, so that the expression for the uncertainty $\delta y_r$ in $y_r$ caused by uncertainties $\delta\alpha$ in $\alpha$ and $\delta\beta$ in $\beta$ is easily found to be

$$\delta y_r = \frac{1}{y_N^{(2)}}\{(y_r^{(1)} y_N^{(2)} - y_N^{(1)} y_r^{(2)})\delta\alpha + y_r^{(2)}\delta\beta\}. \tag{2.68}$$

As in our initial-value problems we see that the *absolute* uncertainty in our answer depends solely on the behaviour of the contributors to the complementary solution of the original problem. If $y_N^{(2)}$ is not very small we have no reason to expect any serious inherent instability. If $y_N^{(2)}$ is very small indeed the absolute uncertainty in $y_r$ is likely to be large even for small $\delta\alpha$ and $\delta\beta$, and we have absolute inherent instability. Again, as in the initial-value problems, whether or not the instability is also relative depends on the size of $y_r$ itself. If $\alpha$ and $\beta$ are such that $\beta - y_N^{(q)}$ in (2.67) is not particularly small, then $A_2$ is very large, $y_r$ is likely to be very large, and there may be little or no relative instability. But if $\beta - y_N^{(q)}$ is also exceptionally small, $y_r$ may be of normal size, $|\delta y_r/y_r|$ may be large and the relative ill-conditioning quite noticeable.

Consider, for example, the problem

$$y_{r+1} - 1.732y_r + y_{r-1} = 1, \quad y_0 = \alpha, \quad y_6 = \beta. \tag{2.69}$$

Table 2.6 gives some rounded results for $\alpha = 0.900$, $\beta = 6.000$ and for these values perturbed slightly to $\alpha = 0.901$, $\beta = 6.001$; and for $\alpha = 0.900$, $\beta = 6.563$ and for the values perturbed to $\alpha = 0.901$, $\beta = 6.564$. The difference in the two sets of results, representing the absolute change, is exactly the same in both cases. With the first set, however, the relative difference is tolerably small, whereas in the second set the relative difference is so large that the solutions of the original problem and the slightly perturbed problem have quite different pictures. Our explanation of the reasons for these results is verified by inspection of the columns for $y_r^{(q)}$ and $y_r^{(2)}$.

If $y_N^{(2)}$ is large, of course, the uncertainty $\delta y_r$ is likely to be quite small, and the problem will be reasonably well conditioned in both absolute and relative senses. Near the end of Section 2.4, for example, we showed that the problem

$$y_{r+1} - 10.1y_r + y_{r-1} = -1.35r, \quad y_0 = \alpha, \quad y_N = \beta \tag{2.70}$$

was inherently stable with respect to small uncertainties $\delta\alpha$ and $\delta\beta$, and the value of $y_N^{(2)}$ is easily seen to be

$$y_N^{(2)} = \tfrac{10}{99}(10^N - 10^{-N}), \tag{2.71}$$

which is certainly large for even moderate-sized $N$. We saw in our earlier work that the corresponding initial-value problem, equation (2.70) with $y_0 = 0$, $y_1 = \tfrac{1}{6}$, is extremely ill conditioned just because it has a rapidly increasing complementary solution, that is just because $y_N^{(2)}$ is large, and our new analysis further illuminates the discussion near the end of Section 2.4 about the possibility

**Table 2.6**

| $r$ | $y_r$ | $y_r$ | $y_r$ | $y_r$ | $y_r^{(q)}$ | $y_r^{(2)}$ |
|---|---|---|---|---|---|---|
| 0 | 0.900 | 0.901 | 0.900 | 0.901 | 0.900 | 0.0000 |
| 1 | 924.306 | 921.026 | 0.765 | $-2.515$ | 0.000 | 1.0000 |
| 2 | 1600.998 | 1595.316 | 1.425 | $-4.257$ | 0.1000 | 1.7320 |
| 3 | 1849.622 | 1843.061 | 2.703 | $-3.858$ | 1.1732 | 1.9998 |
| 4 | 1603.548 | 1597.866 | 4.257 | $-1.425$ | 2.9320 | 1.7317 |
| 5 | 928.723 | 925.443 | 5.670 | $+2.390$ | 4.9050 | 0.9995 |
| 6 | 6.000 | 6.001 | 6.563 | 6.564 | 6.563466 | $-0.0006096$ |

of reformulating some ill-conditioned initial-value problems as well-conditioned boundary-value problems.

A few other comments are relevant. First, the *position* of the second boundary point is involved with the question of ill-conditioning. In Table 2.6, for example, the specification $y_0 = 0.900$, $y_5 = 5.670$ gives results almost identical with those in the third column up to $r = 5$. The slightly perturbed conditions $y_0 = 0.901$, $y_5 = 5.671$ give results which differ at most by about three units in the third decimal place, and this problem is very well conditioned. Its $y_N^{(2)}$ has the quite reasonable value 0.9995 given in Table 2.6.

Second, if $y_N^{(2)} = 0$ it means that the homogeneous problem obtained from (2.64), with $c_r = y_0 = y_N = 0$, has a solution which is not everywhere zero and our very ill-conditioned problem is, in a manner of speaking, very near to this critical situation. The same remark will apply to a problem with more complicated boundary conditions, like

$$y_{r+1} + a_r y_r + b_r y_{r-1} = c_r, \quad p_0 y_0 + q_0 y_1 = r_0, \quad p_N y_N + q_N y_{N-1} = r_N, \quad (2.72)$$

which will be very ill conditioned if we are near to the situation in which the homogeneous problem

$$y_{r+1} + a_r y_r + b_r y_{r-1} = 0, \quad p_0 y_0 + q_0 y_1 = 0, \quad p_N y_N + q_N y_{N-1} = 0 \quad (2.73)$$

has a solution other than $y_r = 0$ for all $r$. Being 'near' to this critical situation may imply that the critical situation is encountered with very small changes in any of the coefficients in (2.73), or even in the value of $N$.

We turn now to a brief consideration of the analogous problem in ordinary differential equations of second order, and consider particularly and immediately the most general linear form with separated boundary conditions given by

$$y'' + f(x)y' + g(x)y = k(x), \quad p_0 y(a) + q_0 y'(a) = r_0, \quad p_1 y(b) + q_1 y'(b) = r_1. \quad (2.74)$$

It should now be clear that this problem is absolutely ill conditioned, say with respect to small changes in $r_0$ and $r_1$, if the homogeneous problem differs only slightly from one in which there is a solution other than $y(x) = 0$ for all $x$ in $(a, b)$. The question of relative ill conditioning is then determined by the non-homogeneous terms in the full problem.

Consider, for example, the problem

$$y'' + y = xe^x, \quad y(0) = 0, \quad 0.5y(1) - 0.78y'(1) = 1. \quad (2.75)$$

The solution of the homogeneous equation is

$$y = A_1 \sin x + A_2 \cos x, \quad (2.76)$$

and this satisfies the first homogeneous boundary condition if $A_2 = 0$. The function $y = A_1 \sin x$ satisfies exactly at $x = 1$ the condition

$$0.5y(1) - 0.5 \tan(1)y'(1) = 0. \quad (2.77)$$

Now $0.5 \tan(1) = 0.7787$, to four figures, so that in comparison with the specified homogeneous boundary condition, the last of (2.75) with zero on the right, we are close to the dangerous situation and expect some absolute ill conditioning.

The general solution of the non-homogeneous differential equation in (2.75) is

$$y = A_1 \sin x + A_2 \cos x + \tfrac{1}{2}(x-1)e^x, \tag{2.78}$$

and it satisfies the specified non-homogeneous boundary conditions in (2.75) if

$$A_2 = \tfrac{1}{2}, \quad A_1 = \frac{1 - 0.25 \cos 1 - 0.39 \sin 1 + 0.39e}{0.5 \sin 1 - 0.78 \cos 1}. \tag{2.79}$$

The denominator of $A_1$ is about $-0.0007$, so that there is definite absolute ill conditioning. If $A_1$ is large, that is if its numerator is not small, then we should not expect serious relative ill conditioning. In fact the solution of (2.75) is very close to

$$y(x) = -2280.2424 \sin x + 0.5 \cos x + \tfrac{1}{2}(x-1)e^x, \tag{2.80}$$

and the solution of a slightly perturbed (2.75), with

$$y(0) = -0.005, \quad 0.5y(1) - 0.78y'(1) = 1.005 \tag{2.81}$$

is the not significantly different

$$y(x) = -2293.9970 \sin x + 0.495 \cos x + \tfrac{1}{2}(x-1)e^x. \tag{2.82}$$

But if the number 1 on the right-hand side of the second boundary condition (in 2.75) is replaced by the number $-0.6$ then $A_1$ is not very large, and the solution is

$$y = 4.4542 \sin x + 0.5 \cos x + \tfrac{1}{2}(x-1)e^x. \tag{2.83}$$

A slight perturbation, to the conditions

$$y(0) = -0.005, \quad 0.5y(1) - 0.78y'(1) = -0.595, \tag{2.84}$$

gives the solution

$$y = -9.3004 \sin x + 0.495 \cos x + \tfrac{1}{2}(x-1)e^x, \tag{2.85}$$

which is completely different in shape from that of (2.83).

The only other relevant remark concerns the possibility of reformulating an ill-conditioned boundary-value problem into one with greater inherent stability. There are various obvious possibilities if we have a little more choice in the specifications of relevant boundary conditions or initial conditions. With the problem (2.69), for example, the reformulation as an initial-value problem, with the specification of both $y_0$ and $y_1$, will produce a very stable alternative. Another retains the problem in boundary-value form but changes the position of the second boundary. For example, as we already noted, the specification of $y_5$ rather than $y_6$ in (2.69) gives a perfectly well-conditioned boundary-value problem.

## 2.7 ADDITIONAL NOTES

*Section 2.1* Not all computing machines have the same circuitry, so that the errors of storage and of arithmetic will vary with different machines. But it is useful to record some representative results for a typical binary computer using floating-point arithmetic with a word length of $t$ binary digits. For such a computer the number $x$ might be stored in the machine as

$$\bar{x} = x(1 + \varepsilon), \qquad |\varepsilon| \leqslant 2^{-t}, \tag{2.86}$$

where the actual value of $\varepsilon$ will depend on $x$ and may be zero. Also any arithmetic operation $*$, where $*$ represents addition, subtraction, multiplication or division, will produce and store, for the numbers $\bar{x}$ and $\bar{y}$ stored in the machine, the value

$$\mathrm{fl}(\bar{x} * \bar{y}) = (\bar{x} * \bar{y})(1 + \varepsilon), \quad |\varepsilon| \leqslant 2^{-t}, \tag{2.87}$$

where $\varepsilon$ now depends on $\bar{x}$ and $\bar{y}$ and may be zero.

Formula (2.86), of course, is relevant to the *inherent instability* of mathematical problems, whereas (2.87) is relevant to any discussion of *induced instability* in the solving technique. Now although this chapter is essentially concerned only with *inherent* instability, the numerical illustrations have been performed on computers with finite word lengths and which therefore *induce* rounding errors in every numerical operation. In initial-value problems such local errors are transmitted in subsequent calculations in precisely the same numerical way as are the physical uncertainties or mathematical errors in the initial values. When the latter are dangerous, so are the former, and any reformulation which mitigates the dangers of uncertainties in initial conditions will also have a corresponding good effect on the transmission of local rounding errors. All the numerical results verify these remarks. Even machines with the same word length, however, may produce somewhat different results due to differences in the rounding hardware, and for various other reasons including the order of the computation. This could be noticeable especially in computations which exhibit significant instability, for example in Tables 2.1, 2.2 and 2.3.

The situation with boundary-value problems is rather different, and in this chapter the arithmetic has been performed so accurately that all the numbers, for example in Tables 2.5 and 2.6, are correct to the number of figures quoted, and any dissatisfaction with the results is due solely to the *inherent* errors. The effect of induced computer errors is later mentioned in the vicinity of equation (2.99), and also in more detail in Chapters 5 and 6.

*Sections 2.2, 2.3* Our problem of sensitivity analysis was not very difficult because we were able to use a differential equation, equation (2.8), whose independent variable had a fixed exact value at the point of interest. Consider, however, the use of the original differential equations (2.6), and consider the problem of determining the range of the projectile, that is the value of $x$ when $y = H$ again.

The difficulty here is that the time at which this will happen is not a fixed exact value but is itself subject to uncertainty caused by uncertainties in the data, here $g$, $V$ and $\alpha$, and we must proceed as follows.

The original problem seeks the value of $x = X$ at the time $t = T$ for which $y = H$. Due to the uncertainties in the data we must replace $X$ by $X + \delta X$, $T$ by $T + \delta T$, and $y = H$ by $y + \delta y = H$. Now

$$y + \delta y = y(T + \delta T, V + \delta V, g + \delta g, \alpha + \delta \alpha)$$

$$= y(T, V, g, \alpha) + \delta T \frac{\partial y}{\partial t} + \delta V \frac{\partial y}{\partial V} + \delta g \frac{\partial y}{\partial g} + \delta \alpha \frac{\partial y}{\partial \alpha}, \tag{2.88}$$

neglecting terms of second and higher orders in $\delta T, \delta V, \delta g$ and $\delta \alpha$, where the derivatives in (2.88) are evaluated at $t = T$ and $y = H$. The first term on the right of (2.88) therefore equals $H$, and $y + \delta y = H$ when

$$\delta T = - q/(\partial y/\partial t), \quad q = \delta V \frac{\partial y}{\partial V} + \delta g \frac{\partial y}{\partial g} + \delta \alpha \frac{\partial y}{\partial \alpha} \tag{2.89}$$

Then also

$$
\left.
\begin{aligned}
X + \delta X &= X(T, V, g, \alpha) + \delta T \frac{\partial X}{\partial t} + p \\
p &= \delta V \frac{\partial X}{\partial V} + \delta g \frac{\partial X}{\partial g} + \delta \alpha \frac{\partial X}{\partial \alpha}
\end{aligned}
\right\}, \tag{2.90}
$$

everything in (2.89) and (2.90) being evaluated at $t = T$. From (2.89) and (2.90) we produce the required result

$$\delta X = p - q \frac{\partial X}{\partial t} \Big/ \frac{\partial y}{\partial t}$$

$$= \delta V \left( \frac{\partial X}{\partial V} - \frac{\dot{x}}{\dot{y}} \frac{\partial y}{\partial V} \right) + \delta g \left( \frac{\partial X}{\partial g} - \frac{\dot{x}}{\dot{y}} \frac{\partial y}{\partial g} \right) + \delta \alpha \left( \frac{\partial X}{\partial \alpha} - \frac{\dot{x}}{\dot{y}} \frac{\partial y}{\partial \alpha} \right), \tag{2.91}$$

where $\dot{x} = \partial x/\partial t$, $\dot{y} = \partial y/\partial t$, with everything evaluated at $t = T$. The derivatives with respect to $V, g$ and $\alpha$ are computed in the usual way from the variational equations, now derived from (2.6) by relevant partial differentiation.

*Section 2.3* Near the end of this section we commented on the fact that the variational equations for non-linear problems do not have the same general form as the original differential equations. In the interests of economy of programming effort we may then occasionally prefer to approximate to the solutions of the variational equations by solving the original equations again with a slight change in the value of the relevant parameter. With respect to (2.18), for example, we might solve (2.18) with $\lambda = \lambda_0$ and again with $\lambda = \lambda_0 + \varepsilon$, and approximate to $y_\lambda = \partial y/\partial \lambda$ by the expression

$$\{\partial y(x)/\partial \lambda\}_{\lambda = \lambda_0} = \{y(x)_{\lambda = \lambda_0 + \varepsilon} - y(x)_{\lambda = \lambda_0}\}/\varepsilon. \tag{2.92}$$

This will often be a good enough approximation for small $\varepsilon$, and avoids the necessity of constructing more than one computer program for solving the differential equations.

*Sections 2.4, 2.5* In these sections on inherent instability all the examples involved linear equations with constant coefficients, because the analytical solutions of such problems are easily obtained and give exact comparison with the computed results of uncertainties. But for equations with non-constant coefficients we can often easily find sufficient information to be able to forecast sufficiently accurately the effects of ill conditioning and reformulation where practicable.

For example, we showed in Section 1.5 that the quantities

$$I_r = \int_0^1 x^r e^{x-1}\, dx, \quad r = 0, 1, 2, \ldots \tag{2.93}$$

satisfy the recurrence relation

$$I_{r+1} + (r+1)I_r = 1. \tag{2.94}$$

But from the work of Section 2.4(i) we would not expect to produce accurate results for large $r$ with the specification of $I_0 = 1 - e^{-1}$. Our expectations are guaranteed by the nature of the complementary solution defined in (2.25). This increases very rapidly in absolute value as $r$ increases, whereas it is easy to see that the $I_r$ of (2.93) decreases as $r$ increases. Correspondingly, the reformulation of the first part of Section 2.4(iv), in which we recur backwards with the specified $I_r = 0$ for some large enough value $r = N$, produces extremely good results for $r$ not too close to $N$.

The other example mentioned in Section 1.5 was the homogeneous second-order recurrence

$$y_{r+1} - \frac{2r}{x} y_r + y_{r-1} = 0, \quad r = 0, 1, 2, \ldots. \tag{2.95}$$

Two independent solutions are the Bessel functions $J_r(x)$ and $Y_r(x)$. As $r$ increases, for fixed $x$, it is known that both $J_r(x)$ and $Y_r(x)$ oscillate with amplitude not greater than unity until $r$ gets greater than $x$, after which $J_r(x)$ decreases monotonically to zero and $Y_r(x)$ increases monotonically to infinity as $r \to \infty$. We shall therefore produce accurate values of $J_r(x)$, by forward recurrence in (2.95) with $y_0 = J_0(x)$, $y_1 = J_1(x)$, only for $r < x$, after which the computed values are increasingly inaccurate. Table 2.7 gives some illustrative results of forward recurrence for $x = 5$.

**Table 2.7**

| $r$ | $\bar{y}_r$ | $r$ | $\bar{y}_r$ | $r$ | $\bar{y}_r$ | $r$ | $\bar{y}_r$ |
|---|---|---|---|---|---|---|---|
| 0 | − 0.177597 | 6 | 0.131049 | 12 | 0.000074 | 18 | − 0.074967 |
| 1 | − 0.327579 | 7 | 0.053376 | 13 | 0.000004 | 19 | − 0.528571 |
| 2 | 0.046565 | 8 | 0.018405 | 14 | − 0.000055 | 20 | − 3.941409 |
| 3 | 0.364831 | 9 | 0.005520 | 15 | − 0.000311 | 21 | − 31.002804 |
| 4 | 0.391232 | 10 | 0.001468 | 16 | − 0.001813 | 22 | − 256.482147 |
| 5 | 0.261141 | 11 | 0.000350 | 17 | − 0.011291 | | |

But an accurate solution can be obtained for $r$ as large as we like by the reformulated method mentioned in the vicinity of equation (2.47). In Table 2.8 we show some results of the backward recurrence starting at $r = N$ for $x = 5$, with $k\bar{y}_r = 0$ at $r = N$ and 1 at $r = N - 1$, and these results scaled to produce $\bar{y}_r$ having the correct value at $r = 0$. It can be shown that if $N$ is large the errors are not very different from $\{J_N(5)/Y_N(5)\} Y_r(5)$, which decrease in size as $r$ decreases with a little oscillation for $r \leqslant 5$. The table gives results for $N = 10$, and we could obviously produce very good results for a larger range of $r$ by starting with a larger value of $N$. In fact with $N \geqslant 15$ we have six-decimal accuracy for all $r < N$.

Note that in the forward recurrence, values will be very accurate certainly for $r < x$, and in this oscillating region there may be a value much larger than that at $r = 0$. Somewhat better results with the backward recurrence method would then be obtained by scaling not to give the correct $y_0$ but the correct $y_r$ of largest absolute value in $r \leqslant x$.

Here we were helped by being able to recognize two independent solutions of (2.95) and their behaviour as $r$ increases, but in general their *nature* is

**Table 2.8**

| $r$ | $y_r$ | $\bar{y}_r$ |
|---|---|---|
| 0 | − 0.177597 | − 0.177597 |
| 1 | − 0.327579 | − 0.327537 |
| 2 | 0.046565 | 0.046582 |
| 3 | 0.364831 | 0.364803 |
| 4 | 0.391232 | 0.391181 |
| 5 | 0.261141 | 0.261088 |
| 6 | 0.131049 | 0.130994 |
| 7 | 0.053376 | 0.053297 |
| 8 | 0.018405 | 0.018239 |
| 9 | 0.005520 | 0.005066 |
| 10 | 0.001468 | 0.000000 |

determinable almost by inspection. For constant $a$ the solutions of the recurrence

$$y_{r+1} - 2ay_r + y_{r-1} = 0 \tag{2.96}$$

are oscillating with unit amplitude if $|a| < 1$, and for $|a| > 1$ they have the forms $p^r$ and $p^{-r}$ with $|p| > 1$. If now the constant $a$ is replaced by a smoothly varying sequence of values $a_r$ the same sort of behaviour takes place. The solutions are oscillatory and of unit constant size if $|a_r| < 1$, and rapidly increase or decrease if $|a_r| > 1$. In our recent example $a_r = r/x$ and no solution is dominant for $r \leqslant x$, but for $r > x$ our $|a_r|$ exceeds unity and one of the solutions becomes dominant and eventually 'swamps' the other.

*Section 2.6* This analysis of the *inherent* instability for linear boundary-value problems includes a technique for performing the computations, exemplified in the equations (2.65), (2.66) and (2.67) and the associated discussion. There may, however, be a significant amount of *induced* instability in this method, particularly in a paradoxical way when the problem is well conditioned, and in many cases we need to use an alternative *stable method*. For problem (2.64), for example, the solution is that of the simultaneous linear equations

$$Ay = b, \tag{2.97}$$

where $A, y$ and $b$ are respectively the matrix and vectors

$$A = \begin{bmatrix} a_1 & 1 & & & \\ b_2 & a_2 & 1 & & \\ & b_3 & a_3 & 1 & \\ & & \cdots & & \\ & & & b_{N-1} & a_{N-1} \end{bmatrix}, \quad y = \begin{bmatrix} y_1 \\ y_2 \\ \cdots \\ y_{N-1} \end{bmatrix}, \quad b = \begin{bmatrix} c_1 - b_1\alpha \\ c_2 \\ \cdots \\ c_{N-1} - \beta \end{bmatrix}, \tag{2.98}$$

and the equations typified by (2.50) are obtained by writing down the recurrence relation for $r = 1, 2, \ldots, N-1$ and taking the known $y_0 = \alpha$ and $y_N = \beta$ to the right-hand side. The unknowns are then just the components of the vector $y$.

The technique outlined in equations (2.65)–(2.67) certainly solves (2.97) exactly with exact arithmetic, which is all we needed for the discussion of *inherent* stability, but to avoid *induced* instability we do need other methods. A classical method is an adapted elimination method, discussed in Chapter 6, for which it can be shown that the computed solution $\bar{y}$ is the *exact* solution of the 'perturbed' problem

$$(A + \delta A)\bar{y} = b + \delta b, \tag{2.99}$$

where the perturbations $\delta A$ and $\delta b$ are respectively small compared with $A$ and $b$. The *method* is therefore stable, giving good results for well-conditioned problems, and the poorness of the solutions for ill-conditioned problems is caused only by the inherent ill-conditioning and not by the method of solution.

The analysis of ill-conditioning can also be performed usefully in terms of this matrix and vector formulation, extending the comments in the vicinity of equation (2.72). In relation to problem (2.64) the critical case arises when the $y_N^{(2)}$ of (2.66) vanishes. Recalling the definition of $y_r^{(2)}$ we note that $y_N^{(2)} = 0$ implies that the homogeneous linear equations

$$Ay^{(2)} = 0 \qquad (2.100)$$

have a solution other than $y^{(2)} \equiv 0$. If this happens the matrix $A$ is *singular*, and equations (2.97) then have no unique solution and no solution at all unless the vector $b$ has special properties. A very small $y_N^{(2)}$ causes the matrix to be nearly singular and its inverse $A^{-1}$ has at least some large elements. Uncertainties $\delta b$ in $b$ in (2.97) produce uncertainties in $y$ given by

$$A\delta y = \delta b, \quad \text{or } \delta y = A^{-1}\delta b, \qquad (2.101)$$

and large elements in $A^{-1}$ produce an ill-conditioned situation. This is at its relative worst if the solution

$$y = A^{-1}b \qquad (2.102)$$

has small values even though $A^{-1}$ has large elements, and this may arise for a vector $b$ close to that for which (2.97) has a solution, though not unique, even in the singular case.

This is perhaps most easily explained in terms of eigenvalues. If the matrix $A$ is symmetric, and has eigenvalues $\lambda_1, \lambda_2, \ldots, \lambda_{N-1}$ and corresponding eigenvectors $x_1, x_2, \ldots, x_{N-1}$, so that

$$Ax_i = \lambda_i x_i, \quad i = 1, 2, \ldots, N-1, \qquad (2.103)$$

then the solution of (2.97) is

$$y = \sum_{i=1}^{N-1} \left( \frac{x_i^{\mathsf{T}} b}{\lambda_i} \right) x_i, \qquad (2.104)$$

where the vectors are normalized so that $x_i^{\mathsf{T}} x_i = 1$. A critical case is when one eigenvalue is zero, say $\lambda_1 = 0$. The matrix is then singular and there is generally no solution. A solution exists only if $x_1^{\mathsf{T}} b = 0$, that is $b$ has rather a special form, but this solution is not unique since any multiple of $x_1$ can be added to any solution in virtue of the fact that $Ax_1 = 0$.

If $\lambda_1$ is small, (2.104) shows there will normally be a large component of $x_1$ in the solution and a sizeable component in the uncertainty

$$\delta y = \sum_{i=1}^{N-1} \left( \frac{x_i^{\mathsf{T}} \delta b}{\lambda_i} \right) x_i. \qquad (2.105)$$

The latter is always likely due to the uncertainty in size and sign of the components of $\delta b$, and this gives large absolute ill-conditioning. The relative ill-conditioning may not be serious, but it will be particularly severe in the special case when $x_1^T b$ is also quite small so that $y$ itself does not have large elements.

As an example of this effect consider the results of Table 2.6. The first two columns show large absolute but small relative ill-conditioning for (2.69) with $\alpha = 0.900$, $\beta = 6.000$, while the next two columns show the same absolute but catastrophically large relative ill conditioning for problem (2.69) with $\alpha = 0.900$, $\beta = 6.563$. The matrix is $5 \times 5$ and given by

$$A = \begin{bmatrix} -1.732 & 1 & & & \\ 1 & -1.732 & 1 & & \\ & 1 & -1.732 & 1 & \\ & & 1 & -1.732 & 1 \\ & & & 1 & -1.732 \end{bmatrix}, \quad (2.106)$$

and its smallest eigenvalue and corresponding normalized eigenvector are approximately

$$\lambda_1 = -0.0000508, \quad x_1^T = (0.288675, 0.5, 0.577350, 0.5, 0.288675). \quad (2.107)$$

The vectors $b$ in (2.97) are

$$b_1^T = (0.1, 1, 1, 1, -5); \quad b_2^T = (0.1, 1, 1, 1, -5.563) \quad (2.108)$$

corresponding to the two cases, and we find

$$x_1^T b_1 = 0.16284, \quad x_1^T b_2 = 0.00032. \quad (2.109)$$

So in the first problem the solution is large, with a large multiple of $x_1$ of something like 3000, while the second solution is small with only some $6x_1$.

There is a similar though slightly more complicated analysis for unsymmetric matrices. There also exist analogous theories for linear boundary-value problems in ordinary differential equations, which for example would explain the statements in Section 1.3 relevant to problems with no solutions or no unique solutions.

---

## EXERCISES

1. The number $x$ is stored in the computing machine as $\bar{x} = x(1 + \varepsilon)$, where $|\varepsilon| \leqslant 2^{-t}$. For $n$ small compared with $t$ show that for $x > 0$ the inherent relative error in the resulting $n$th power of $x$ is approximately $n^2$ times that of the resulting $n$th root of $x$.

2. With respect to the projectile problem, ignoring the dart board, show from equation (2.9) that $y = H$ again when $x = (V^2 \sin 2\alpha)/g$, and show that the

uncertainty in this value due to uncertainties in $g$, $V$ and $\alpha$ is

$$\delta x = -\frac{V^2}{g^2}\sin 2\alpha\,\delta g + \frac{2V\sin 2\alpha}{g}\delta V + \frac{2V^2\cos 2\alpha}{g}\delta\alpha.$$

For the data of equation (2.11) determine whether the problem for $x$ is more or less ill conditioned than that of (2.9) for $y$.

3. The result (2.91) is important because all the relevant quantities can be computed by numerical solution of the original equations and the variational equations. Here these can also be solved analytically. Do this to show that $y = H$ again when $t = T = (2V\sin\alpha)/g$, and that (2.91) reproduces the result given in Exercise 2.

4. Show that $I_r = \int_0^1 x^{2r}e^{-x^2}\,dx$ decreases monotonically as $r$ increases. Prove that $I_r$ satisfies the first-order recurrence relation

$$I_{r+1} - (r + \tfrac{1}{2})I_r = -\tfrac{1}{2}e^{-1}.$$

Show that the problem of computing $I_r$ is inherently ill conditioned if the only given condition associated with the recurrence relation is the specification of $I_0 = \int_0^1 e^{-x^2}\,dx \simeq 0.7468$. Verify this by performing some forward recurrence with $\bar{I}_0 = 0.746$ and $\bar{I}_0 = 0.747$. Estimate the approximate value of $r$ at which these two sets of results will differ by at least 1.0.

5. Suggest an accurate method for computing $I_r$ defined in Exercise 4, and perform some calculations to confirm the inherent stability of your reformulated problem.

6. Show that the general solution of the recurrence relation

$$y_{r+1} - 10y_r + 8.19y_{r-1} = 0$$

is $y_r = A(9.1)^r + B(0.9)^r$. Prove that you will fail to get the solution $y_r = (0.9)^r$ if the specified conditions are the values $y_0 = 1$ and $y_1 = 0.9$ stored in a binary computer. Verify this by some computer calculations.

7. Discuss the method of Section 2.4(iv) in the vicinity of equation (2.47) to obtain reasonable accuracy for $y_r = (0.9)^r$ defined in Exercise 6. Verify the accuracy of your method by some suitable computation.

8. For the problem of the type of Exercise 6, in which the general solution is $y_r = A_1 p_1^r + A_2 p_2^r$, where $p_1$ and $p_2$ are real and $|p_1| \gg |p_2|$, show that with the method of Exercise 7 aimed at the computation of $p_2^r$ the relative error in your computed $y_r$ is very close to $(p_2/p_1)^{N-r}$.

9. Verify that the solution of the initial-value problem

$$y_{r+1} - \frac{2r}{e}y_r + y_{r-1} = -2r\sin(r\pi/2), \quad y_0 = 0, \quad y_1 = e,$$

is $y_r = e \sin(r\pi/2)$. By performing some relevant computations demonstrate that you will fail to get this solution with the specification of the machine's values of $y_0 = 0, y_1 = e$, or of $y_{20} = 0, y_{19} = -e$.

10. Use a reputable computer linear algebraic equation solver to find solutions of the recurrence relation in Exercise 9 with the boundary conditions $y_0 = 0$, $y_{20} = 0$, and then with $y_0 = 0$, $y_{10} = 0$, demonstrating the inherent stability of this reformulation.

11. If you have a reputable computer routine for solving second-order ordinary differential equations of boundary-value type, use it to solve the problem

$$y'' + k^2 y = 1, \quad y(0) = 0, \quad y(1) = 0,$$

first for $k = 3.141$ and next for $k = 3.142$. Why are the two solutions very different in form?

12. Show that the problem

$$y'' + k^2 y = xe^x, \quad y(0) = 1, \quad y'(\pi) - ky(\pi) = 3,$$

will manifest a considerable amount of inherent instability if $k$ is near to $\frac{1}{4}$. Verify your conclusion by obtaining numerical solutions, first with $k = 0.249$ and second with $k = 0.251$.

13. Consider Exercise 10 of Chapter 1. What would be the limit of $y_{r+1}/y_r$, if the recurrence relation were used with computer arithmetic?

# 3

# Initial-value problems: one-step methods

## 3.1 INTRODUCTION

In this chapter we start our discussion of numerical methods, concentrating on the initial-value problem

$$y' = f(x, y), \quad y(a) = \alpha. \tag{3.1}$$

Our aim will be to compute approximate values of the solution at selected discrete points

$$x_0 = a, \quad x_r = x_{r-1} + h_r, \quad r = 1, 2, \ldots, \tag{3.2}$$

where $h_r$ is the *local interval size* or *local step length*. Quite often we shall take $h_r$ to be a constant $h$, independent of $r$ at least for a large range of successive values of $r$, and then (3.2) can be written

$$x_0 = a, \quad x_r = x_0 + rh, \quad r = 1, 2, \ldots. \tag{3.3}$$

We shall denote by $Y_r$ the computed approximate value of $y(x_r) = y_r$, and of course we have no hesitation in taking $Y_0 = y_0 = \alpha$.

The methods of this and the next chapter are step-by-step methods, in which we compute first $Y_1$ as an approximation to $y(x_1) = y_1$, then $Y_2$, $Y_3$, and so on. In this chapter we concentrate on the so-called *one-step* or *single-step* methods, in which the value of $Y_{r+1}$ is obtained using only the already computed $Y_r$ and other information easily obtained at the point $x_r$, in various ways which will be described in detail in the appropriate contexts.

In the major part of this chapter we concentrate on a simple class of finite-difference methods, examining their accuracy, stability, application to single first-order equations and to a system of such equations, both linear and non-linear, and methods of improving the accuracy of first approximations.

The final two sections discuss two one-step methods which are not of finite-difference type and are essentially relevant to the use of Taylor series at the discrete points. Accuracy and stability of these methods are also considered.

## 3.2 THREE POSSIBLE ONE-STEP METHODS
## (FINITE-DIFFERENCE METHODS)

One-step methods may not have any real context with finite differences, but the three methods of this section have relations with multi-step methods, which are realistically described as finite-difference methods and which are treated in the next chapter. Even in the one-step methods, however, one technique for improving the approximations does use finite-difference ideas and techniques.

### (i) The forward or explicit Euler method

The simplest of all possible methods uses the formula

$$Y_{r+1} = Y_r + hf(x_r, Y_r) \tag{3.4}$$

to obtain $Y_{r+1}$ from a knowledge of $Y_r$. Starting with $Y_0 = y(a) = \alpha$ the computation is almost trivial, depending on the complexity of the function $f(x, y)$. One can interpret the relation (3.4) in various ways. For example, $(Y_{r+1} - Y_r)/h$ might be regarded as a simple approximation to the derivative of $y$ at the point $x_r$, which is $f(x_r, y_r)$ and which we approximate as well as we can with $f(x_r, Y_r)$. An alternative and more helpful idea is to think of (3.4) as representing approximately the first two terms of the Taylor-series expansion of the solution $y(x)$ about the point $x_r$, given by

$$y(x_r + h) = y(x_r) + hy'(x_r) + \frac{h^2}{2!} y''(x_r) + \cdots. \tag{3.5}$$

This idea will help us later in the analysis of error of the method. Note that both our interpretations assume that $Y_r$ is close to $y_r$.

The word 'forward' in the title is obvious, and the use of the word 'explicit' denotes that we can calculate $Y_{r+1}$ from (3.4) without having to solve any type of equation. The advantage of this becomes obvious when we consider our second method.

### (ii) The backward or implicit Euler method

If we centre the Taylor series (3.5) at the point $x_{r+1}$, and write

$$y(x_r) = y(x_{r+1} - h) = y(x_{r+1}) - hy'(x_{r+1}) + \frac{h^2}{2!} y''(x_{r+1}) - \cdots, \tag{3.6}$$

we easily produce the backward Euler formula

$$Y_{r+1} = Y_r + hf(x_{r+1}, Y_{r+1}). \tag{3.7}$$

The word 'backward' is obvious, and the method is *implicit* because $Y_{r+1}$, the quantity we are trying to find, appears on both sides of (3.7) and we have to solve an equation to get it. If $f(x, y)$ is linear, of the form

$$f(x, y) = p(x)y + q(x),\tag{3.8}$$

then the solution of (3.7) is quite easy, and indeed

$$Y_{r+1} = \frac{Y_r + hq(x_{r+1})}{1 - hp(x_{r+1})}.\tag{3.9}$$

But if $f(x, y)$ is non-linear, for example of the form

$$f(x, y) = xy^5 + p(x)y + q(x),\tag{3.10}$$

then we obtain $Y_{r+1}$ only by solving the non-linear equation

$$Y_{r+1} = Y_r + h(x_{r+1}Y_{r+1}^5 + p_{r+1}Y_{r+1} + q_{r+1}).\tag{3.11}$$

No 'formula' will do this for us, and we must use some sort of iterative method. We shall mention some possibilities later, but the solution of (3.11) for $Y_{r+1}$ obviously involves far more time and effort than the corresponding computation for the explicit Euler method, which merely requires the *evaluation* of (3.11) with all the suffixes $r + 1$ on the right-hand side replaced by the suffix $r$.

### (iii) The trapezoidal rule method

Our last technique of this class is in a sense the average of the forward and backward Euler methods, embodied in the formula

$$Y_{r+1} = Y_r + \tfrac{1}{2}h\{f(x_{r+1}, Y_{r+1}) + f(x_r, Y_r)\}.\tag{3.12}$$

It is called the trapezoidal rule method because of its close relation to the formula

$$y_{r+1} - y_r = \int_{x_r}^{x_{r+1}} y'(x)\,dx \sim \tfrac{1}{2}h\{y'(x_{r+1}) + y'(x_r)\},\tag{3.13}$$

which is the well-known trapezoidal rule for integration. This technique is also obviously implicit and shares this particular disadvantage, as far as computing economy is concerned, with the backward Euler method.

### 3.3 ERROR ANALYSIS: LINEAR PROBLEMS

Some aspect of the quality of any technique is provided by what is called the *local truncation error*, which is the difference between the left- and right-hand sides of the relevant formula (3.4), (3.7) or (3.12) when the correct solution is used instead of the approximate solution. With obvious notation we have

$$\left.\begin{aligned}
T_{r+1} &= y_{r+1} - (y_r + hy_r') && \text{(forward Euler)}\\
T_{r+1} &= y_{r+1} - (y_r + hy_{r+1}') && \text{(backward Euler)}\\
T_{r+1} &= y_{r+1} - \{y_r + \tfrac{1}{2}h(y_r' + y_{r+1}')\} && \text{(trapezoidal rule)}
\end{aligned}\right\},\tag{3.14}$$

where we have temporarily replaced $f(x, y)$ by its equivalent $y'(x)$.

The local truncation error is not, of course, the global error of our final result, which we define as

$$e_r = y(x_r) - Y_r = y_r - Y_r, \tag{3.15}$$

but we can get some idea of the relation between $e_r$ and $T_r$ by analysing a simple but important case, when

$$f(x, y) = \lambda y + g(x) \tag{3.16}$$

for constant $\lambda$. Now the computation of $Y_{r+1}$ will involve a rounding error $\varepsilon_{r+1}$, so that for the forward Euler method we must write

$$Y_{r+1} = Y_r + h(\lambda Y_r + g_r) - \varepsilon_{r+1}. \tag{3.17}$$

The first of (3.14) is

$$y_{r+1} = y_r + h(\lambda y_r + g_r) + T_{r+1}, \tag{3.18}$$

and subtraction of (3.17) from (3.18) gives the recurrence relation

$$e_{r+1} = (1 + h\lambda)e_r + T_{r+1} + \varepsilon_{r+1} \tag{3.19}$$

for the required error $e_r$.

As we noticed in the sensitivity analysis of Chapter 2, we may not know exactly or be able to store exactly the number $Y_0 = y(x_0) = y(a) = \alpha$, so that we must take $e_0$ to be $\varepsilon_0$, and then subsequent errors are related to the local truncation and rounding errors according to the formulae

$$\left.\begin{aligned}
e_0 &= \varepsilon_0 \\
e_1 &= (1 + h\lambda)\varepsilon_0 + T_1 + \varepsilon_1 \\
e_2 &= (1 + h\lambda)^2\varepsilon_0 + (1 + h\lambda)(T_1 + \varepsilon_1) + T_2 + \varepsilon_2 \\
&\text{etc.}
\end{aligned}\right\} \tag{3.20}$$

The corresponding analysis for the backward Euler method gives

$$e_{r+1} = (1 - h\lambda)^{-1}e_r + (1 - h\lambda)^{-1}(T_{r+1} + \varepsilon_{r+1}), \tag{3.21}$$

so that

$$\left.\begin{aligned}
e_0 &= \varepsilon_0 \\
e_1 &= (1 - h\lambda)^{-1}\varepsilon_0 + (1 - h\lambda)^{-1}(T_1 + \varepsilon_1) \\
e_2 &= (1 - h\lambda)^{-2}\varepsilon_0 + (1 - h\lambda)^{-2}(T_1 + \varepsilon_1) + (1 - h\lambda)^{-1}(T_2 + \varepsilon_2) \\
&\text{etc.}
\end{aligned}\right\} \tag{3.22}$$

For the trapezoidal rule method we find

$$e_{r+1} = (1 - \tfrac{1}{2}h\lambda)^{-1}(1 + \tfrac{1}{2}h\lambda)e_r + (1 - \tfrac{1}{2}h\lambda)^{-1}(T_{r+1} + \varepsilon_{r+1}), \tag{3.23}$$

so that

$$
\begin{aligned}
e_0 &= \varepsilon_0 \\
e_1 &= (1 - \tfrac{1}{2}h\lambda)^{-1}(1 + \tfrac{1}{2}h\lambda)\varepsilon_0 + (1 - \tfrac{1}{2}h\lambda)^{-1}(T_1 + \varepsilon_1) \\
e_2 &= (1 - \tfrac{1}{2}h\lambda)^{-2}(1 + \tfrac{1}{2}h\lambda)^2\varepsilon_0 + (1 - \tfrac{1}{2}h\lambda)^{-2}(1 + \tfrac{1}{2}h\lambda)(T_1 + \varepsilon_1) \\
&\quad + (1 - \tfrac{1}{2}h\lambda)^{-1}(T_2 + \varepsilon_2)
\end{aligned} \right\} \quad (3.24)
$$

etc.

Now in considering these expressions the only quantity at our disposal is the interval $h$ between our successive discrete points, sometimes called 'mesh points'. The choice of $h$ affects both the size of the local truncation error and also the number of steps needed to reach a fixed point $x = X$. The truncation error is best examined by using the Taylor series as in (3.5). Immediately we see from (3.5) and the first of (3.14) that

$$
T_{r+1} = \tfrac{1}{2}h^2 y''(x_r) + \tfrac{1}{6}h^3 y'''(x_r) + \cdots \quad \text{(forward Euler)}, \quad (3.25)
$$

and similar manipulation with (3.5) and the rest of (3.14) gives

$$
\begin{aligned}
T_{r+1} &= -\tfrac{1}{2}h^2 y''(x_r) - \tfrac{1}{3}h^3 y'''(x_r) + \cdots \quad \text{(backward Euler)} \\
T_{r+1} &= \qquad\qquad\qquad -\tfrac{1}{12}h^3 y'''(x_r) + \cdots \quad \text{(trapezoidal rule)}
\end{aligned} \right\} \quad (3.26)
$$

For sufficiently small $h$ the first term in the truncation error will dominate, and we say that the $T_{r+1}$ in (3.25) and the first of (3.26) are of order $h^2$, and that in the second of (3.26) is of order $h^3$. For a fixed small $h$ we should obviously expect to get better answers with a technique with smaller truncation errors, and particularly with local truncation errors of higher orders, so that here we should expect the two Euler methods to be comparable and the trapezoidal rule method to be significantly better.

With any method, moreover, we should expect the result to improve as the interval $h$ is made smaller. This is not quite so obvious, since even though a smaller $h$ gives a smaller $T_r$, it also increases the number of steps to reach the point $X$, this number being $(X - a)/h$. Nevertheless, we should intuitively expect that a local truncation error $O(h^p)$ would produce, after a number of steps proportional to $h^{-1}$, a global error $O(h^{p-1})$, so that the global error for both Euler methods would be $O(h)$ and that of the trapezoidal rule method $O(h^2)$. These results have been proved rigorously in more advanced mathematical literature.

Notice, however, that the individual rounding errors do not get smaller as $h$ decreases, and with an increasing number of steps to reach a given point their accumulated effect is likely to be greater as $h$ decreases. Normally, of course, the local truncation error is much larger than a rounding error, so that the latter is not in general a serious problem.

Notice also that the first term in all our error equations, caused by an uncertainty in $y(a)$ or an error in its stored value, measures the technique's estimate of the sensitivity error discussed in Chapter 2. Moreover, as the interval becomes smaller and approaches zero, we see that for the three methods we have considered the sensitivity errors at the point $X$ for which $X - a = nh$ are respectively

$$
\left.
\begin{aligned}
e_n &= (1 + h\lambda)^n \varepsilon_0 = (1 + h\lambda)^{(X-a)/h} \varepsilon_0 \quad &\text{(forward Euler)} \\
e_n &= (1 - h\lambda)^{-n} \varepsilon_0 = (1 - h\lambda)^{-(X-a)/h} \varepsilon_0 \quad &\text{(backward Euler)} \\
e_n &= \left(\frac{1 + \frac{1}{2}h\lambda}{1 - \frac{1}{2}h\lambda}\right)^n \varepsilon_0 = \left(\frac{1 + \frac{1}{2}h\lambda}{1 - \frac{1}{2}h\lambda}\right)^{(X-a)/h} \varepsilon_0 \quad &\text{(trapezoidal rule)}
\end{aligned}
\right\}, \quad (3.27)
$$

and in every case it is known mathematically that as $h \to 0$

$$
e_n = e(X) \to e^{\lambda(X-a)} \varepsilon_0. \tag{3.28}
$$

This is precisely the result proved in Chapter 2, the sensitivity error being the appropriate multiple of the complementary solution which is here $e^{\lambda x}$.

The actual size of $h$ which produces a reasonably small local truncation error depends, of course, on the solution to the problem. If the latter is dominated by the complementary solution we need to consider what sort of interval will give a small local truncation error for the function $e^{\lambda x}$, and (3.25) and (3.26) indicate that $\lambda h$ must be quite small in absolute value, so that corresponding to a large $|\lambda|$ we need a small $h$. This is also reasonably intuitively obvious, a function with very rapid change, not only in its value but in all its derivatives, needing 'sampling' at quite adjacent points for its proper appreciation.

The solution to the problem

$$
y' = y - (1+x)^{-2} - (1+x)^{-1}, \quad y(0) = 2, \tag{3.29}
$$

is

$$
y = e^x + (1+x)^{-1}. \tag{3.30}
$$

**Table 3.1**

| | | \multicolumn{2}{c}{*FE*} | | \multicolumn{2}{c}{*BE*} | | \multicolumn{2}{c}{*TR*} |
|------|----------|-----------|------------|-----------|------------|-----------|------------|
| $x$ | $y$ | $h=0.1$ | $h=0.05$ | $h=0.1$ | $h=0.05$ | $h=0.1$ | $h=0.05$ |
| 0.0 | 2.00000 | 0.00000 | 0.00000 | 0.00000 | 0.00000 | 0.00000 | 0.00000 |
| 0.1 | 2.01426 | 0.01426 | 0.00723 | −0.01512 | −0.00745 | 0.00034 | 0.00009 |
| 0.2 | 2.05474 | 0.02829 | 0.01439 | −0.03038 | −0.01491 | 0.00058 | 0.00015 |
| 0.3 | 2.11909 | 0.04278 | 0.02181 | −0.04649 | −0.02274 | 0.00074 | 0.00019 |
| 0.4 | 2.20611 | 0.05826 | 0.02978 | −0.06404 | −0.03122 | 0.00086 | 0.00022 |
| 0.5 | 2.31539 | 0.07520 | 0.03854 | −0.08354 | −0.04061 | 0.00093 | 0.00023 |
| 0.6 | 2.44712 | 0.09403 | 0.04830 | −0.10551 | −0.05115 | 0.00096 | 0.00024 |
| 0.7 | 2.60199 | 0.11515 | 0.05927 | −0.13046 | −0.06308 | 0.00097 | 0.00024 |
| 0.8 | 2.78110 | 0.13900 | 0.07170 | −0.15894 | −0.07665 | 0.00094 | 0.00024 |
| 0.9 | 2.98592 | 0.16603 | 0.08582 | −0.19153 | −0.09215 | 0.00088 | 0.00022 |
| 1.0 | 3.21828 | 0.19674 | 0.10188 | −0.22888 | −0.10986 | 0.00078 | 0.00020 |

Table 3.1 illustrates the numerical solution obtained by our three methods at two different intervals, $h = 0.1$ and $h = 0.05$. The column headed $y$ is the true solution, and the other columns give the errors obtained with the forward Euler method (FE), the backward Euler method (BE) and the trapezoidal rule method (TR), the intermediate values at the smaller interval being omitted.

We see: (a) that in all cases the errors are smaller with the smaller interval; (b) that the two Euler methods give very similar errors but with opposite signs and with the former slightly superior; (c) for both Euler methods the errors with $h = 0.05$ are about half of those at the same values of $x$ with $h = 0.1$; (d) the trapezoidal rule results are considerably better at both intervals; and (e) the halving of the interval in this case has divided the errors by a much larger factor very nearly equal to four.

Point (a) is our intuitive expectation. Point (b) would be expected from (3.25) and the first of (3.26), which show that the local truncation errors are of opposite sign, with those of the forward Euler method rather smaller in this example, and from the coefficients in (3.20) and (3.22) which multiply these local errors and which are very nearly equal for $h\lambda = 0.1$ and 0.05. Point (c) follows from our expectation that the order of the error is a multiple of $h$ in the Euler methods, so that if $h$ is halved so approximately is the error. Point (d) stems from the second of (3.26), which shows that the local truncation error of the trapezoidal rule method is significantly smaller than that of the Euler methods, and from equation (3.24) which shows that the multipliers are about the same as those of the other methods for small $h$. Point (e) follows from the fact that we now expect the final error to be of order $h^2$, so that a halving of $h$ approximately divides the error by four.

### 3.4 ERROR ANALYSIS AND TECHNIQUES FOR NON-LINEAR PROBLEMS

For a non-linear problem the analysis is rather more complicated, but there is no essential change in the conclusions. The expressions (3.25) and (3.26) for the local truncation errors of our three methods are of course unchanged, but for a general $f(x, y)$ we have to replace (3.17), for example, by

$$Y_{r+1} = Y_r + hf(x_r, Y_r) - \varepsilon_{r+1}. \tag{3.31}$$

We then write the first of (3.14) in the form

$$y_{r+1} = y_r + hf(x_r, y_r) + T_{r+1}, \tag{3.32}$$

and subtraction of (3.31) from (3.32) gives

$$e_{r+1} = e_r + h\{f(x_r, y_r) - f(x_r, Y_r)\} + T_{r+1} + \varepsilon_{r+1}. \tag{3.33}$$

Now if $y_r - Y_r = e_r$ is sufficiently small we can approximate to the second term on the right of (3.33) by the expression

$$f(x_r, y_r) - f(x_r, Y_r) \sim (y_r - Y_r)\frac{\partial f}{\partial y}(x_r, y_r), \tag{3.34}$$

and from (3.33) we produce for the error the approximate recurrence relation

$$e_{r+1} = \left\{1 + h\frac{\partial f}{\partial y}(x_r, y_r)\right\}e_r + T_{r+1} + \varepsilon_{r+1}. \tag{3.35}$$

This is a generalization of equation (3.19), in which $\partial f/\partial y$ is just the constant $\lambda$. In the equations corresponding to (3.20) the coefficients are not now powers of $(1 + h\lambda)$ but products like $(1 + h\partial f/\partial y_1)(1 + h\partial f/\partial y_2)$, where $\partial f/\partial y_1$ means $\partial f/\partial y$ evaluated at $(x_1, y_1)$, etc. The equations corresponding to (3.21) and (3.22) for the backward Euler method, and to (3.23) and (3.24) for the trapezoidal rule, are also obtained by replacing each successive $h\lambda$ by the appropriate $h\partial f/\partial y_r$.

It is worth noting, in passing, that for a more general linear equation than (3.16), in which $\lambda$ is replaced by some function $k(x)$, the analysis for the linear case follows as before, but equations (3.19), (3.21) and (3.23) are then replaced by the respective equations

$$\left.\begin{aligned}
e_{r+1} &= (1 + hk_r)e_r + T_{r+1} + \varepsilon_{r+1} \\
e_{r+1} &= (1 - hk_{r+1})^{-1}e_r + (1 - hk_{r+1})^{-1}(T_{r+1} + \varepsilon_{r+1}) \\
e_{r+1} &= (1 - \tfrac{1}{2}hk_{r+1})^{-1}(1 + \tfrac{1}{2}hk_r)e_r + (1 - \tfrac{1}{2}hk_{r+1})^{-1}(T_{r+1} + \varepsilon_{r+1})
\end{aligned}\right\}, \tag{3.36}$$

and again there are products rather than powers in the equations corresponding to (3.20), (3.22) and (3.24).

Table 3.2 shows, with the notation of Table 3.1, the results of our techniques

**Table 3.2**

| | | FE | | BE | | TR | |
|---|---|---|---|---|---|---|---|
| $x$ | $y$ | $h = 0.1$ | $h = 0.05$ | $h = 0.1$ | $h = 0.05$ | $h = 0.1$ | $h = 0.05$ |
| 0.0 | 1.00000 | 0.00000 | 0.00000 | 0.00000 | 0.00000 | 0.00000 | 0.00000 |
| 0.1 | 0.90909 | 0.00909 | 0.00477 | −0.01103 | −0.00525 | 0.00047 | 0.00012 |
| 0.2 | 0.83333 | 0.01821 | 0.00953 | −0.02207 | −0.01048 | 0.00091 | 0.00023 |
| 0.3 | 0.76923 | 0.02726 | 0.01426 | −0.03314 | −0.01570 | 0.00133 | 0.00034 |
| 0.4 | 0.71429 | 0.03616 | 0.01891 | −0.04427 | −0.02089 | 0.00173 | 0.00044 |
| 0.5 | 0.66667 | 0.04482 | 0.02346 | −0.05547 | −0.02604 | 0.00212 | 0.00053 |
| 0.6 | 0.62500 | 0.05318 | 0.02787 | −0.06676 | −0.03115 | 0.00249 | 0.00063 |
| 0.7 | 0.58824 | 0.06120 | 0.03214 | −0.07816 | −0.03620 | 0.00285 | 0.00072 |
| 0.8 | 0.55556 | 0.06883 | 0.03625 | −0.08970 | −0.04120 | 0.00320 | 0.00081 |
| 0.9 | 0.52632 | 0.07607 | 0.04018 | −0.10139 | −0.04615 | 0.00353 | 0.00089 |
| 1.0 | 0.50000 | 0.08291 | 0.04393 | −0.11327 | −0.05104 | 0.00386 | 0.00098 |

applied to the non-linear problem

$$y' = y^3 - (1 + x)^{-3} - (1 + x)^{-2}, \quad y(0) = 1, \tag{3.37}$$

whose solution is $y = (1 + x)^{-1}$. The comments following Table 3.1 are virtually unchanged for the solution of this non-linear problem.

We digress to consider two iterative techniques for solving the non-linear equations which arise at each stage in our implicit methods, and we consider especially the trapezoidal rule method. For simplicity we write (3.37) in the form

$$y' = y^3 + g(x), \tag{3.38}$$

and the trapezoidal rule formula gives for the computation of $Y_{r+1}$ the equation

$$Y_{r+1} = Y_r + \tfrac{1}{2}h(Y_{r+1}^3 + g_{r+1} + Y_r^3 + g_r), \tag{3.39}$$

with $Y_r$ known. One of the best iterative methods for solving a non-linear equation is that of Newton, which finds a zero of $\phi(Y)$ as the limit of the iterative sequence defined by

$$Y^{(s+1)} = Y^{(s)} - \phi(Y^{(s)})/\phi'(Y^{(s)}), \quad Y^{(0)} \text{ specified.} \tag{3.40}$$

If $Y^{(0)}$ is sufficiently close to a simple zero of $\phi(Y)$, and if there is no other zero very near to this, then the sequence will converge, and usually quite rapidly.

The guarantee of convergence and the number of iterations needed to produce a result to the required accuracy depend on the availability of a good first approximation, and this can be obtained quite easily by using the explicit forward Euler method for this purpose. Consider, for example, the first step in the solution of (3.37), for which with $h = 0.1$ equation (3.39) is

$$Y_1 = 0.05Y_1^3 + 0.871112, \tag{3.41}$$

rounded to six decimal places. Equation (3.40) gives

$$Y_1^{(s+1)} = Y_1^{(s)} - \{Y_1^{(s)} - 0.05Y_1^{(s)3} - 0.871112\}/(1 - 0.15Y_1^{(s)2}). \tag{3.42}$$

The required $Y_1^{(0)}$ obtained from the forward Euler method is

$$Y_1^{(0)} = 1 + 0.1(1^3 - 2) = 0.9, \tag{3.43}$$

and substitution in (3.42) gives the next estimate $Y_1^{(1)}$ of about 0.90861, which is in error by only one unit in the last figure. The next step produces the correct five-decimal value whose accuracy is also confirmed by this sequence.

Another method for solving $\phi(Y) = 0$ expresses this equation in an equivalent form

$$Y = \psi(Y), \tag{3.44}$$

and uses the iteration given by

$$Y^{(s+1)} = \psi(Y^{(s)}), \quad Y^{(0)} \text{ specified.} \tag{3.45}$$

Equation (3.39) is already of this type, so that we can use the iteration (3.45) with a $Y^{(0)}$ computed as before from a simple explicit formula. This is here given by (3.43) for (3.45), and the successive iterative steps give values 0.9, 0.90756, 0.90849, 0.90860 and 0.90862. A few more steps are needed than for the Newton method, but no differentiation is involved and the method is commonly economically more satisfactory.

This is an example of what is called a *predictor–corrector* method, the first approximation being obtained from an explicit formula, hence called the *predictor*, and successive improved values by direct substitution in the equation we are trying to solve, hence called the *corrector*. Of course we 'predict' only once, but 'correct' as many times as is necessary to achieve the required accuracy. We shall learn more about predictor-corrector methods in the next chapter, finding that with multi-step methods our explicit predicting formula, often chosen to have the same order as the corrector, can give a much better first approximation than that of our recent example.

## 3.5 INDUCED INSTABILITY: PARTIAL INSTABILITY

The next interesting question is how our methods compare when applied to the particular type of linear problem which we discussed in Chapter 2 in connection with inherent instability. Do they, in this particular context, exhibit any significant form of *induced instability*, producing answers in some sense poor even in well-conditioned problems? The particular problem we consider relates to the differential equation

$$y' = \lambda y - (1 + x)^{-2} - \lambda(1 + x)^{-1}, \qquad (3.46)$$

whose general solution is

$$y = Ae^{\lambda x} + (1 + x)^{-1}, \qquad (3.47)$$

and if we take the initial condition $y(0) = 1$ the constant $A$ is zero and the problem has the same solution $y = (1 + x)^{-1}$ for all values of $\lambda$. The arithmetic is different for different $\lambda$, however, and equations (3.19) to (3.24) indicate how successive local truncation errors and rounding errors are treated by the three methods for various $\lambda$ and a constant interval $h$. This we take to be fairly small so that the local truncation errors of the true solution are also reasonably small. In fact, with $h = 0.1$ the local truncation error for the forward Euler method decreases steadily from about 0.0091 in the first step to about 0·0014 in the last step.

We consider first the case $\lambda > 0$, the analysis of Chapter 2 showing that relevant problems of this class suffer from both absolute and relative ill-conditioning, increasingly as $\lambda$ increases. For small $h\lambda > 0$ equations (3.19) to (3.24) reveal that in all our three methods local truncation and rounding errors are multiplied by factors exceeding unity in successive steps. (We shall use the term *multiplying factor* frequently in what follows.) With $\lambda = \frac{1}{2}$ and $h = 0.1$ these factors for the forward Euler (FE) method, the backward Euler (BE) method and the trapezoidal rule (TR) method are respectively $1 + h\lambda$, $(1 - h\lambda)^{-1}$ and $(1 + \frac{1}{2}h\lambda)/(1 - \frac{1}{2}h\lambda)$, the first being 1.05 and the other two very close to this number. This is perfectly satisfactory since the complementary solution which will contribute to the complete solution for any other initial value is an increasing exponential, and in our case the relevant inherent instability in the problem is also correctly reflected in each formula.

For all large $h\lambda$ the FE method at least takes some account of the complementary solution, with a positive multiplying factor increasing with $h\lambda$, but the other methods behave rather differently. The BE factor exceeds unity and increases as $h\lambda$ increases from 0 to near 1, but at $h\lambda = 1$ the method fails. For $1 < h\lambda < 2$ the factor is negative and exceeds unity, but with decreasing size as $h\lambda$ increases; and for $h\lambda > 2$ the factor is negative but continues to decrease in size from 1 to 0 as $h\lambda \rightarrow \infty$. The TR factor exceeds unity and increases as $h\lambda$ increases from 0 to near 2, at which point the method fails. For $h\lambda > 2$ the factor is negative and larger than unity in size, the size decreasing and tending to 1 as $h\lambda \rightarrow \infty$.

These remarks reinforce our suggestions, in the paragraph preceding Table 3.1, about the size of the interval needed for an accurate solution in the presence of a contributing complementary solution of increasing exponential type. Unless $h\lambda$ is sufficiently small the multiplying factors of the two implicit methods do not exhibit the correct behaviour. The FE method will also need a small interval since its factor $1 + h\lambda$ effectively comes from the first two terms of the Taylor series for $e^{h\lambda}$, and the neglected terms are large for large $h\lambda$. The local truncation errors are also large for large $h\lambda$, and any computed result must be quite inaccurate. We have both inherent and also *induced instability*. For example, with $\lambda = 20$ and $y(0) = 1.0001$ the true solution of our problem is

$$y = 0.0001e^{20x} + (1 + x)^{-1}, \tag{3.48}$$

and at $x = 1$ its value is approximately $4.852 \times 10^4$. With $h = 0.1$ the FE method gives the value $-233.622$ at $x = 1$ and the BE method the value 0.5042.

Returning to our problem in which there is no contributing exponential solution, we give some computed results in Table 3.3. The column labelled $y$ is the solution $(1 + x)^{-1}$, and the other columns give the errors of solutions computed by the various methods for the various values of $\lambda$ quoted and for $h = 0.1$.

The results for $\lambda = \frac{1}{2}$ confirm our expectations for the FE and BE methods. The local truncation errors are of much the same size but of opposite sign, the methods have virtually the same multiplying factor and they finish up with much the same size of global error with opposite signs. For the TR method, with results for $\lambda = \frac{1}{2}$ not given in the table, the local truncation error is much smaller and the results correspondingly better, the maximum error being 0.00209 at $x = 1$.

**Table 3.3**

| | | FE | | BE | | | TR | |
|---|---|---|---|---|---|---|---|---|
| $x$ | $y$ | $(\lambda = \frac{1}{2})$ | $(\lambda = 10)$ | $(\lambda = \frac{1}{2})$ | $(\lambda = 20)$ | $(\lambda = 40)$ | $(\lambda = 100)$ | $(\lambda = 200)$ |
| 0.0 | 1.00000 | 0.00000 | 0.00000 | 0.00000 | 0.00000 | 0.00000 | 0.00000 | 0.00000 |
| 0.1 | 0.90909 | 0.00909 | 0.00909 | −0.00870 | 0.00826 | 0.00275 | −0.00010 | −0.00005 |
| 0.2 | 0.83333 | 0.01643 | 0.02507 | −0.01580 | −0.00195 | 0.00119 | 0.00008 | 0.00002 |
| 0.3 | 0.76923 | 0.02260 | 0.05548 | −0.02182 | 0.00688 | 0.00125 | −0.00018 | −0.00005 |
| 0.4 | 0.71429 | 0.02795 | 0.11519 | −0.02710 | −0.00296 | 0.00089 | 0.00023 | 0.00005 |
| 0.5 | 0.66667 | 0.03275 | 0.23377 | −0.03187 | 0.00613 | 0.00076 | −0.00037 | −0.00007 |
| 0.6 | 0.62500 | 0.03717 | 0.47032 | −0.03629 | −0.00353 | 0.00061 | 0.00053 | 0.00008 |
| 0.7 | 0.58824 | 0.04132 | 0.94295 | −0.04048 | 0.00569 | 0.00052 | −0.00081 | −0.00010 |
| 0.8 | 0.55556 | 0.04531 | 1.88781 | −0.04452 | −0.00388 | 0.00043 | 0.00121 | 0.00012 |
| 0.9 | 0.52632 | 0.04920 | 3.77725 | −0.04848 | 0.00541 | 0.00037 | −0.00182 | −0.00015 |
| 1.0 | 0.50000 | 0.05305 | 7.55588 | −0.05242 | −0.00410 | 0.00032 | 0.00272 | 0.00018 |

For large values of $\lambda$, as we prophesied, things are quite different. For $\lambda = 10$ the explicit FE method soon introduces an increasing solution which becomes quite dominant, with $1 + h\lambda = 2$ and almost exactly this factor multiplying each successive error. Note that the accurate increasing solution $e^{10x}$ would produce the factor $e^{10h} = e \sim 2.7$, so that the FE method, again as we expected, exhibits a certain amount of induced instability from this cause alone.

The BE method fails for $\lambda = 10$, but for $\lambda = 20$ we are just at the point where errors stop accumulating with factors exceeding unity, and in fact the global error is just the sum of local errors with alternating signs. With $\lambda = 40$ there is at each stage a substantial dampening of previous errors, and at some point the local truncation error plays a sufficiently dominating role to keep the global error of constant sign. In this example the alternating signs actually make the global error smaller than the local truncation error, which is as much as $-0.0013$ in the last step.

It therefore appears that for $\lambda > 20$ the BE method tries increasingly powerfully to suppress the existence (as in the results following equation (3.48)) or the inherent introduction (as in the results of Table 3.3) of the relevant exponentially increasing term. The method obviously has induced instabilities, which the results following equation (3.48) indicate, but in a real sense this suppression effect is important, giving quite accurately the results we seek even in ill-conditioned problems which lack the increasing exponential part of the general solution. The FE method obviously fails to do this, increasingly as $\lambda$ increases. As an extra striking example, with $y(0) = 1$, $\lambda = 100$ and $h = 0.05$, the BE method has a maximum error of $6.3 \times 10^{-5}$ at $x = 1$, and the FE method has its maximum error of $1.7 \times 10^{12}$ at the same point.

The TR method restricts the growth of the increasing exponential term but does not completely suppress it. With $\lambda = 100$ the early errors are quite small because the local truncation errors are quite small. The multiplying factor, however, is the non-trivial $-1.5$, and this has the dominating effect in the second half of the range. With $\lambda = 200$ the alternating signs are more potent in cancelling out the early errors, and with a multiplying factor of only $-11/9$ the global errors do not get very large in this range of $x$. Despite its much smaller local truncation error, however, the TR method is not always superior to the BE method for the case $\lambda > 0$. For the example mentioned at the end of the previous paragraph the TR method produces a maximum error at $x = 1$ of about $2.74 \times 10^{-3}$, much greater than that of the BE method.

We consider next the case $\lambda < 0$, for which the original problem is well conditioned and any errors are due to induced instability. For small $|\lambda|$ the behaviour of the methods is quite similar to that of the previous case and should produce reasonably satisfactory results. The FE method should continue to work well up to $\lambda = -10$, at which point the relevant factor $1 + h\lambda$ is zero with $h = 0.1$, and the global error at any point is just the local truncation error at that point plus a few rounding errors. As $-\lambda$ increases, however, $1 + h\lambda$ becomes negative and larger than unity in absolute value for $|h\lambda| > 2$. The factor $1 + h\lambda$ then ceases to have much in common with the true factor $e^{\lambda h}$, $\lambda < 0$, but it also has the decisive failure of exceeding unity in absolute value. The error therefore increases, rapidly for large $|h\lambda|$, and with alternating sign.

For the BE method the relevant factor is $(1 - h\lambda)^{-1}$, and with $\lambda < 0$ this is always positive, always $< 1$ and decreases as $|\lambda|$ increases. Any errors are therefore suppressed even more ruthlessly than in the case $\lambda > 0$. For the TR method the factor is $(1 + \frac{1}{2}h\lambda)/(1 - \frac{1}{2}h\lambda)$, which is also less than one in absolute value for $\lambda < 0$ and is negative for $|h\lambda| > 2$. For large $|\lambda|$ the factor is not very different from $-1$, but for any negative $\lambda$ errors do not propagate increasingly as $x$ increases, and the TR method for $\lambda < 0$ is significantly better than it is for $\lambda > 0$.

The results of Table 3.4 verify these predictions. For $\lambda = -100$ the errors in the FE method are soon multiplied by the factor $-9$, and the unlisted entries in the relevant column follow this trend. For this $\lambda$ the BE method gives very good results, with $1 - h\lambda = 11$ so that previous errors make very little contribution to the global error at any point and even the local truncation error is divided by this factor. For $\lambda = -100$ the TR method has no such large factor, but the factor is smaller than unity in size and negative. Moreover, the local truncation errors are small and they contribute with alternating signs, all of which contribute to the production of a very accurate result.

When the complementary solution is not an increasing exponential the relevant multiplying factor of any good method must not exceed unity in absolute value. For one-step methods this can happen only for an interval greater than a particular size, and when it does happen we say that the method suffers from *partial instability*. For $\lambda < 0$ the FE method has partial instability if $h > 2/|\lambda|$, while the BE method and TR method are both free from partial instability.

For an equation of type (3.46), with a negative value of $\lambda$, the initial condition

**Table 3.4**

| | | FE | | | BE | | TR |
|---|---|---|---|---|---|---|---|
| $x$ | $y$ | $(\lambda = -\frac{1}{2})$ | $(\lambda = -10)$ | $(\lambda = -100)$ | $(\lambda = -\frac{1}{2})$ | $(\lambda = -100)$ | $(\lambda = -100)$ |
| 0.0 | 1.00000 | 0.00000 | 0.00000 | 0.00000 | 0.00000 | 0.00000 | 0.00000 |
| 0.1 | 0.90909 | 0.00909 | 0.00909 | 0.00909 | −0.00787 | −0.00075 | 0.00007 |
| 0.2 | 0.83333 | 0.01552 | 0.00689 | −0.07493 | −0.01351 | −0.00064 | 0.00000 |
| 0.3 | 0.76923 | 0.02009 | 0.00534 | 0.67972 | −0.01756 | −0.00051 | 0.00003 |
| 0.4 | 0.71429 | 0.02331 | 0.00423 | −6.1133 | −0.02046 | −0.00040 | 0.00000 |
| 0.5 | 0.66667 | 0.02555 | 0.00340 | 55.0229 | −0.02251 | −0.00033 | 0.00002 |
| 0.6 | 0.62500 | 0.02705 | 0.00278 | | −0.02392 | −0.00027 | 0.00000 |
| 0.7 | 0.58824 | 0.02799 | 0.00230 | | −0.02484 | −0.00022 | 0.00001 |
| 0.8 | 0.55556 | 0.02852 | 0.00192 | | −0.02539 | −0.00019 | 0.00000 |
| 0.9 | 0.52632 | 0.02871 | 0.00162 | | −0.02564 | −0.00016 | 0.00001 |
| 1.0 | 0.50000 | 0.02866 | 0.00139 | | −0.02568 | −0.00013 | 0.00000 |

will, of course, only rarely suppress completely the decreasing exponential complementary solution. Quite commonly, however, in what are called *stiff* equations, we might have a rapidly decreasing exponential term associated with a much more slowly varying function, such as

$$y = e^{-100x} + (1 + x)^{-1}, \qquad (3.49)$$

which satisfies (3.46) with $y(0) = 2$ and $\lambda = -100$. The rapidly decreasing exponential, of course, needs a very small interval to produce accurate results, but certainly by the time we reach $x = 0.1$ this term contributes only $e^{-10} \sim 0.00005$ to the true solution, and in successive steps we should be able with confidence to obtain the succeeding dominating $(1 + x)^{-1}$ at quite a large interval. For example, starting with the correct value of (3.49) at $x = 0.1$ and with interval $h = 0.1$ we produce at $x = 1$ the satisfactory values 0.50013 and 0.50000 with the BE and TR methods respectively, whereas the FE method produces the remarkably unsatisfactory value $-2.9 \times 10^5$.

To conclude our discussion of partial instability we look at the non-linear case of equation (3.1). In the analysis of Section 3.4 we observed that in the non-linear case the term $1 + h \partial f / \partial y$ is the natural analogue of the $1 + h\lambda$ of the linear case. An interesting example, analogous to that of (3.46), is provided by the differential equation

$$y' = \lambda y^3 - \lambda(1 + x)^{-3} - (1 + x)^{-2}, \qquad (3.50)$$

which with $y(0) = 1$ has the solution $y = (1 + x)^{-1}$ for all $\lambda$. The critical quantity $h \partial f / \partial y$ is just

$$h \frac{\partial f}{\partial y} = 3h\lambda y^2 = 3h\lambda(1 + x)^{-2}, \qquad (3.51)$$

**Table 3.5**

| | | FE | | | BE | | TR | |
|---|---|---|---|---|---|---|---|---|
| $x$ | $y$ | $(\lambda=-10)$ | $(\lambda=-20)$ | $(\lambda=-25)$ | $(\lambda=-10)$ | $(\lambda=-100)$ | $(\lambda=-10)$ | $(\lambda=-100)$ |
| 0.0 | 1.00000 | 0.00000 | 0.00000 | 0.00000 | 0.00000 | 0.00000 | 0.00000 | 0.00000 |
| 0.1 | 0.90909 | 0.00909 | 0.00909 | 0.00909 | −0.00237 | −0.00032 | 0.00018 | 0.00003 |
| 0.2 | 0.83333 | −0.00634 | −0.02866 | −0.03982 | −0.00281 | −0.00030 | 0.00012 | −0.00001 |
| 0.3 | 0.76923 | 0.01231 | 0.10024 | 0.18299 | −0.00278 | −0.00028 | 0.00011 | 0.00003 |
| 0.4 | 0.71429 | −0.00497 | −0.20706 | −0.44700 | −0.00264 | −0.00026 | 0.00009 | −0.00001 |
| 0.5 | 0.66667 | 0.00609 | 0.63171 | 2.5606 | −0.00249 | −0.00024 | 0.00008 | 0.00002 |
| 0.6 | 0.62500 | 0.00082 | 0.04198 | −15.1601 | −0.00234 | −0.00022 | 0.00007 | 0.00000 |
| 0.7 | 0.58824 | 0.00216 | −0.04766 | 9817.1 | −0.00221 | −0.00021 | 0.00006 | 0.00001 |
| 0.8 | 0.55556 | 0.00185 | 0.06144 | | −0.00208 | −0.00020 | 0.00006 | 0.00000 |
| 0.9 | 0.52632 | 0.00177 | −0.03859 | | −0.00198 | −0.00019 | 0.00005 | 0.00001 |
| 1.0 | 0.50000 | 0.00169 | 0.03175 | | −0.00188 | −0.00018 | 0.00005 | 0.00000 |

which does, of course, vary with $x$. Partial stability is here in essence a local matter, and indeed for some problems $h\partial f/\partial y$ could have different signs in different ranges of $x$ even for constant $\lambda$. Here its sign is that of $\lambda$ for all $x$.

The main interest, as in the linear case, is when $\lambda$ is negative, and we observe that in this case the FE method will be free from partial instability only if

$$h \leqslant 2(1+x)^2/3|\lambda| \tag{3.52}$$

and therefore sufficiently, for positive $x$, if $h \leqslant 2/3|\lambda|$. The BE and TR methods will both be free from partial instability.

Table 3.5 gives some results for various values of negative $\lambda$, extending the results of Table 3.2 for $\lambda = 1$. The case $\lambda = -10$ for the FE method just violates the stability restriction (3.52) for $h = 0.1$, and errors begin to grow. But as $x$ grows, the right of equation (3.52) increases, (3.52) is satisfied and the errors decrease. For $\lambda = -20$ the early growth is still greater, but (3.52) is satisfied for $h = 0.1$ for $(1+x)^2 \leqslant 3$ and there is still some decrease for larger $x$. For $\lambda = -25$ the early errors grow very rapidly and soon become so great that equation (3.34) no longer holds and the size of the jump from 0.6 to 0.7 cannot be predicted by the theory stemming from (3.34) (hence the termination of the listing at that point). Indeed, such a spectacular jump could occur only in a non-linear problem. The BE results show no signs of instability, and neither do the TR results which are superior due to the smaller local truncation error of this method.

### 3.6 SYSTEMS OF EQUATIONS

Our three methods can all be applied with no essential modifications to obtain approximate solutions to a system of simultaneous first-order equations typified by the two equations

$$\frac{dx}{dt} = f(x, y, t), \qquad \frac{dy}{dt} = g(x, y, t), \tag{3.53}$$

with specified values of $x(a)$ and $y(a)$ at the same starting point $t = a$. The explicit forward Euler method has no real extra computational problem, but the implicit methods involve significantly more work, especially in non-linear problems. For example, the trapezoidal rule method for (3.53) gives rise to the equations

$$\left. \begin{array}{l} X_{r+1} = X_r + \tfrac{1}{2}h\{f(X_r, Y_r, t_r) + f(X_{r+1}, Y_{r+1}, t_{r+1})\} \\ Y_{r+1} = Y_r + \tfrac{1}{2}h\{g(X_r, Y_r, t_r) + g(X_{r+1}, Y_{r+1}, t_{r+1})\} \end{array} \right\}, \tag{3.54}$$

and the computation of $X_{r+1}$ and $Y_{r+1}$ involves the solution of two simultaneous non-linear equations. With $n$ differential equations in (3.53) each step of the trapezoidal rule method and also of the backward Euler method requires the solution of $n$ simultaneous non-linear equations. An extension of the Newton method is possible (and is described in later chapters), but it involves the computation of $n^2$ partial derivatives and the solution of $n$ simultaneous linear equations in each step of the iteration. The labour involved increases very rapidly with $n$.

Despite this, implicit methods may be worth while, especially in situations in which the forward Euler method exhibits induced (partial) instability. For a single equation we discussed partial instability with respect to (3.1) and (3.16). For the simultaneous set the analogous equation is

$$\frac{dy}{dt} = Ay + f(t), \tag{3.55}$$

where $y$ and $f$ are each vectors with $n$ components, $A$ is an $(n \times n)$ matrix of constant terms, and the $n$ components of $y$ are all specified at $t = a$. In many cases there is a non-singular matrix $M$ such that

$$M^{-1}AM = D, \tag{3.56}$$

where $D$ is a diagonal matrix whose diagonal elements are the eigenvalues of the matrix $A$. Then with

$$z = M^{-1}y, \qquad g(t) = M^{-1}f(t) \tag{3.57}$$

we can write (3.55) in the form

$$\frac{dz}{dt} = Dz + g(t), \tag{3.58}$$

where $z$ and $g(t)$ are also vectors with $n$ components. But the matrix $D$ is now diagonal, the equations (3.55) have effectively been 'uncoupled', and the component $z_i$ of $z$ is now obtainable directly from the single equation

$$\frac{\mathrm{d}z_i}{\mathrm{d}t} = \lambda_i z_i + g_i(t), \tag{3.59}$$

where $\lambda_i$ is the $i$th eigenvalue of the matrices $D$ and $A$.

The stability of the numerical method for the system of equations therefore depends on that of the method applied to (3.59) for every $i$, and this involves all the eigenvalues of the matrix $A$. If all these are real and negative, we have in particular the restriction on partial stability for the forward Euler method that

$$h < \frac{2}{\max |\lambda_i|}, \quad i = 1, 2, \ldots, n. \tag{3.60}$$

The implicit methods are, as before, free from partial instability.

We should, of course, not usually deliberately find the matrix $M$ and carry out the decoupling process which gives $n$ individual equations (3.59). Instead we should solve (3.55) in its original form, which for the implicit methods would require the solution of $n$ simultaneous linear equations at each step. If all the data in (3.55) are real we need no complex arithmetic. But even a real $A$ may have complex eigenvalues and we have not considered the question of partial instability for such a case. The most common problem, in analogy with that of real eigenvalues, is when we have a complex eigenvalue

$$\lambda = a + ib, \tag{3.61}$$

and the real part $a$ is large and negative. For the forward Euler method partial stability requires that

$$|1 + h\lambda| \leqslant 1 \Rightarrow (1 + ah)^2 + b^2 h^2 \leqslant 1. \tag{3.62}$$

This states that $h$ should be chosen so that, for each eigenvalue $\lambda$, $h\lambda$ lies in the disc in the complex plane with radius 1 and centre $-1$. The determination of an $h$ for which (3.62) is true for all eigenvalues is clearly not a trivial problem, requiring the prior computation of all the eigenvalues of the matrix $A$.

Our two implicit methods retain their freedom from partial instability, since both $|1 - h\lambda|^{-1}$ and $|(1 + \frac{1}{2}h\lambda)/(1 - \frac{1}{2}h\lambda)|$ are less than unity if $\lambda$ is defined by (3.61) and $a$ is negative.

## 3.7 IMPROVING THE ACCURACY

The results obtained with a particular interval $h$ will almost certainly have some sort of error, and we would like to be able to find a reasonable estimate for this and indeed to produce more accurate results without effectively changing the method. One method that is used from time to time performs the same computations with successively decreasing sizes of interval, observing the rate of convergence to the true results and especially accelerating this rate by rather easy operations. In the next section we discuss this method with some numerical examples. Its title, 'The deferred approach to the limit', hardly needs any explanation.

Another method, now called the method of deferred correction, improves the first result by using it to get an approximation to neglected local truncation errors and hence correcting the results with an iterative method without changing the size of the interval. This method, however, is hardly ever used with initial-value problems, being reserved for the corresponding solution of boundary-value problems. We shall therefore deal with it in more detail in Chapter 6, giving in the additional notes in Section 3.9 some indications of how the method could be used in the current context.

### The deferred approach to the limit

We have already inferred, by inspection of Tables 3.1 and 3.2, that for sufficiently small $h$ the global errors are proportional to $h$ for the Euler methods and to $h^2$ for the trapezoidal rule method. In successive solutions in which the interval is successively halved we would therefore expect the gobal error, at mesh points in common in the successive solutions, to decrease at each stage by a factor two for the Euler methods and a factor four for the TR methods. Of course we know the global errors in Tables 3.1 and 3.2 only because we have the analytical solutions, but they are easily assessed. For at a point $x = X$ which is a mesh point for all selected intervals, and if $Y(X, h)$ means the approximate value obtained at $X$ with interval $h$, we have the approximations

$$y(X) - Y(X, h) = Ah, \qquad y(X) - Y(X, \tfrac{1}{2}h) = \tfrac{1}{2}Ah \qquad (3.63)$$

for the Euler methods, so that the current global error is estimated by

$$y(X) - Y(X, \tfrac{1}{2}h) = Y(X, \tfrac{1}{2}h) - Y(X, h). \qquad (3.64)$$

In Table 3.1, at $X = 1$ with $h = 0.1$, $h = 0.05$, it is not difficult to see that this value is 0.09486 for the FE method, so that we should have to take something like 13 more interval halvings (since $2^{13} = 8192$), to reduce the global error at this point to about one unit in the fifth decimal.

With the trapezoidal rule method the right-hand sides of (3.63) are respectively $Ah^2$ and $\tfrac{1}{4}Ah^2$, and (3.64) becomes

$$y(X) - Y(X, \tfrac{1}{2}h) = \tfrac{1}{3}\{Y(X, \tfrac{1}{2}h) - Y(X, h)\}. \qquad (3.65)$$

At the same $X$-place in Table 3.1 this value is about 0.00019, so that only two more interval halvings are needed to reduce the error to about $10^{-5}$.

In all relevant cases, however, the rate of convergence can be improved with the use of a little extra knowledge and an almost trivial amount of extra computation. For the Euler methods it can be shown that if the differential equation and its solution are sufficiently smooth then the error at $X$ depends on the interval $h$ according to the formula

$$y(X) = Y(X, h) + hu_1(X) + h^2u_2(X) + \cdots, \tag{3.66}$$

where the $u_i(X)$ are certain functions independent of $h$ whose nature need not concern us. Now if we repeat the computation with interval $\frac{1}{2}h$ we can eliminate the term in $u_1(X)$ and find

$$y(X) = 2Y(X, \tfrac{1}{2}h) - Y(X, h) - \tfrac{1}{2}h^2u_2(X) + \cdots. \tag{3.67}$$

Whereas $Y(X, \frac{1}{2}h)$ was our previous estimate for $y(X)$, with error $O(h)$, we now have the better estimate

$$Y(X, \tfrac{1}{2}h, h) = Y(X, \tfrac{1}{2}h) + \{Y(X, \tfrac{1}{2}h) - Y(X, h)\} \tag{3.68}$$

with error $O(h^2)$.

We can extend this idea as far as we like. Taking the interval $\frac{1}{4}h$, and rewriting (3.67) appropriately, we can now eliminate the term in $u_2(X)$ and find a still better estimate

$$Y(X, \tfrac{1}{4}h, \tfrac{1}{2}h, h) = Y(X, \tfrac{1}{4}h, \tfrac{1}{2}h) + \tfrac{1}{3}\{Y(X, \tfrac{1}{4}h, \tfrac{1}{2}h) - Y(X, \tfrac{1}{2}h, h)\} \tag{3.69}$$

with error $O(h^3)$. The behaviour of our new sequences of values $Y(X, h)$, $Y(X, \frac{1}{2}h, h)$, $Y(X, \frac{1}{4}h, \frac{1}{2}h, h)\ldots$ gives most of the information we need about the correct solution.

To show how it works, consider the values at $x = X = 1$ of the approximations to the solution of problem (3.29) for the forward Euler method. For $h = 0.2, 0.1$ and 0.05 they are 2.84996, 3.02154 and 3.11640. The way our process accelerates the convergence to the correct answer, which is 3.21828 to five decimals, is indicated in Table 3.6. Our best approximation has an error of 0.00097, a result as good as we would obtain with the unrefined method for $h$ as small as 0.0004.

The same kind of operation can be performed with the trapezoidal rule results with even more economy of effort. The very first approximation now has global error $O(h^2)$, and formula (3.66) is here replaced by

$$y(X) = Y(X, h) + h^2v_2(X) + h^4v_4(X) + \cdots, \tag{3.70}$$

**Table 3.6**

| $h$ | $Y(1, h)$ | $Y(1, \tfrac{1}{2}h, h)$ | $Y(1, \tfrac{1}{4}h, \tfrac{1}{2}h, h)$ |
|---|---|---|---|
| 0.2 | 2.84996 | | |
| | | 3.19312 | |
| 0.1 | 3.02154 | | 3.21731 |
| | | 3.21126 | |
| 0.05 | 3.11640 | | |
| Error | $O(h)$ | $O(h^2)$ | $O(h^3)$ |

the symmetry of the TR formula effectively suppressing the odd powers of $h$ in (3.66). The elimination of $v_2(X)$, from results at intervals $h$ and $\frac{1}{2}h$, gives corresponding to (3.68) the expression

$$Y(X, \tfrac{1}{2}h, h) = Y(X, \tfrac{1}{2}h) + \tfrac{1}{3}\{Y(X, \tfrac{1}{2}h) - Y(X, h)\}, \qquad (3.71)$$

which differs from $y(X)$ by an amount $O(h^4)$, and the elimination of both $v_2(X)$ and $v_4(X)$ from (3.70), from approximations at intervals $h, \frac{1}{2}h, \frac{1}{4}h$ gives the result

$$Y(X, \tfrac{1}{4}h, \tfrac{1}{2}h, h) = Y(X, \tfrac{1}{4}h, \tfrac{1}{2}h) + \tfrac{1}{15}\{Y(X, \tfrac{1}{4}h, \tfrac{1}{2}h) - Y(X, \tfrac{1}{2}h, h)\}, \qquad (3.72)$$

with error $O(h^6)$ compared with $y(X)$. Table 3.7 gives the results for the TR method for $x = X = 1.0$ corresponding to those of Table 3.6 for the FE method, showing now how very quickly the process converges to the correct result 3.21828 to five decimals.

It is worth noting that the method of the deferred approach to the limit must be used with some care, particularly in the early part of tables like Table 3.6 and Table 3.7. To some extent we are relying on *consistency* to guarantee accurate results, and occasionally we can be deceived by 'inaccurate consistency'. Consider, for example, the solution by the TR method of the problem

$$y' = -y + k\cos kx + \sin kx, \quad y(0) = 0, \quad k = 1.2454. \qquad (3.73)$$

At $x = 2$, with $h = 1$ and $0.5$, the first few results are those shown in Table 3.8.

**Table 3.7**

| $h$ | $Y(1, h)$ | $Y(1, \tfrac{1}{2}h, h)$ | $Y(1, \tfrac{1}{4}h, \tfrac{1}{2}h, h)$ |
|---|---|---|---|
| 0.2 | 3.21533 | | |
| | | 3.21822 | |
| 0.1 | 3.21750 | | 3.21827 |
| | | 3.21827 | |
| 0.05 | 3.21808 | | |
| Error | $O(h^2)$ | $O(h^4)$ | $O(h^6)$ |

**Table 3.8**

| $h$ | $Y(2, h)$ | $Y(2, \tfrac{1}{2}h, h)$ |
|---|---|---|
| 1.0 | 0.6081 | |
| | | 0.6081 |
| 0.5 | 0.6081 | |
| Error | $O(h^2)$ | $O(h^4)$ |

Here we have some consistency, but the correct answer at $x = 2$ is 0.6058 and not 0.6081. Experience would tell us that the interval 0.5 is too large to give adequate representation of terms like $\cos 1.2454x$ and $\sin 1.2454x$ and the complementary function $e^{-x}$ which might be part of the answer. Extending the table a little will provide a later consistency, and this is much more likely to be accurate.

It is worth remarking that in Tables 3.6 and 3.7 the computed $Y(1, h)$ have been rounded to the figures quoted, and successive numbers have been computed with further roundings to five decimal places. Exercise 11 considers the errors from this source in various parts of the table. Normally we should compute everything to the word length of the machine, rounding only the final results to the required number of figures. It should also be noted that not all machines with a word length of $t$ digits would produce exactly the same results. This depends on several factors, including the precise rounding arrangements of the machine and also the order in which the arithmetic is performed.

### 3.8 MORE ACCURATE ONE-STEP METHODS

The one-step methods we have so far considered, without the correcting devices of the last section, have a fixed order of accuracy. We now consider two methods of a different class, which are also one-step but which at any stage take into account much more computable information at the previous mesh point, and which as a result have as much accuracy, effectively at any interval, as we are prepared to work for. Both methods are closely related to the existence and use of the Taylor series.

### (i) The Taylor series method

The first method, in fact, merely computes the Taylor series at successive mesh points and uses the information to extend successively the step-by-step process. Consider, for example, the problem given by

$$y' = x - y^2, \quad y(1) = 1. \tag{3.74}$$

The Taylor series at $x = 1$ gives for the value of $y_1 = y(1 + h)$ the series

$$y_1 = y(1) + hy'(1) + \frac{h^2}{2!} y''(1) + \frac{h^3}{3!} y'''(1) + \frac{h^4}{4!} y^{iv}(1) + \cdots$$

$$= y_0 + hy_0' + \frac{h^2}{2!} y_0'' + \frac{h^3}{3!} y_0''' + \frac{h^4}{4!} y_0^{iv} + \cdots \tag{3.75}$$

in our usual notation. The value of $y_0$ is specified, so that we can compute $y_0'$ directly from the differential equation, and so far we have done nothing that was not included in our previous methods.

Now, however, we have to do different things, obtaining higher derivatives by successive differentiation of the original differential equation. We have

$$y'' = 1 - 2yy', \quad y''' = -2(y')^2 - 2yy'', \quad y^{iv} = -6y'y'' - 2yy''', \quad \text{etc.} \quad (3.76)$$

and we easily compute, for example, the values

$$y_0' = 0, \quad y_0'' = 1, \quad y_0''' = -2, \quad y_0^{iv} = 4, \quad \text{etc.,} \quad (3.77)$$

so that

$$y_1 = 1 + \frac{h^2}{2!} - \frac{2h^3}{3!} + \frac{4h^4}{4!} + \cdots. \quad (3.78)$$

Except in the neighbourhood of singularities the Taylor series will converge for a significant range of $h$, and we have to decide whether to use a reasonably large $h$ and perform much differentiation or to limit $h$ and hence the number of terms needed in (3.78) to achieve any accuracy we require. With $h = 0.2$ we have to five decimals

$$y_1 = 1 + 0.02 - 0.00267 + 0.00027 + \cdots, \quad (3.79)$$

and the terms are decreasing sufficiently rapidly to guarantee an accurate four-decimal result without computing any more derivatives. With $h = 1$, on the other hand, the term in $y^{vii}$ is still contributing some $-0.07$ to the solution and many more terms are needed to guarantee a four-decimal result.

To progress beyond $x = 1.2$, with $h = 0.2$ and a similar amount of work at each step, we decide that we have a new problem, with the differential equation in (3.74) but with the initial condition $y(1.2) = \alpha$, where $\alpha$ is the number already computed from (3.79). Here we use the Taylor series at $x = 1.2$, with $y(1.2)$ given, $y'(1.2)$ calculable from the differential equation and higher derivatives from successively differentiated forms thereof. At the point $x = 1{\cdot}2$, for example, we compute the values

$$y = 1.01760, \quad y' = 0.16449, \quad y'' = 0.66523, \quad y''' = -1.40799,$$
$$y^{iv} = 2.20900, \quad \text{etc.,} \quad (3.80)$$

giving the series

$$y(1.2 + h) = 1.01760 + h(0.16449) + \frac{h^2}{2}(0.66523) - $$

$$- \frac{h^3}{6}(1.40799) + \frac{h^4}{24}(2.20900) + \cdots$$

$$= 1.01760 + 0.03290 + 0.01330 - 0.00188 + 0.00015 + \cdots \quad (3.81)$$

with $h = 0.2$, again good enough for four-place accuracy. An interesting check here is to take $h = -0.2$ at this stage, hoping to reproduce the computed value at the previous point, here the known value at $x = 1$, thereby confirming the accuracy of our work. Indeed, with the adjusted signs (3.81) produces the excellent four-decimal value 1.0000.

The principle of the method and even its numerical detail are really quite obvious for quite a variety of problems. The treatment of simultaneous first-order equations with specified initial values is perfectly straightforward, and the treatment of higher-order equations is a fairly obvious extension. For a second-order equation with specified function and first derivative at the starting point we merely use the Taylor series for both function and derivative at the next mesh point. For example, with

$$y'' = xy^2, \quad y(0) = 1, \quad y'(0) = 0 \tag{3.82}$$

we can treat the problem as a pair of simultaneous first-order equations

$$y' = z, \quad z' = xy^2, \quad y(0) = 1, z(0) = 0, \tag{3.83}$$

or treat it directly as follows. From (3.82) we can calculate $y''$ directly, then $y'''$, $y^{iv}$, etc., from successive differentiation of the differential equation, and for example

$$y''' = y^2 + 2xyy', y^{iv} = 4yy' + 2x(y')^2 + 2xyy'',\ldots. \tag{3.84}$$

We then calculate both the function and the first derivative at the next mesh point from the series

$$\left.\begin{aligned} y(x_r + h) &= y(x_r) + hy'(x_r) + \frac{h^2}{2!} y''(x_r) + \cdots \\ y'(x_r + h) &= y'(x_r) + hy''(x_r) + \frac{h^2}{2!} y'''(x_r) + \cdots \end{aligned}\right\}, \tag{3.85}$$

and start the whole process again from the point $x_{r+1} = x_r + h$ with specified value and first derivative.

The error analysis for the special linear problem (3.16) follows the lines of that of the forward Euler method, and the quantity corresponding to the $(1 + h\lambda)$ of (3.19) and (3.20) for that method is

$$p = 1 + h\lambda + \frac{h^2}{2!} \lambda^2 + \cdots + \frac{h^n}{n!} \lambda^n, \tag{3.86}$$

assuming that we compute and use up to the $n$th derivative in the Taylor series. In particular, for a negative $\lambda$ we are concerned with the partial stability property of the method for stiff equations, and we decide that we have such stability only if $|p| \leqslant 1$. For $n = 1$, of course, we have exactly the forward Euler method, for $n = 2$ we find the restriction $|h\lambda| \leqslant 2$, and for $n = 4$ we have $|h\lambda| \leqslant 2.78$ approximately.

### (ii) Runge–Kutta methods

The main disadvantage of the Taylor series method is the need for successive differentiation, which may be substantial for large non-linear systems, and this is avoided with the Runge–Kutta schemes, which for the differential equation (3.1) replace the computation of successive derivatives by successive evaluations of the function $f(x, y)$ for different arguments.

For example, the second derivative of $y$ defined by (3.1) is

$$y'' = \frac{\partial f}{\partial x} + \frac{\partial f}{\partial y}\frac{dy}{dx} = \frac{\partial f}{\partial x} + f\frac{\partial f}{\partial y}, \tag{3.87}$$

and we can approximate to this function by using the truncated Taylor series, with $f$ occasionally written for $f(x, y)$, given by

$$f(x + h, y + hf) \sim f(x, y) + h\left(\frac{\partial f}{\partial x} + f\frac{\partial f}{\partial y}\right) = f(x, y) + hy''. \tag{3.88}$$

Then we have the similarly truncated Taylor series

$$y(x + h) \sim y(x) + hy'(x) + \frac{h^2}{2!}y''(x)$$

$$\sim y(x) + hf(x, y) + \frac{h}{2}\{f(x + h, y + hf) - f(x, y)\}$$

$$\sim y(x) + \frac{h}{2}\{f(x, y) + f(x + h, y + hf)\}. \tag{3.89}$$

The corresponding *second-order* Runge–Kutta method is usually represented by the formulae

$$k_1 = hf(x, y), \quad k_2 = hf(x + h, y + k_1), \quad y(x + h) \sim y(x) + \tfrac{1}{2}(k_1 + k_2). \tag{3.90}$$

Like the second-order Taylor series the local truncation error is $O(h^3)$, though the precise nature of this error is completely different in form from that of the Taylor series. In fact it can be written in the form

$$T_{r+1} = -\tfrac{1}{12}h^3\left(y_r''' - 3\frac{\partial f_r}{\partial y}y_r''\right) + O(h^4), \tag{3.91}$$

so that the local truncation error depends explicitly not only on the solution but also on the differential equation.

It is possible to construct Runge–Kutta methods of higher orders, and probably the one most used is a fourth-order method given by

$$\left.\begin{array}{l} k_1 = hf(x, y), \quad k_2 = hf(x + \tfrac{1}{2}h, y + \tfrac{1}{2}k_1), \quad k_3 = hf(x + \tfrac{1}{2}h, y + \tfrac{1}{2}k_2) \\ k_4 = hf(x + h, y + k_3), \quad y(x + h) \sim y(x) + \tfrac{1}{6}(k_1 + 2k_2 + 2k_3 + k_4) \end{array}\right\}, \tag{3.92}$$

with local truncation error $O(h^5)$. We say 'a' fourth-order method since there are several such methods, each involving four evaluations of $f(x, y)$ corresponding to the evaluation of four derivatives in the fourth-order Taylor series method. Beyond this order the formulae become more complicated, and more than one extra function evaluation may be needed to give an improvement in the order. For example, six evaluations are needed to give a method of order five.

The methods are not applicable directly to equations of higher order, but they can be applied to a corresponding system of first-order equations, and for example the formulae corresponding to (3.92) for the pair of first-order equations

$$y' = f(x, y, z), \quad z' = g(x, y, z) \tag{3.93}$$

are given by

$$
\left.
\begin{aligned}
k_1 &= hf(x, y, z) & m_1 &= hg(x, y, z) \\
k_2 &= hf(x + \tfrac{1}{2}h, y + \tfrac{1}{2}k_1, z + \tfrac{1}{2}m_1) & m_2 &= hg(x + \tfrac{1}{2}h, y + \tfrac{1}{2}k_1, z + \tfrac{1}{2}m_1) \\
k_3 &= hf(x + \tfrac{1}{2}h, y + \tfrac{1}{2}k_2, z + \tfrac{1}{2}m_2) & m_3 &= hg(x + \tfrac{1}{2}h, y + \tfrac{1}{2}k_2, z + \tfrac{1}{2}m_2) \\
k_4 &= hf(x + h, y + k_3, z + m_3) & m_4 &= hg(x + h, y + k_3, z + m_3) \\
y(x + h) &\sim y(x) + \tfrac{1}{6}(k_1 + 2k_2 + 2k_3 + k_4) \\
z(x + h) &\sim z(x) + \tfrac{1}{6}(m_1 + 2m_2 + 2m_3 + m_4)
\end{aligned}
\right\}.
$$

$$\tag{3.94}$$

The answer to the question of partial stability is, of course, precisely the same as that of the Taylor series method for the same order. For the equation $y' = \lambda y$ we have from (3.92) the expressions

$$
\left.
\begin{aligned}
k_1 &= h\lambda y, \quad k_2 = h\lambda(y + \tfrac{1}{2}h\lambda y) = h\lambda y + \tfrac{1}{2}h^2\lambda^2 y \\
k_3 &= h\lambda(y + \tfrac{1}{2}k_2) = h\lambda y + \tfrac{1}{2}h^2\lambda^2 y + \tfrac{1}{4}h^3\lambda^3 y \\
k_4 &= h\lambda(y + k_3) = h\lambda y + h^2\lambda^2 y + \tfrac{1}{2}h^3\lambda^3 y + \tfrac{1}{4}h^4\lambda^4 y
\end{aligned}
\right\}, \tag{3.95}
$$

and then from the last of (3.92) we have

$$y(x + h) \sim (1 + h\lambda + \tfrac{1}{2}h^2\lambda^2 + \tfrac{1}{6}h^3\lambda^3 + \tfrac{1}{24}h^4\lambda^4)y, \tag{3.96}$$

which is exactly the fourth-order Taylor series formula.

It follows that explicit Runge–Kutta methods are somewhat inefficient for stiff equations. We have seen that with one-step finite-difference methods only the implicit BE and TR methods are completely free from partial instability, and it is natural to investigate the possibility of finding some kinds of implicit Runge–Kutta methods which are either completely stable or have a less restrictive interval size than the explicit formula. The trapezoidal rule formula can itself be regarded as an implicit Runge–Kutta formula of second order, with the simple rewriting

$$k_1 = hf(x, y), \quad k_2 = hf(x + h, y + \tfrac{1}{2}k_1 + \tfrac{1}{2}k_2), \quad y(x + h) \sim y(x) + \tfrac{1}{2}(k_1 + k_2). \tag{3.97}$$

Here the number $k_2$ rather than $y(x + h)$ is obtained implicitly from the second of (3.97), but the arithmetic is essentially identical and (3.97) has no stability problems.

The full theory of generalized implicit Runge–Kutta methods is rather elaborate, and here we mention just a few examples to illustrate the important points. We could write a more general second-order method than (3.97) in the form

$$k_1 = hf\{x + (\alpha + \beta)h, y + \alpha k_1 + \beta k_2\}, \quad k_2 = hf\{x + (\gamma + \delta)h, y + \gamma k_1 + \delta k_2\}$$
$$y(x + h) \sim y(x) + c_1 k_1 + c_2 k_2$$

$$(3.98)$$

in which the coefficients are chosen to give best accuracy and best partial stability and to satisfy other organizational conveniencies. A particular form of (3.98), produced 30 years ago, has $\alpha = \frac{1}{4}$, $\beta = \frac{1}{4} + \frac{1}{6}\sqrt{3}$, $\gamma = \frac{1}{4} - \frac{1}{6}\sqrt{3}$ and $\delta = \frac{1}{4}$, with $c_1 = c_2 = \frac{1}{2}$. Unlike the explicit case there is no obvious connection between the order of the formula and the number of the $k_i$ functions involved, and indeed when we apply the last formula to $y' = \lambda y$ we obtain the simple result

$$y(x + h) = \left(\frac{1 + \frac{1}{2}h\lambda + \frac{1}{2}h^2\lambda^2}{1 - \frac{1}{2}h\lambda + \frac{1}{2}h^2\lambda^2}\right)y(x), \tag{3.99}$$

expansion by the binomial theorem giving an agreement with (3.96) up to the term in $h^4\lambda^4$, so that this is a fourth-order method. For all $\lambda < 0$ the multiplying factor in (3.99) is smaller than unity in absolute value so that the method has unrestricted stability.

Such methods, however, involve rather a large amount of computation. In (3.98) both $k_1$ and $k_2$ occur implicitly in both the equations which define them, and for a system of $N$ simultaneous first-order differential equations we should have to solve at each step a system of $2N$ simultaneous non-linear algebraic equations to determine the relevant $k_1$ and $k_2$. A more attractive fourth-order method is typified by the scheme given here for a single equation by

$$k_1 = hf(x, y), \quad k_2 = hf(x + \frac{1}{2}h, y + \frac{1}{4}k_1 + \frac{1}{4}k_2),$$
$$k_3 = hf(x + h, y + k_2), \quad y(x + h) \sim y(x) + \frac{1}{6}(k_1 + 4k_2 + k_3)$$

$$(3.100)$$

This type of method is said to be 'semi-explicit' or 'diagonally implicit', since it requires the solution of only one set of equations for the unknown $k_2$. The other $k_i$ are defined explicitly.

We also mention in Chapter 9 an extension of the Runge–Kutta approach to 'block' methods, which deal essentially with two or more adjacent intervals as a block. These have also been investigated for some 30 years, but the most significant research in implicit Runge–Kutta and block methods is more recent, and we give some further discussion and references to this in Chapter 9. At least one of these references gives a good list of earlier relevant publications.

We should make one final point about the explicit Runge–Kutta methods. In the Taylor series method inspection of the successive computed derivatives gives us a good indication of what accuracy we have achieved with a specified order and interval, and in particular whether the latter can be increased or must be decreased to obtain the required accuracy with this order of method. There is no such indication with the explicit Runge–Kutta formulae, and what we need is some reasonably simple method of estimating the error and of adjusting the interval accordingly. An early and still popular method of this kind, due to Merson, uses a fourth-order method, somewhat different from that of (3.92) but which permits the computation of an error estimate with just one more evaluation of $f(x, y)$. The formulae are

$$k_1 = hf(x, y), \quad k_2 = hf(x + \tfrac{1}{3}h, y + \tfrac{1}{3}k_1), \quad k_3 = hf(x + \tfrac{1}{3}h, y + \tfrac{1}{6}k_1 + \tfrac{1}{6}k_2)$$
$$k_4 = hf(x + \tfrac{1}{2}h, y + \tfrac{1}{8}k_1 + \tfrac{3}{8}k_3), \quad k_5 = hf(x + h, y + \tfrac{1}{2}k_1 - \tfrac{3}{2}k_3 + 2k_4) \tag{3.101}$$

and in the original notation of Merson the numbers

$$y_4 = y_0 + \tfrac{1}{2}k_1 - \tfrac{3}{2}k_3 + 2k_4, \quad y_5 = y_0 + \tfrac{1}{6}k_1 + \tfrac{2}{3}k_4 + \tfrac{1}{6}k_5 \tag{3.102}$$

both differ from $y(x + h)$ by errors $O(h^5)$. In fact if $f(x, y)$ is linear in both $x$ and $y$ it can be shown that the respective dominant error terms are $-\tfrac{1}{120}h^5 y^{(v)}$ and $-\tfrac{1}{720}h^5 y^{(v)}$, so that in this special linear case a good estimate for the error in $y_5$ is

$$y(x + h) - y_5 \sim \tfrac{1}{5}(y_5 - y_4) = \tfrac{1}{30}(-2k_1 + 9k_3 - 8k_4 + k_5). \tag{3.103}$$

This error in the accepted $y_5$ is a measure of the local truncation error $T_1$ in the first step, though in subsequent steps it is not the global error and it is used as a control for the local step length for a required accuracy.

Consider, for example, the first step, at $h = 0.2$, of the solution of problem (3.29). We find

$$k_1 = 0, \quad k_2 = 0.036719, \quad k_3 = 0.037943, \quad k_4 = 0.055738, \quad k_5 = 0.105357 \tag{3.104}$$

and

$$y_4 = 2.05456, \quad y_5 = 2.05472. \tag{3.105}$$

With $y(0.2) = 2.05474$ we have $10^{-5}$ (2, 18) for the respective errors in $y_5$ and $y_4$, and $y_5 + \tfrac{1}{5}(y_5 - y_4)$ is in error by only one unit in the fifth decimal. The estimates here are quite reliable, even though $f(x, y)$ is not linear in $x$ in this region.

They may not be quite so reliable for problems non-linear in $y$, but they are usually reasonably satisfactory. Consider, for example, the calculation of the first step of problem (3.37), with $h = 0.2$. We find

$$k_1 = -0.2, \quad k_2 = -0.177969, \quad k_3 = -0.176042, \quad k_4 = -0.165342,$$
$$k_5 = -0.138870, \tag{3.106}$$

and

$$y_4 = 0.83338, \quad y_5 = 0.83329. \tag{3.107}$$

With $y(0.2) = 0.83333$ we have $10^{-5}(4, -5)$ for the respective errors in $y_5$ and $y_4$, and the estimated error in $y_5$, from equation (3.103), is $-2.10^{-5}$, which has the wrong sign but the right order, here of very small magnitude.

## 3.9 ADDITIONAL NOTES

*Section 3.4* In the predictor–corrector methods which use the TR formula as corrector we used an explicit one-step formula for the predictor whose local truncation error was one order poorer than that of the corrector. This is inevitable in the very first step. We could do better in subsequent steps, although with a predictor of explicit form we should have to use a two-step formula to produce a local truncation error $O(h^3)$. We shall return to this point in the next chapter.

*Sections 3.5, 3.6* It is clear that for stiff problems only the implicit variety of the one-step class of methods has any value. It will turn out in the next chapter that this is true also of the class of multi-step methods, and indeed implicit methods are essential, particularly for stiff simultaneous equations of type (3.55), where the eigenvalues of $A$ have negative real parts which vary greatly in magnitude.

This is also true of non-linear simultaneous equations of type (3.53), whose stiffness is determined by the eigenvalues of the Jacobian matrix involved in the Newton iteration. We mention this again in Section 4.4 and also in Chapter 8 in the remarks on computer programs for stiff equations.

*Section 3.7* To some extent in this chapter, and certainly in later chapters, we

**Table 3.9**

| Forward differences | | | | Backward differences | | | | Central differences | | | |
|---|---|---|---|---|---|---|---|---|---|---|---|
| $y_0$ | | | | $y_0$ | | | | $y_0$ | | | |
| | $\Delta y_0$ | | | | $\nabla y_1$ | | | | $\delta y_{1/2}$ | | |
| $y_1$ | | $\Delta^2 y_0$ | | $y_1$ | | $\nabla^2 y_2$ | | $y_1$ | | $\delta^2 y_1$ | |
| | $\Delta y_1$ | | $\Delta^3 y_0$ | | $\nabla y_2$ | | $\nabla^3 y_3$ | | $\delta y_{3/2}$ | | $\delta^3 y_{3/2}$ |
| $y_2$ | | $\Delta^2 y_1$ | | $y_2$ | | $\nabla^2 y_3$ | | $y_2$ | | $\delta^2 y_2$ | |
| | $\Delta y_2$ | | $\Delta^3 y_1$ | | $\nabla y_3$ | | $\nabla^3 y_4$ | | $\delta y_{5/2}$ | | $\delta^3 y_{5/2}$ |
| $y_3$ | | $\Delta^2 y_2$ | | $y_3$ | | $\nabla^2 y_4$ | | $y_3$ | | $\delta^2 y_3$ | |
| | $\Delta y_3$ | | | | $\nabla y_4$ | | | | $\delta y_{7/2}$ | | |
| $y_4$ | | | | $y_4$ | | | | $y_4$ | | | |

shall need to use finite differences not infrequently, and it is convenient to record some relevant elementary facts at this stage.

Table 3.9 shows a sample of the difference table for a function $y(x)$ tabulated at equidistant points $x_0, x_1, x_2,\ldots$. After the first column successive orders of differences are obtained by subtraction, for example

$$\Delta y_0 = y_1 - y_0, \quad \nabla^2 y_3 = \nabla y_3 - \nabla y_2, \quad \delta^3 y_{3/2} = \delta^2 y_2 - \delta^2 y_1, \quad (3.108)$$

and in the three tables numbers in the same positions are identical.

The differences of a polynomial of degree $n$ are all zero in orders greater than $n$, but if a tabulated polynomial has possible rounding errors of $\pm \varepsilon$ in its last figure the differences of order $n + 1$ may have corresponding errors of amount $\pm 2^{n+1}\varepsilon$, with alternating signs. A more general function whose differences first show this sort of oscillation in the $(n + 1)$th order is then regarded as behaving like a polynomial of degree $n$, and differences of orders greater than $n$ are ignored for all purposes. The degree of the 'approximating polyomial' for a general function may, of course, vary slowly throughout a range of the independent variable. The realistic degree may also depend on the number of figures we are concerned with in our final results, that is the required precision of our answers.

The difference table can then be used for various purposes. The first of these is *interpolation*, the calculation of $y(x)$ at a point somewhere in the range but not at a mesh point. The second is *integration* or *quadrature*, the evaluation of the definite integral of $y(x)$ between two particular points, which is commonly a requirement of some physical problems. The third is *differentiation*, the evaluation of first or higher derivatives of $y(x)$ at mesh points.

Interpolation may be needed for various purposes, one such purpose being relevant to the improved accuracy obtained with the method of the deferred approach to the limit. As we have observed, the better results are obtained only at points which are mesh points at all the intervals involved, and better values at other points can then be obtained by interpolation. For example, if the values $Y_{r-1}, Y_r, Y_{r+1}$ and $Y_{r+2}$ at equal intervals $h$ have errors of order $h^3$ then the estimate

$$Y_{r+\frac{1}{2}} = \tfrac{1}{16}(- Y_{r-1} + 9Y_r + 9Y_{r+1} - Y_{r+2}) \qquad (3.109)$$

also has error $O(h^3)$. More generally, and at a more general point $x_r + ph, p < 1$, we could use the formula

$$Y(x_r + ph) = Y_r + p\delta Y_{r+\frac{1}{2}} + \tfrac{1}{4}p(p - 1)(\delta^2 Y_r + \delta^2 Y_{r+1})$$
$$+ \tfrac{1}{6}p(p - 1)(p - \tfrac{1}{2})\delta^3 Y_{r+\frac{1}{2}} + \cdots, \qquad (3.110)$$

at points for which enough central differences can be calculated. Near the ends of the range we may have to use less satisfactory forward or backward difference formulae, given respectively by

$$Y(x_r + ph) = Y_r + p\Delta Y_r + \tfrac{1}{2}p(p-1)\Delta^2 Y_r + \tfrac{1}{6}p(p-1)(p-2)\Delta^3 Y_r + \cdots \quad (3.111)$$

and

$$Y(x_r - ph) = Y_r - p\nabla Y_r + \tfrac{1}{2}p(p-1)\nabla^2 Y_r - \tfrac{1}{6}p(p-1)(p-2)\nabla^3 Y_r + \cdots. \quad (3.112)$$

We shall ignore quadrature formulae, since they are not particularly relevant to our methods for solving ordinary differential equations, and they can be found in most numerical analysis texts. Formulae for at least the first derivative are needed here, however, and we mention candidates like

$$\left.\begin{aligned}
hy_r' &= (\mu\delta - \tfrac{1}{6}\mu\delta^3 + \tfrac{1}{30}\mu\delta^5 - \cdots)y_r \\
hy_r' &= (\Delta + \tfrac{1}{2}\Delta^2 - \tfrac{1}{6}\Delta^3 + \tfrac{1}{12}\Delta^4 - \tfrac{1}{20}\Delta^5 + \cdots)y_{r-1} \\
hy_r' &= (\Delta - \tfrac{1}{2}\Delta^2 + \tfrac{1}{3}\Delta^3 - \tfrac{1}{4}\Delta^4 + \tfrac{1}{5}\Delta^5 - \cdots)y_r
\end{aligned}\right\}, \quad (3.113)$$

where in the first of these the symbol $\mu$ means 'mean' or average, and for example

$$\mu\delta^3 y_r = \tfrac{1}{2}(\delta^3 y_{r-\frac{1}{2}} + \delta^3 y_{r+\frac{1}{2}}), \quad (3.114)$$

the right-hand side containing the numbers in the difference table.

The right-hand sides of (3.113) can be expressed in terms of the mesh values in what is generally called the Lagrangian form. With direct inclusion of differences of orders up to the fourth, for example, we can write

$$\left.\begin{aligned}
hy_r' &= \tfrac{1}{12}(-y_{r+2} + 8y_{r+1} - 8y_{r-1} + y_{r-2}) + \cdots \\
hy_r' &= \tfrac{1}{12}(y_{r+3} - 6y_{r+2} + 18y_{r+1} - 10y_r - 3y_{r-1}) + \cdots \\
hy_r' &= \tfrac{1}{12}(-3y_{r+4} + 16y_{r+3} - 36y_{r+2} + 48y_{r+1} - 25y_r) + \cdots
\end{aligned}\right\}. \quad (3.115)$$

It is intuitively obvious that if we want to draw an accurate tangent we would like the relevant point to be surrounded by a number of mesh points on each side. The central-difference formula, the first of (3.113), satisfies this requirement very well. The second of (3.113) involves many points on one side but only one on the other, and the third involves no points at all on one side of the tangent position. Correspondingly, we see from the explicit terms in (3.115) that the errors in $hy_r'$ caused by a rounding error of half a unit in each tabulated $y$ value can be as large respectively as $\tfrac{9}{12}$, $\tfrac{19}{12}$ and $\tfrac{64}{12}$ units. Moreover, in (3.115) the multiples of the neglected differences in (3.113) vary from very small values with central differences to quite large values with the second of the forward-difference formulae. Clearly the first of (3.113) is preferable when such a formula is possible.

We can now give a brief account of the second method quoted in Section 3.7 for the correction of our first approximate results in the solution of initial-value problems of first order. Now called 'the method of deferred correction', it can be used with virtually all techniques for which we have reasonable expressions for the local truncation error, but to be specific we consider here the forward

Euler method. Here the true solution to problem (3.1) satisfies the equations

$$y_{r+1} = y_r + hf(x_r, y_r) + T_{r+1}, \quad y_0 = \alpha, \tag{3.116}$$

and our first approximation satisfies (3.4) with $Y_0 = \alpha$.

At a small enough interval we would expect that the discrete $Y_r$ values themselves represent a smooth function not too different from $y(x)$, and we should be able to compute a 'local truncation error' for $Y_r$ which is not too different from that for $y(x_r)$. If $Y_r^{(1)}$ is the $Y_r$ of (3.4), and $T_r^{(1)}$ its 'local truncation error', we might expect to compute a better solution from the formulae

$$Y_{r+1}^{(2)} = Y_r^{(2)} + hf(x_r, Y_r^{(2)}) + T_{r+1}^{(1)}, \quad Y_0^{(2)} = \alpha. \tag{3.117}$$

Next we compute the local truncation error for $Y_r^{(2)}$, and take another step in the iteration which is given comprehensibly by the formulae

$$Y_{r+1}^{(s+1)} = Y_r^{(s+1)} + hf(x_r, Y_r^{(s+1)}) + T_{r+1}^{(s)}, \quad Y_0^{(s+1)} = \alpha, \quad s = 1, 2, \ldots. \tag{3.118}$$

It is reasonable to suppose that the $Y_r^{(s)}$ so obtained will converge to $y_r$ as $s$ increases, and indeed convergence is commonly quite rapid, acceptable solutions being produced for quite small values of $s$.

In terms of central differences the local truncation error for the FE method is easily obtained from the first of (3.14), the first of (3.113), and the fact that

$$y_{r+1} - y_r - \mu\delta y_r = \delta y_{r+\frac{1}{2}} - \tfrac{1}{2}(\delta y_{r+\frac{1}{2}} + \delta y_{r-\frac{1}{2}}) = \tfrac{1}{2}\delta^2 y_r, \tag{3.119}$$

and we obtain

$$T_{r+1}^{(s)} = (\tfrac{1}{2}\delta^2 + \tfrac{1}{6}\mu\delta^3 - \tfrac{1}{30}\mu\delta^5 + \tfrac{1}{140}\mu\delta^7 - \cdots)Y_r^{(s)}. \tag{3.120}$$

This form is commonly called the *difference correction*, with notation $c(Y_r^{(s)})$.

In the original version of the method the first approximation $Y_r^{(1)}$ is computed for all points in the range of interest, but then extended beyond both ends of the range to permit the computation of (3.120) near 'boundary' points. For the same subsequent purpose $Y_r^{(2)}$ has to be extended correctly at both ends, which requires the computation of $c(Y_r^{(1)})$ at a few external points. The total number of external values which have to be computed and corrected depends on the final value of $s$ in (3.118) and on the number of differences to be retained in (3.120). The extension should be performed with some care, incidentally, using equation (3.118) for all values of $r$, whether positive or negative. The difference correction then has the same form at every point and is a smooth function of $r$, this smoothness being of some value in the rate of convergence of the process. Note, of course, that if $r = -1$ equation (3.118) reads

$$Y_0^{(s+1)} = Y_{-1}^{(s+1)} + hf(x_{-1}, Y_{-1}^{(s+1)}) + T_0^{(s)}, \tag{3.121}$$

and this is now implicit, with a little more computation in the non-linear case, for the computation of $Y_{-1}^{(s+1)}$.

The number of differences to be retained in the computation of the difference corrections was originally decided by direct inspection of the table of differences of the relevant mesh values $Y_r^{(s)}$, the decision being a function of the behaviour of the differences, their size, and the number of accurate figures required in the final results. The last value of the iteration number $s$ was determined when in successive iterative steps the $Y_r$ had stopped changing to the required accuracy.

It is, however, difficult and time-consuming to ask a modern computer, rather than the human eye and brain, to investigate the difference table and decide which orders can be neglected. We should probably now vary this technique in line with modern research applied to the corresponding treatment of boundary-value problems which we discuss in Chapter 6. This research suggests that for our initial-value techniques we could use successive segments of the difference-correction expression in successive difference corrections. For the FE method we should take with obvious notation the expressions

$$
\left.
\begin{aligned}
c^{(1)}(Y_r^{(1)}) &= \tfrac{1}{2}\delta^2 Y_r^{(1)} \\
c^{(2)}(Y_r^{(2)}) &= (\tfrac{1}{2}\delta^2 + \tfrac{1}{6}\mu\delta^3)Y_r^{(2)} \\
c^{(3)}(Y_r^{(3)}) &= (\tfrac{1}{2}\delta^2 + \tfrac{1}{6}\mu\delta^3)Y_r^{(3)} \\
c^{(4)}(Y_r^{(4)}) &= (\tfrac{1}{2}\delta^2 + \tfrac{1}{6}\mu\delta^3 - \tfrac{1}{30}\mu\delta^5)Y_r^{(4)} \\
c^{(5)}(Y_r^{(5)}) &= (\tfrac{1}{2}\delta^2 + \tfrac{1}{6}\mu\delta^3 - \tfrac{1}{30}\mu\delta^5)Y_r^{(5)}
\end{aligned}
\right\},
\tag{3.122}
$$

the third and fifth of these being repeated (though applied to different functions) to take account of fourth and sixth differences even though their coefficients are zero. The theory suggests that with this choice the global errors are $O(h^s)$ in $Y_r^{(s)}$, and when the technique is applied to the problem of equation (3.29) at interval $h = 0.05$, some first results for which are given in Table 3.1, the maximum errors of $Y_r^{(1)}$, $Y_r^{(2)}$, $Y_r^{(3)}$, $Y_r^{(4)}$ and $Y_r^{(5)}$ are respectively $10^{-5}(10188, 234, 54, 1, 0)$ a reasonably satisfactory result.

The corresponding technique for the TR method is, of course, considerably superior. The difference corrections are of a superior order, with much smaller coefficients, and at each stage we gain two orders of global accuracy with

$$
\left.
\begin{aligned}
c^{(1)}(Y_r^{(1)}) &= -\tfrac{1}{12}\delta^3 Y_{r+\frac{1}{2}}^{(1)} \\
c^{(2)}(Y_r^{(2)}) &= (-\tfrac{1}{12}\delta^3 + \tfrac{1}{120}\delta^5)Y_{r+\frac{1}{2}}^{(2)} \\
c^{(3)}(Y_r^{(3)}) &= (-\tfrac{1}{12}\delta^3 + \tfrac{1}{120}\delta^5 - \tfrac{1}{840}\delta^7)Y_{r+\frac{1}{2}}^{(3)}
\end{aligned}
\right\},
\tag{3.123}
$$

the global errors in $Y_r^{(s)}$ being $O(h^{2s})$. Applied to the non-linear problem (3.37) this method gives, at interval $h = 0.1$, respective maximum errors in $Y_r^{(1)}$, $Y_r^{(2)}$ and $Y_r^{(3)}$ of magnitude $10^{-5}(386, 7, 0)$, an extremely good result.

Three further remarks are relevant. First, it is now unlikely that one would compute the difference tables, much less examine them, and the difference corrections would usually be calculated from the appropriate Lagrangian forms. Second, in computing external values it is dangerous to get too close to a possible singularity, for example at $x = -1$ in our current illustrations. If this appears to be needed to compute difference corrections then the interval should be reduced, at least in the neighbourhood of the starting point $x = 0$ in these cases. Third, at some interval the process may not give convergence to the required accuracy if the function is varying too rapidly in any region, again possibly near $x = 0$ in our illustrations, or may not converge satisfactorily with the use of a specified number of differences and maximum specified $s$. In both cases we should again use a smaller interval.

Section 3.8(i). The Taylor-series method is used quite commonly for the first few steps of a process in which subsequent steps are performed using very accurate finite-difference formulae described in the next chapter. Its drawback as a method in its own right is the need for a significant amount of differentiation. Some automatic routines are available but they are not easily transported from one machine range to another.

Section 3.8(ii). A fourth-order Runge–Kutta method was established and used in the very early days of automatic computation. Such methods have retained their popularity and value, for a large range of problems, for at least 30 years.

The Merson method for assessing the accuracy is more recent, and is not quite satisfactory since the formulae (3.101)–(3.103) assume that $f(x, y)$ is a linear function of $x$ and $y$ over the region concerned. Moreover, the error estimate, as here in the nonlinear case, is not always larger than the true error, and this is clearly unsatisfactory. In Chapter 8 we mention some other possibilities for error control.

### Final evaluation

The main purpose of this chapter is to introduce *ideas* using simple formulae, and which in subsequent chapters are extended to produce very accurate and economic methods. Of the methods mentioned here the explicit nature of the forward Euler method by no means compensates for its low accuracy and poor stability properties. The backward Euler method is not very accurate, but it is not at all bad for the solution of stiff equations when the non-stiff part of the solution can be approximated well at quite a large interval. The trapezoidal rule method has much better accuracy than the others and is quite superior except possibly for very stiff systems.

The deferred approach and deferred correction are rarely used for improving the accuracy of solutions of initial-value problems, due to the availability of Runge–Kutta methods and the next chapter's more accurate multi-step methods. They are used, however, for boundary-value problems, particularly with the trapezoidal rule, and we comment again on this point in Chapter 6. ·

The Taylor series method can be very good indeed, and some routines exist for specific computers which perform the whole thing automatically. These, however, are not everywhere available, and Runge–Kutta methods have generally taken over. In fact it is probable that for all problems which are not stiff the fourth-order Runge–Kutta method of equation (3.94), in virtue of its simplicity and reasonable accuracy, is used far more often than any other method for the solution of initial-value problems.

---

## EXERCISES

1. The problem $y' - 2\lambda xy = -2\lambda x(1+x)^{-1} - (1+x)^{-2}$, $y(0) = 2$, has the solution $y = e^{\lambda x^2} + (1+x)^{-1}$. Produce the equivalent of Table 3.1 for this problem for $\lambda = 1$.

2. For the problem of Exercise 1 take various positive and negative values of $\lambda$, with $y(0) = 1$, and produce the equivalents of Tables 3.3 and 3.4. Show, and verify by computation, that for the FE method $h = 0.1$ will give partial stability for $\lambda = -10$ up to $x = 1.0$, but that successively smaller intervals will be needed for larger ranges of $x$.

3. Perform the first step of the solution of problem (3.37) using (3.4) as predictor and (3.7) as corrector by (a) the Newton technique typified by (3.40) and the succeeding paragraph, and (b) the predictor-corrector method typified by (3.45) and the succeeding sentence. Take both methods to convergence, to five decimal places, with $h = 0.1$.

4. Consider the finite-difference formula $Y_{r+1} = Y_r + h(2Y'_r - Y'_{r+1})$ for the solution of problem (3.1). Show that its local truncation error is about three times that of the forward Euler method, and that its interval limit for partial stability is only one-third of that of the FE method. Partially check this result by performing some computations with this formula for problem (3.46) with $y(0) = 1$, $\lambda = -10$, and $h = 0.1$.

5. For the pair of simultaneous equations

$$\left. \begin{array}{l} y' = -12.5y + 6.3z + f_1(x) \\ z' = 5y - 2.4z + f_2(x) \end{array} \right\}$$

show that $h = 2/15$ is the largest step-length for partial stability for the forward Euler method and for the second-order Taylor series method.

6. For the problem of Exercise 5 perform some computation with different intervals for the fourth-order Runge–Kutta method defined in equations (3.94), with $f_1(x) = -(1+x)^{-2} + 18.8(1+x)^{-1}$, $f_2(x) = (1+x)^{-2} - 7.4(1+x)^{-1}$, and $y(0) = 2$, $z(0) = 1$, showing that the computations reveal partial instability for intervals exceeding 0.2 approximately.

7. Show that the exact solution of the problem of Exercise 6 is

$$y = (1+x)^{-1} + e^{0.1x}, \quad z = -(1+x)^{-1} + 2e^{0.1x}.$$

Perform some computations using the trapezoidal rule method at intervals $h = 0.4, 0.2, 0.1, 0.05$ and $0.025$ up to $x = 0.8$. Use the deferred approach to the limit to obtain the best possible approximations at $x = 0.8$, and compare them with the correct values.

8. For problem (3.29) show that successive derivatives $y^{(r)}(x)$ are obtained, for $r > 1$, from the four-term recurrence relation

$$(1 + x)^2 y^{(r+1)} + \{2r(1 + x) - (1 + x)^2\} y^{(r)} + r(r - 3 - 2x) y^{(r-1)} - - r(r - 1) y^{(r-2)} = 0.$$

(*Hint*: multiply the differential equation by $(1 + x)^2$ and differentiate $r$ times using the relevant theorem of Leibnitz.)

9. For the problem (3.29) compute $Y(0.4)$ with intervals 0.4, 0.2 and 0.1 using the second-order Taylor series method. Say what you think would be the global error for this method, on the lines of equation (3.66). Using your predictions find better values for your estimate of $Y(0.4)$. Compare these with the $O(h^3)$ error obtained by two steps of the deferred approach to the limit for the value $Y(0.4)$ obtained with the forward Euler method for $h = 0.2, 0.1$ and 0.05, the last two of which are effectively contained in Table 3.1.

10. It could be argued that in results like those of Table 3.1 the numbers computed with any of the methods could get too far from the truth for the deferred approach to the limit to give good results, for example if there is a singularity not too far from the range of interest. It may then be preferable to use the deferred approach to get the required accuracy at each step before proceeding to the next. Suppose, for example, that we want accurate results at $x = 0.2(0.2)1.0$. For the first step we would use intervals 0.2, 0.1, 0.05 and smaller intervals if necessary, then use the deferred approach to get an accurate result at $x = 0.2$ before proceeding, with the same technique, to $x = 0.4$. For the trapezoidal rule method, for example, we would calculate $Y(0.2, h)$ to be 2.05247, 2.05416 and 2.05459 for $h = 0.2, 0.1$ and 0.05 respectively, and apart from a possible small rounding error would be content with the $O(h^6)$ result 2.05473.

  Carry out this process for a few more steps, comparing your results at $x = 1$ with those obtained in Table 3.7 at $x = 1$.

11. In Table 3.7 suppose that the entries with error $O(h^2)$ have rounding errors of up to five units in the first neglected figure, and that further rounding errors of this amount are made in the computation of the numbers in the next two columns. Show that the rounding error in the numbers with errors $O(h^6)$ could be as much as two units in the fifth figure after the decimal point.

12. For the differential equation $y' = f(x)$ show that the fourth-order Runge–Kutta formulae (3.92) are equivalent to the approximate integration of $f(x)$ with Simpson's rule.

13. Write down what you think will be the Merson-type equations for the errors in the Runge–Kutta solution of the simultaneous first-order equations $y' = f(x, y, z)$, $z' = g(x, y, z)$. Perform some computations for the problem of Exercise 6 with $h = 0.1$, and compare the calculated errors with the true errors.

14. Verify the partial stability of the implicit Runge–Kutta method, defined by equations (3.98) and the coefficients following, by performing some relevant computations on the problem given by (3.46) with $\lambda = -100$, $h = 0.1$ and $y(0) = 1$. What is the interval restriction for the stability of the explicit Runge–Kutta method (3.92) applied to this problem?

15. For the problem of equation (3.74) show that the substitution $y = z'/z$ produces the linear differential equation $z'' = xz$. This equation being homogeneous, we can take $z(1)$ to be anything we like, say $z(1) = 1$, and the condition on $y$ then gives $z'(1) = 1$ and we have enough conditions for this second-order initial-value problem. Solve this with the substitution $z' = t$, giving the pair of simultaneous first-order equations $z' = t$, $t' = xz$, with initial conditions $z(1) = 1$, $t(1) = 1$, performing a few steps with all the methods of this chapter up to $x = 2$.

# 4

# Initial-value problems: multi-step methods

## 4.1 INTRODUCTION

The one-step finite-difference methods of the previous chapter, with the possible exception of the trapezoidal rule method, are basically rather too uneconomic for general use. Their local truncation error is of rather low order, with the result that to achieve good accuracy we need to use either: (a) a rather small step-by-step interval, involving many steps to cover a specified range; or (b) one of our correcting devices which involve some additional computation and more computer programming. The special one-step methods of the previous chapter, the Taylor series method and the explicit Runge–Kutta methods, do not share these disadvantages, but they involve extra computer storage, the Taylor series method involves a possibly large amount of non-automatic differentiation, and with the Runge–Kutta methods we have to compute possibly rather complicated expressions several times in each step. Moreover, both these methods have rather poor partial stability properties.

In this chapter we discuss more accurate finite-difference methods, which at their best will succeed in covering a given range with an economic number of steps and with an accuracy reasonably well determined and obtained without special correcting devices. In the discussion we pay particular attention to their stability, the need for special starting procedures, and the use and analysis of predictor–corrector methods in non-linear problems. Unlike in the one-step methods a constant interval is here very convenient, and is assumed for most of the chapter, but we also need to mention techniques for changing the size of the interval in regions in which the size of the local truncation error makes this desirable.

## 4.2 MULTI-STEP FINITE-DIFFERENCE FORMULAE

For the differential equation (3.1) the one-step finite-difference formula with local truncation error of the highest order is the trapezoidal rule, with error $O(h^3)$. This is easy to prove by considering the most general one-step finite-difference formula

$$Y_{r+1} + \alpha_1 Y_r = h(\beta_0 Y'_{r+1} + \beta_1 Y'_r), \quad Y' = f(x, Y). \tag{4.1}$$

Its local truncation error is

$$T_{r+1} = y_{r+1} + \alpha_1 y_r - h(\beta_0 y'_{r+1} + \beta_1 y'_r), \tag{4.2}$$

and expansion in Taylor series about the point $x_r$ gives

$$T_{r+1} = (1 + \alpha_1)y_r + (1 - \beta_0 - \beta_1)hy'_r + (\tfrac{1}{2} - \beta_0)h^2 y''_r + (\tfrac{1}{6} - \tfrac{1}{2}\beta_0)h^3 y'''_r +$$
$$+ (\tfrac{1}{24} - \tfrac{1}{6}\beta_0)h^4 y_r^{iv} + \cdots. \tag{4.3}$$

We can expect to make the first three terms zero because there are three coefficients to be specified in (4.1), and we find

$$\alpha_1 = -1, \quad \beta_0 = \beta_1 = \tfrac{1}{2}, \quad T_{r+1} = -\tfrac{1}{12}h^3 y'''_r - \tfrac{1}{24}h^4 y_r^{iv} + \cdots \tag{4.4}$$

which, as we have seen, is the trapezoidal rule.

To get a $T_{r+1}$ of higher order we need more coefficients in (4.1), that is more points involving $Y$ and/or $Y'$. We can get such formulae in many ways. For example, let us approximate to the dominant term $-\tfrac{1}{12}h^3 y'''_r$ in $T_{r+1}$ in (4.4) by using the known finite-difference formula

$$h^3 y'''_r = h \frac{d^2}{dx^2}(h^2 y'_r) \simeq h\delta^2 y'_r = h(y'_{r+1} - 2y'_r + y'_{r-1}), \tag{4.5}$$

with the central-difference notation of Chapter 3. The 'extended' trapezoidal rule formula then becomes

$$Y_{r+1} - Y_r = \tfrac{1}{12}h(5Y'_{r+1} + 8Y'_r - Y'_{r-1}). \tag{4.6}$$

The local truncation error in the approximation (4.5) is $O(h^5)$, so that the dominant local error in (4.6) is $-\tfrac{1}{24}h^4 y_r^{iv}$, the dominant term in (4.4) not used in (4.6), and this result is easily checked by expanding the truncation error for (4.6) in a Taylor series about the point $x_r$.

Alternatively we could use the approximation

$$h^3 y'''_r \sim \Delta^3 y_{r-1} = y_{r+2} - 3y_{r+1} + 3y_r - y_{r-1}, \tag{4.7}$$

in the forward-difference notation of Chapter 3. The new 'extended' trapezoidal rule formula is then

$$Y_{r+1} - Y_r = \tfrac{1}{2}h(Y'_{r+1} + Y'_r) - \tfrac{1}{12}(Y_{r+2} - 3Y_{r+1} + 3Y_r - Y_{r-1}), \tag{4.8}$$

and a little manipulation, and with $r$ replaced by $r-1$ everywhere, produces the result, corresponding to (4.6), given by

$$Y_{r+1} + 9Y_r - 9Y_{r-1} - Y_{r-2} = 6h(Y'_r + Y'_{r-1}). \tag{4.9}$$

Here we are a little lucky. The local error involved in (4.7) cancels out the $h^4$

term in (4.4), and the Taylor series analysis shows that the local truncation error of (4.9) is dominated by the term $\frac{1}{10}h^5 y_r^{(v)}$.

Formulae (4.6) and (4.9) are special cases of the more general $k$-step formula

$$Y_{r+1} + \alpha_1 Y_r + \alpha_2 Y_{r-1} + \cdots + \alpha_k Y_{r-k+1}$$
$$= h(\beta_0 Y'_{r+1} + \beta_1 Y'_r + \beta_2 Y'_{r-1} + \cdots + \beta_k Y'_{r-k+1}). \qquad (4.10)$$

It is called a *k-step formula* because either $Y$ or the function $f(x, Y)$ is involved at $k$ successive mesh points prior to the point $x_{r+1}$ at which we seek to compute the next value $Y_{r+1}$.

We must be concerned with the *order* of our formulae. If the local truncation error is $O(h^{p+1})$ then, in analogy with the results of our one-step methods, we expect the global error to be $O(h^p)$, and we say that our method is of order $p$. So (4.6) gives a *two-step implicit* method of order 3 and (4.9) gives a *three-step explicit* method of order 4. But (4.6) may not be the most accurate two-step implicit method. This would be obtained by using another term on both sides of (4.1) and performing a similar analysis. There are five coefficients which we compute so that the contributions to the truncation error of the terms in $y_r$ and its first four derivatives are all zero, and we find the result

$$Y_{r+1} - Y_{r-1} = \tfrac{1}{3}h(Y'_{r+1} + 4Y'_r + Y'_{r-1}). \qquad (4.11)$$

This *two-step fourth-order method* is clearly analogous to the *Simpson rule* for integration. Note that the coefficient of $Y_r$ on the left of (4.11) turns out to be zero.

We might ask, in the same spirit, whether (4.9) is the most accurate three-step explicit method, noting that this would have the form

$$Y_{r+1} + \alpha_1 Y_r + \alpha_2 Y_{r-1} + \alpha_3 Y_{r-2} = h(\beta_1 Y'_r + \beta_2 Y'_{r-1} + \beta_3 Y'_{r-2}), \qquad (4.12)$$

with an extra term on the right-hand side of (4.9). Again we use the Taylor series approach to calculate the $\alpha$ and $\beta$ coefficients, and find

$$Y_{r+1} + 18Y_r - 9Y_{r-1} - 10Y_{r-2} = h(9Y'_r + 18Y'_{r-1} + 3Y'_{r-2}), \qquad (4.13)$$

with dominant local error $\frac{1}{20}h^6 y_r^{vi}$. This is smaller than that of (4.9), as we must expect since the correct specification of six coefficients should produce at least $O(h^6)$ for the local truncation error. The 'luck' which gave (4.9) is only partial!

So we could produce any number of multi-step methods, explicit, implicit and of various orders, and we need to investigate the relative uses and advantages of the various possibilities. Economy and convenience are two criteria, but by far the most important refers to their stability properties, which we now proceed to discuss.

## 4.3 CONVERGENCE, CONSISTENCY AND ZERO STABILITY

In the last chapter we suggested that with a useful method we should expect that as the interval is reduced the approximate solution gets steadily closer to the correct solution. This is the question of *convergence*, that at a fixed point $x = X$, such that $X = x_0 + nh$, we should expect our computed result to tend to the correct result as $h \to 0$, that is as $n \to \infty$, such that $nh = X - x_0$.

Two factors affect this type of convergence. The first is that the finite-difference formula relevant to the method should be *consistent*, that is should be sufficiently accurate with respect to its local truncation error. We have frequently observed that if our formula has local truncation error $O(h^{p+1})$ the corresponding global error would be expected to be $O(h^p)$, and for convergence this must clearly $\to 0$ as $h \to 0$. So, for integer $p$, we must have $p \geqslant 1$ so that the local truncation error is at least $O(h^2)$. For the general $k$-step formula (4.10) we easily find that the two required conditions for consistency are

$$1 + \alpha_1 + \alpha_2 + \cdots + \alpha_k = 0 \qquad (4.14)$$

and

$$1 - \alpha_2 - 2\alpha_3 - \cdots - (k-1)\alpha_k = \beta_0 + \beta_1 + \cdots + \beta_k. \qquad (4.15)$$

The second factor relating to convergence is that of *zero stability*, and this requires an investigation of the solutions of (4.10) as $h \to 0$, which, as we know, are of the form $A_i q_i^r$ or $(A_i + B_i r) q_i^r$, where the $q_i$, $i = 1, 2, \ldots, k$, are the roots of the polynomial equation

$$q^k + \alpha_1 q^{k-1} + \cdots + \alpha_{k-1} q + \alpha_k = 0, \qquad (4.16)$$

and the $B_i r q_i^r$ term appears in the case of two equal roots. We should expect that one root is $q_1 = 1$, representing the approximate solution corresponding to the single true solution of the differential equation, and equation (4.14), the first of the consistency equations, in fact guarantees its existence in (4.16). The other $k - 1$ roots are *parasitic* or *spurious*, having no connection with the differential equation. They must therefore be made to play an insignificant or at least a non-dominating role in the solution for $h = 0$, and this can happen only if no other $q_i$, $i = 2, 3, \ldots, k$, has modulus exceeding unity, and also that no roots with unit modulus are coincident. With these two criteria both any exponentially increasing solution and any linear growth are suppressed, the method is said to have *zero stability* or *strong stability*, and in the literature it has been proved that *zero stability and consistency imply convergence*.

A one-step method, of course, gives rise to no spurious solutions, and consistency alone should be enough for convergence. With multi-step methods we have to be more careful, and several of the formulae we have mentioned are not convergent, the lack of zero stability usually being the important reason. Numerical results can be quite spectacular. Although a concentration on $h = 0$ may appear to be a rather meaningless mathematical concept, it is known that if any $|q_i| > 1$ for $h = 0$ then it is true that for at least a small region of $h > 0$ the corresponding spurious solution of

$$(1 - h\lambda\beta_0)Y_{r+1} + (\alpha_1 - h\lambda\beta_1)Y_r + \cdots + (\alpha_k - h\lambda\beta_k)Y_{r-k+1} = 0, \qquad (4.17)$$

which is (4.10) applied to the test equation

$$y' = \lambda y, \tag{4.18}$$

will have a factor exceeding unity in absolute value.

Consider, for example, the use of equation (4.9). The formula is consistent, but the relevant equation

$$q^3 + 9q^2 - 9q - 1 = 0 \tag{4.19}$$

has roots $q_1 = 1$, $q_2 = -5 + \sqrt{24}$, $q_3 = -5 - \sqrt{24}$, and $q_3$ is the dangerous parasite. The behaviour of the method is illustrated in Table 4.1, which relates to the problem

$$y' = -y - (1+x)^{-2} + (1+x)^{-1}, \quad y(0) = 1, \quad y = (1+x)^{-1}, \tag{4.20}$$

used in various contexts in the previous chapter. We have used the known solution to provide the three values needed to start off the step-by-step process. The behaviour of the error, as the interval is reduced, exhibits the expected lack of convergence in a quite startling way. At interval $h$ the complementary solution is a sum of terms $\sum_{s=1}^{3} A_s p_s^r$, where the $p_s$ are roots of the cubic equation

$$p^3 + (9 + 6h)p^2 - (9 - 6h)p - 1 = 0, \tag{4.21}$$

and the table reveals that for small $h$ the dominant spurious solution is not very different from the $q_3$ of (4.19), about $-10$ in value. To reach a given point, with the smaller interval, the errors are multiplied by about $-10$ many times more, and the results then get steadily and catastrophically worse as the interval is reduced.

In contrast, the relevant zeros of the polynomial corresponding to the $h = 0$ part of formula (4.6) are 1 and 0, and the method is consistent with local truncation error $O(h^4)$ and expected global error $O(h^3)$. Table 4.2 gives the

**Table 4.1**

| $x$ | $1/(1+x)$ | Error ($h = 0.2$) | Error ($h = 0.1$) |
|-----|-----------|-------------------|-------------------|
| 0.0 | 1.00000 | 0.00000 | 0.00000 |
| 0.1 | 0.90909 | | 0.00000 |
| 0.2 | 0.83333 | 0.00000 | 0.00000 |
| 0.3 | 0.76923 | | -0.00005 |
| 0.4 | 0.71429 | 0.00000 | 0.00048 |
| 0.5 | 0.66667 | | -0.00506 |
| 0.6 | 0.62500 | -0.00085 | 0.05249 |
| 0.7 | 0.58824 | | -0.54591 |
| 0.8 | 0.55556 | 0.00832 | 5.6766 |
| 0.9 | 0.52632 | | -59.0285 |
| 1.0 | 0.50000 | -0.09164 | 613.812 |

**Table 4.2**

| $x$ | $1/(1 + x)$ | Error $(h = 0.2)$ | Error $(h = 0.1)$ |
|-----|-------------|-------------------|-------------------|
| 0.0 | 1.00000 | 0.00000 | 0.00000 |
| 0.1 | 0.90909 |         | 0.00000 |
| 0.2 | 0.83333 | 0.00000 | − 0.00006 |
| 0.3 | 0.76923 |         | − 0.00009 |
| 0.4 | 0.71429 | − 0.00056 | − 0.00010 |
| 0.5 | 0.66667 |         | − 0.00011 |
| 0.6 | 0.62500 | − 0.00071 | − 0.00011 |
| 0.7 | 0.58824 |         | − 0.00011 |
| 0.8 | 0.55556 | − 0.00071 | − 0.00011 |
| 0.9 | 0.52632 |         | − 0.00010 |
| 1.0 | 0.50000 | − 0.00065 | − 0.00009 |

results for this method corresponding to those of Table 4.1, with the two required initial values taken from the true solution. Here we clearly have convergence, with the error being quite reasonably proportional to $h^3$.

Though zero stability and consistency are enough for convergence there is a sort of battle between them, and it turns out that the local truncation error must be of a much smaller order than we might reasonably expect. An explicit $k$-step formula, with $2k$ arbitrary constants, can be constructed with local truncation error $O(h^{2k})$, and a corresponding implicit formula could have local truncation error $O(h^{2k+1})$. An important theorem, however, states that no zero-stable $k$-step method can have local truncation error exceeding $O(h^{k+2})$ when $k$ is odd and $O(h^{k+3})$ when $k$ is even.

Only for $k = 1$ and $k = 2$ are our expectations to some extent justified. The successful one-step trapezoidal rule method has local truncation error $O(h^3)$, and the two-step Simpson rule method (4.11) has local truncation error $O(h^5)$, both of which are 'optimal' with respect to the theorem. These cases, however, are quite exceptional. There is therefore a tendency in modern computer programs to use methods like that embodied in (4.6), with $Y_{r+1} - Y_r$ on the left-hand side so that the spurious roots at $h = 0$ are as small as possible, and with the right-hand side arranged to give reasonable local truncation error and also the other main requirement for stability which we consider in the next section.

## 4.4 PARTIAL AND OTHER STABILITIES

The other main stability requirement is a natural extension of what we discussed in connection with one-step methods in the previous chapter and, as there, it is best discussed in relation to the global error, the difference between the true solution and the numerical solution. Corresponding to an equation like (3.19) we have, in connection with formula (4.10) applied to the differential equation

$$y' = \lambda y + g(x), \tag{4.22}$$

the recurrence relation for successive global errors given by

$$(1 - h\lambda\beta_0)e_{r+1} + (\alpha_1 - h\lambda\beta_1)e_{r+2} + \cdots + (\alpha_k - h\lambda\beta_k)e_{r-k+1} = T_{r+1} + \varepsilon_{r+1}, \tag{4.23}$$

where $T_{r+1}$ is the local truncation error and $\varepsilon_{r+1}$ a local rounding error. If we take $T_{r+1} + \varepsilon_{r+1}$ to be a constant $K$, then in virtue of (4.14) the general solution of (4.23) is

$$e_r = A_1 p_1^r + A_2 p_2^r + \cdots + A_k p_k^r - K \left/ \sum_{s=0}^{k} h\lambda\beta_s \right. , \tag{4.24}$$

where the $p_i$, $i = 1, 2, \ldots, k$, are roots of the polynomial equation

$$(1 - h\lambda\beta_0)p^k + (\alpha_1 - h\lambda\beta_1)p^{k-1} + \cdots + (\alpha_k - h\lambda\beta_k) = 0, \tag{4.25}$$

and where there are extra terms $rB_i p_i^r$ on the right of (4.24) for any double roots $p_i$.

As in the one-step case we are concerned that the errors do not increase in size in the direction of integration. In this respect a method is said to be *absolutely stable*, for a given value of $\lambda h$, if for this value no roots of (4.25) lie outside the unit circle and any root on the boundary of the circle is a simple root. An interval $[\mu_1, \mu_2]$ is said to be an *interval of absolute stability* if the method is absolutely stable for all values of $\lambda h$ in $[\mu_1, \mu_2]$.

If the method is zero stable it is obvious from these definitions that it is absolutely stable when $\lambda h = 0$. A method is said to be *partially unstable* if it has a finite interval of absolute stability which includes the origin. This is most relevant of course for $\lambda < 0$, $h > 0$, and if the region of absolute stability includes the whole of the negative axis, so that the method is absolutely stable for all $h\lambda < 0$, then the method has unrestricted stability.

For the zero-stable implicit method (4.6), for example, we are concerned with the two roots of the quadratic equation

$$(1 - \tfrac{5}{12}h\lambda)p^2 - (1 + \tfrac{8}{12}h\lambda)p + \tfrac{1}{12}h\lambda = 0. \tag{4.26}$$

The root corresponding to the true solution has $p_1 = 1$ for $h = 0$, and this root very properly increases for $\lambda > 0$ and decreases for $\lambda < 0$. Here partial stability depends on the spurious root $p_2$, and Table 4.3 gives some relevant values of $p_1$ and $p_2$. It is clear that the interval $-6 \leqslant \lambda h \leqslant 0$ is an interval of absolute

**Table 4.3**

| $h\lambda$ | $+1$ | $0$ | $-1$ | $-2$ | $-3$ | $-4$ | $-5$ | $-6$ | $-7$ |
|---|---|---|---|---|---|---|---|---|---|
| $p_1$ | 2.81 | 1.00 | 0.39 | 0.22 | 0.18 | 0.16 | 0.15 | 0.14 | 0.13 |
| $p_2$ | 0.05 | 0.00 | $-0.15$ | $-0.41$ | $-0.62$ | $-0.78$ | $-0.91$ | $-1.00$ | $-1.07$ |

stability for the method, which therefore becomes partially unstable when $h > 6/|\lambda|$.

As another example consider the most accurate explicit two-step method with $Y_{r+1} - Y_r$ on the left-hand side, which is easily found to be

$$Y_{r+1} - Y_r = h(\tfrac{3}{2} Y_r' - \tfrac{1}{2} Y_{r-1}').  \tag{4.27}$$

It is not difficult to show that for negative real $\lambda$ in (4.22) we have partial stability with (4.27) only if $h \leqslant 1/|\lambda|$.

For the more general equation

$$y' = f(x, y)  \tag{4.28}$$

the partial stability requirement is essentially a local matter, as we mentioned in the last chapter, and the criterion may change in successive steps. As before, here we replace $\lambda$ by some value of $\partial f/\partial y$, perhaps a reasonably average value over the range of $x$ involved in the formula.

For the system of $m$ linear first-order equations with constant coefficients, given by

$$y' = Ay + f(x),  \tag{4.29}$$

where $y$ and $f$ are vectors with $m$ components and $A$ is a matrix of order $m$, partial stability requires consideration of each of the $m$ eigenvalues of the matrix $A$. If the equations are linear but the coefficients of $A$ depend on $x$ then we take these to be locally constant in every step. Here, as we mentioned in the last chapter, the eigenvalues of $A$ may be complex even when $A$ is real, so that we have to consider the case when $\lambda = a + ib$ for negative $a$.

Finally, if we consider (4.28) as a system of first-order non-linear equations, with $y$ and $f$ vectors with $m$ components, then we have to replace $\lambda$ by every eigenvalue of the *Jacobian matrix* of $f(x, y)$, evaluated at some point local to the interval. For example, with the three equations

$$\left.\begin{aligned}
y_1' &= f_1(x, y_1, y_2, y_3) \\
y_2' &= f_2(x, y_1, y_2, y_3) \\
y_3' &= f_3(x, y_1, y_2, y_3)
\end{aligned}\right\},  \tag{4.30}$$

the relevant Jacobian matrix is

$$J = \begin{bmatrix}
\partial f_1/\partial y_1 & \partial f_1/\partial y_2 & \partial f_1/\partial y_3 \\
\partial f_2/\partial y_1 & \partial f_2/\partial y_2 & \partial f_2/\partial y_3 \\
\partial f_3/\partial y_1 & \partial f_3/\partial y_2 & \partial f_3/\partial y_3
\end{bmatrix}.  \tag{4.31}$$

Of course the evaluation of the eigenvalues of a large matrix, repeated at each step of the integration, is a major operation and will not usually be performed. We may be able to apply some other criterion, but in general the simple analysis of the various possible formulae is perhaps most useful for producing some good information about the relative performance of these methods with regard to partial stability.

Several important facts of this kind are known. First, implicit methods of order $p$ are significantly superior to explicit methods of the same order, not only having a smaller coefficient $C$ of the local truncation error term $Ch^{p+1}$, but also allowing a much larger interval of absolute stability. Second, implicit methods like (4.6), with $Y_{r+1} - Y_r$ on the left, have very good partial stability properties, the maximum interval of absolute stability decreasing slightly as the order increases. Third, the class of formulae given by

$$Y_{r+1} + \alpha_1 Y_r + \cdots + \alpha_k Y_{r-k+1} = h\beta_0 Y'_{r+1} \tag{4.32}$$

is even better in this respect, and is commonly used for stiff equations in which the ratio of the largest negative $\lambda$ to the smallest negative $\lambda$ is very large indeed, whereas we need a reasonably large bound on $h|\lambda|$ to make the method economic. All explicit methods are quite uneconomic in this context.

In many numerical analysis books there is a discussion of the region of absolute stability of a method, which is the region of the complex plane such that if $\lambda h$ is within this region then the roots of (4.25) have the required restrictions in size and simplicity. The ideal method for solving stiff systems would have as stability region the whole of the negative half-plane, but unfortunately only a one-step method can have this property. For stiff systems in practice it is usually sufficient for the stability region to include all complex numbers $z = x + iy$, where $x \leqslant -A$, and also those for which $-A \leqslant x \leqslant 0$ and $-\alpha \leqslant y \leqslant \alpha$ for some suitable positive numbers $A$ and $\alpha$. This region is shown in Fig. 4.1, and a method with such a region is called *stiffly stable*.

The reason why this is satisfactory is that if $\lambda$ has a fairly small negative real part and a large imaginary part then $e^{\lambda x}$ will oscillate rapidly with small decaying amplitude as $x$ increases. The numerical method will then need a small $h$ to control the size of the local truncation error, and this will usually be small

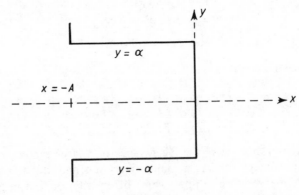

**Figure 4.1**

enough to guarantee stability. But when the real part is large and negative and the corresponding rapidly decreasing $e^{\lambda x}$ has become negligible, we need to be able to increase the step size without our method becoming unstable.

For example, formulae of class (4.32) are truncated forms of the backward-difference formula for a first derivative, given by

$$hy'_{r+1} = (\nabla + \tfrac{1}{2}\nabla^2 + \tfrac{1}{3}\nabla^3 + \cdots)y_{r+1}. \tag{4.33}$$

In the last chapter we preferred central-difference formulae to 'sloping' difference formulae, but when stability is much more important than size of local truncation error, these formulae are very useful. For real negative $\lambda$ they have unrestricted partial stability for orders up to $k = 6$, and for complex $\lambda$ they are stiffly stable up to $k = 6$ but not beyond this. The corresponding stability region of Fig. 4.1 for $k = 6$ has $A$ close to 6 and $\alpha$ to 0.5.

The formulae like (4.6), also quoted as having good partial stability properties, are also easily derived from the truncated backward-difference formula, effectively the 'inverse' of (4.33), given by

$$y_{r+1} - y_r = h(1 - \tfrac{1}{2}\nabla - \tfrac{1}{12}\nabla^2 - \tfrac{1}{24}\nabla^3 - \cdots)y'_{r+1}. \tag{4.34}$$

They are not, in fact, as valuable for very stiff systems as the class of equations (4.32), but they are generally very useful with small local truncation errors. In the next section we shall find this to be a useful corrector formula, and its standard predictor companion is derived from the corresponding explicit formula given by

$$y_{r+1} - y_r = h(1 + \tfrac{1}{2}\nabla + \tfrac{5}{12}\nabla^2 + \tfrac{3}{8}\nabla^3 + \cdots)y'_r. \tag{4.35}$$

Equations (4.34) and (4.35) are the basis of the well-known Adams–Bashforth method.

We must mention one final stability concept which is particularly relevant to the zero-stable Simpson rule method given by equation (4.11). The equation (4.25) is here

$$(1 - \tfrac{1}{3}h\lambda)p^2 - \tfrac{4}{3}h\lambda p - (1 + \tfrac{1}{3}h\lambda) = 0. \tag{4.36}$$

It is possible to show that with good accuracy the roots are close to

$$p_1 = e^{h\lambda}, \quad p_2 = -e^{-\tfrac{1}{3}h\lambda} \tag{4.37}$$

when $h|\lambda|$ is small. One of these always has magnitude greater than unity, so that the method is not absolutely stable for any $\lambda h$ except with $h = 0$. We therefore introduce the concept of *relative stability*. Every useful method has the one true root, which we have usually called $p_1$, which is here represented by the first of (4.37). The method is said to be *relatively stable* for a value of $h\lambda$ such that $|p_j| \leqslant |p_1|$ for $j = 2, 3, \ldots, k$. This means that none of the parasitic solutions increases faster than the true solution; and a region is said to be a *region of relative stability* if the method is relatively stable for all values of $h\lambda$ in this region.

With these definitions it follows that the whole positive line is a region of relative stability for the Simpson rule method and the whole negative line is a region of relative instability. It implies that the method is perfectly satisfactory for the solution of (4.22) with $\lambda > 0$, but care is neded with $\lambda < 0$. The method is convergent, so that we should get any required accuracy at any positive value of $x$ with a small enough interval. But the larger this value of $x$, the more guarding figures we should need to use in our arithmetic to prevent storage errors in initial values, and rounding and truncation errors in successive steps, from introducing too large a multiple of the spurious solution which is now the second of (4.37) with negative $\lambda$.

## 4.5 PREDICTOR–CORRECTOR METHODS

In the last chapter we mentioned the use of predictor–corrector methods for the solution of the non-linear recurrence relations stemming from non-linear differential equations. In the one-step case our corrector was typically the second-order one-step trapezoidal rule formula and the predictor the first-order one-step explicit Euler formula. For multi-step methods there are some advantages in using corrector and predictor formulae of the same order. In particular the latter produces a quite close approximation for the start of the iteration in the corrector.

Consider, for example, the corrector formula (4.6), which we solve according to the iteration

$$Y_{r+1}^{(s+1)} = Y_r + \tfrac{1}{12}h\{5f(x_{r+1}, Y_{r+1}^{(s)}) + 8Y_r' - Y_{r-1}'\}, \tag{4.38}$$

where the first estimate $Y_{r+1}^{(0)}$ is obtained from the explicit predictor of the same order, given by

$$Y_{r+1}^{(0)} = Y_r + \tfrac{1}{12}h(23Y_r' - 16Y_{r-1}' + 5Y_{r-2}'). \tag{4.39}$$

In (4.38) we have written $f(x_{r+1}, Y_{r+1}^{(s)})$ for the equivalent $Y_{r+1}^{(s)\prime}$ for a reason which becomes obvious almost immediately. Now if we record (4.38) with $s$ changed to $s-1$, and subtract one equation from the other, we see that

$$Y_{r+1}^{(s+1)} - Y_{r+1}^{(s)} = \tfrac{5}{12}h\frac{\partial f}{\partial Y}(Y_{r+1}^{(s)} - Y_{r+1}^{(s-1)}), \tag{4.40}$$

where $\partial f/\partial Y$ is evaluated for $Y$ somewhere between $Y_{r+1}^{(s)}$ and $Y_{r+1}^{(s-1)}$. It follows that the iteration is very likely to converge if

$$\tfrac{5}{12}h\left|\frac{\partial f}{\partial Y}\right| < 1, \tag{4.41}$$

and the smaller the quantity on the left, the faster the convergence.

Consider, for example, the problem

$$y' = y^3 - (1 + x)^{-3} - (1 + x)^{-2}, \quad y(0) = 1, \tag{4.42}$$

already used in the previous chapter for a similar purpose. Starting with the correct values at $x = 0.3$, $0.4$ and $0.5$, we predict from (4.39) the value at $x = 0.6$ to be 0.6248539, and successive estimates from the corrector (4.38) are 0.6250056, 0.6250130 and 0.6250134. The convergence is quite fast. The only available useful estimate for $\frac{5}{12}h(\partial f/\partial Y)$ from (4.40) is

$$\frac{0.6250130 - 0.6250056}{0.6250056 - 0.6248539} \sim \tfrac{1}{20}, \tag{4.43}$$

whilst its correct value at $x = 0.6$ is

$$\tfrac{5}{12}(0.1)(3)(0.625)^2 \sim \tfrac{1}{20}. \tag{4.44}$$

But the mere fact that the process converges makes it likely that (4.41) is satisfied, and since the criterion for partial stability of (4.6) is $h|\partial f/\partial Y| < 6$ we are confident that our method is stable at this interval, even without computing (4.43). In almost all cases, moreover, except for very stiff equations, convergence of the iteration is very fast since the interval will need to be well below the partial stability requirement to guarantee a sufficiently small local truncation error. This is exemplified by (4.43) and the fact that the error in this step is 0.000013, so that a smaller interval is needed for even five-decimal-digit accuracy.

The predictor–corrector method is less satisfactory for very stiff equations, where with a large $|\partial f/\partial Y|$ in (4.41) the criterion for the convergence of the iteration may require a quite unacceptably small value of $h$. Of course we have to · maintain partial stability, but for this purpose we may be prepared to use a method of no great order but with no very restrictive bound on $h$ for stability, for example even the trapezoidal rule which has no such interval restriction at all. In that event the predictor–corrector iteration is unlikely to converge, and we must then use something like the Newton method which will certainly converge if the first approximation is good enough.

For a system of equations, however, this involves quite a lot of computation. For two equations, for example, of the form

$$\left. \begin{aligned} y' &= f(x, y, z) \\ z' &= g(x, y, z) \end{aligned} \right\}, \tag{4.45}$$

the trapezoidal rule method will involve the solution of the pair of non-linear algebraic equations given by

$$\left. \begin{aligned} Y_{r+1} &= \tfrac{1}{2}hf(x_{r+1}, Y_{r+1}, Z_{r+1}) + \phi \\ Z_{r+1} &= \tfrac{1}{2}hg(x_{r+1}, Y_{r+1}, Z_{r+1}) + \psi \end{aligned} \right\}, \tag{4.46}$$

where $\phi$ and $\psi$ are known. This we can write conveniently as

$$F_1(Y, Z) = 0, \quad F_2(Y, Z) = 0, \quad Y = Y_{r+1}, \quad Z = Z_{r+1}. \tag{4.47}$$

The Newton method makes corrections $\delta Y$ and $\delta Z$ to first approximations $Y^{(0)}$ and $Z^{(0)}$ by solving the linear equations

$$\left. \begin{aligned} \frac{\partial F_1}{\partial Y} \delta Y + \frac{\partial F_1}{\partial Z} \delta Z + F_1 = 0 \\ \frac{\partial F_2}{\partial Y} \delta Y + \frac{\partial F_2}{\partial Z} \delta Z + F_2 = 0 \end{aligned} \right\}, \tag{4.48}$$

where $F_1, F_2$ and the partial derivatives are evaluated at $Y^{(0)}, Z^{(0)}$. The matrix of coefficients in (4.48) is the Jacobian matrix of the $F_i$, $i = 1, 2$, and for a set of $n$ equations, with $F_1, F_2, \ldots, F_n$ in (4.47) and with $n$ unknowns, the matrix is $(n \times n)$ and its respective elements are the partial derivatives of all the $F_i$ with respect to all the unknowns.

## 4.6 ERROR ESTIMATION AND CHOICE OF INTERVAL

In the last chapter we mentioned a method which gave an estimate of the error produced in one step by a fourth-order Runge–Kutta method, which gives the chance of adjusting the interval so that the local error and, with stability, the global error is as small as we please. We clearly need some similar device for our multi-step methods, and this is provided with a little extra work applied to the predictor–corrector methods, again with the same order of local truncation error in both the predictor and the corrector. Consider, for example, the predictor (4.39) and the corrector (4.6), and assume that there have been no errors in the values of $Y$ and $Y'$ at points $r, r - 1$ and $r - 2$, so that $Y_i = y_i$ for $i < r + 1$. Then the correct value at $r + 1$ satisfies

$$y_{r+1} = y_r + \tfrac{1}{12} h(23y'_r - 16y'_{r-1} + 5y'_{r-2}) + T^{(p)}_{r+1}, \tag{4.49}$$

where $T^{(p)}_{r+1}$ is the local truncation error of the predictor, and the error of the predicted value $Y^{(0)}_{r+1}$ at this point is then just

$$e^{(p)}_{r+1} = y_{r+1} - Y^{(0)}_{r+1} = T^{(p)}_{r+1}. \tag{4.50}$$

The result for the corrector is a little more complicated. The correct value at $r + 1$ satisfies the equation

$$y_{r+1} = y_r + \tfrac{1}{12} h(5y'_{r+1} + 8y'_r - y'_{r-1}) + T^{(c)}_{r+1}, \tag{4.51}$$

where $T^{(c)}_{r+1}$ is the local truncation error of the corrector, and one application of the corrector, which is all that is needed for this purpose, gives the estimate

$$Y^{(1)}_{r+1} = y_r + \tfrac{1}{12} h(5Y^{(0)\prime}_{r+1} + 8y'_r - y'_{r-1}). \tag{4.52}$$

Subtraction of (4.52) from (4.51) gives

$$e_{r+1}^{(c)} = y_{r+1} - Y_{r+1}^{(1)} = \tfrac{5}{12} h \frac{\partial f}{\partial y_{r+1}} e_{r+1}^{(p)} + T_{r+1}^{(c)}, \tag{4.53}$$

so that

$$e_{r+1}^{(c)} = T_{r+1}^{(c)} + \text{terms of higher order in } h. \tag{4.54}$$

The dominant term in $T_{r+1}^{(p)}$ is easily found to be $\tfrac{3}{8} h^4 y_r^{iv}$, and that in $T_{r+1}^{(c)}$ is $-\tfrac{1}{24} h^4 y_r^{iv}$, so that to this order in $h$ we see that

$$y_{r+1} - Y_{r+1}^{(1)} = -\tfrac{1}{10}(Y_{r+1}^{(1)} - Y_{r+1}^{(0)}), \tag{4.55}$$

a simple and simply calculated expression for the local error. Although the notation suggests that this is the global error, it has this property only if the previous values are correct. Otherwise it is just an estimate for the local error, giving no indication of the global error but a very useful criterion for the successful choice of the current step-size $h$.

For the numerical results following equation (4.42) we then estimate the error in this step to be $-\tfrac{1}{10}(0.625006 - 0.624854) = -0.000015$. The error is in fact $-0.000013$, so that this is quite a reasonable estimate. If we wanted six-decimal accuracy it would suggest that, with a third-order method, an interval slightly less than 0.05 would give a good result, whereas if only three-decimal accuracy is needed we could at least double the original interval.

## 4.7 STARTING THE COMPUTATION

Unlike the one-step methods the multi-step methods need several $Y_i$ and $Y_i'$ for $i < r + 1$, and in the early steps of the computation these do not exist. Probably the easiest way of providing this information is the use of a Runge–Kutta method, say of fourth order and with a Merson-type routine which estimates the error and therefore the size of the interval for required accuracy. Since we want to use a constant interval in the successive computation with multi-step methods we should want to use the same interval in all the Runge–Kutta steps which may not be well calculable until all such computations have been performed. When the multi-step routines come into operation they may be more accurate than the chosen Runge–Kutta routine, and the error control procedure of the previous section may come into operation quite soon.

In modern computer routines there is an increasing tendency to use at the start a set of multi-step methods of different orders, particularly the so called implicit Adams-formulae of type (4.34) with $Y_{r+1} - Y_r$ on the left, which we have already stated have good partial stability properties. The first few of these, with the order of the formula attached, are

$$Y_{r+1} - Y_r = \tfrac{1}{2}h(Y'_{r+1} + Y'_r), \qquad O(h^2)$$
$$Y_{r+1} - Y_r = \tfrac{1}{12}h(5Y'_{r+1} + 8Y'_r - Y'_{r-1}), \qquad O(h^3)$$
$$Y_{r+1} - Y_r = \tfrac{1}{24}h(9Y'_{r+1} + 19Y'_r - 5Y'_{r-1} + Y'_{r-2}), \qquad O(h^4)$$

$$\left.\right\} \qquad (4.56)$$

The interval needed in the first step will probably have to be rather small, and the same interval is likely to be used for convenience in a few successive steps until we reach the formula of maximum order we are prepared to take. After that our error-control process will come into operation, and the interval may be increased or decreased or the order varied in a fairly automatic manner, described in more detail in Chapter 8, in an attempt to achieve the required accuracy most economically.

### 4.8 CHANGING THE INTERVAL

Finally we must consider how best to increase or decrease the interval size when this becomes necessary or convenient. For example, the main problem we have used for illustration in this and the previous chapter has the solution $(1 + x)^{-1}$. For $x$ close to $-1$ this function and its derivatives are large and change rapidly with $x$, and small finite-difference intervals are required. As $x$ goes away from $-1$ in either direction the function gets increasingly 'smoother', and increasingly larger intervals can be used for its tabulation.

With one-step methods there is no problem, since the formulae can be used without effective change for any value of $h$. But with multi-step methods like (4.10) the coefficients $\alpha_i$ and $\beta_i$ we have used depend on the fact that $h = x_{r+1} - x_r = x_r - x_{r-1} = \cdots$ for all terms in the formula. If we use unequal intervals we have to recalculate the coefficients and indeed the size of the local truncation error.

The easiest situation is when the interval is doubled or halved in a particular step and then kept constant for subsequent steps. The first of these possibilities is shown in Fig. 4.2.

Here

$$x_r - x_{r-1} = x_{r-1} - x_{r-2} = \cdots = \tfrac{1}{2}h, \qquad (4.57)$$

and

$$x_{r+1} - x_r = x_{r+2} - x_{r+1} = \cdots = h. \qquad (4.58)$$

Clearly we can use the standard formula (4.10) to compute $Y_{r+1}$ by using the

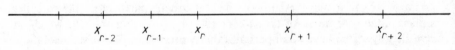

**Figure 4.2**

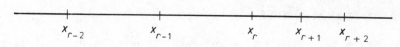

**Figure 4.3**

previous values at $x_r, x_{r-2}, \ldots$, with an interval $h$, and successive $Y_{r+2}$, etc., at a further interval of size $h$, are computed with no change in the formula.

Halving the interval needs some modification, which we illustrate with formula (4.6). With respect to Fig. 4.3 we have halved the interval in proceeding from $x_r$ to $x_{r+1}$, and subsequently maintained this new constant interval.

We now need a special formula to compute $Y_{r+1}$, and the standard use of Taylor series shows this to be

$$Y_{r+1} = Y_r + \frac{h}{36}(16Y'_{r+1} + 21Y'_r - Y'_{r-1}), \tag{4.59}$$

where

$$h = x_{r+1} - x_r = \tfrac{1}{2}(x_r - x_{r-1}). \tag{4.60}$$

The local truncation error of (4.59) is dominated by the term $-\frac{5}{72}h^4 y_r^{iv}$, the same order as that of (4.6) but with a larger coefficient.

Following the use of (4.59) for $Y_{r+1}$, the value at $x_{r+2}$ is here computed by the original two-step formula, though for a three-step case another special formula would be needed for $Y_{r+2}$, and so on. The formulae and their derivation are quite simple and are not in any real sense uneconomic.

They may, however, have rather too drastic an effect, and we may prefer a smoother change in interval size. Formulae for quite general intervals can, of course, be obtained by our standard Taylor series analysis. Corresponding to formula (4.6) for the equal-interval case, for example, we need to compute the coefficients in

$$Y_{r+1} - Y_r = \gamma_0 Y'_{r+1} + \gamma_1 Y'_r + \gamma_2 Y'_{r-1}, \tag{4.61}$$

where $x_{r+1} - x_r = h_{r+1}$, $x_r - x_{r-1} = h_r$. We find

$$\gamma_0 = \frac{h_{r+1}(2h_{r+1} + 3h_r)}{6(h_{r+1} + h_r)}, \quad \gamma_1 = \frac{h_{r+1}(h_{r+1} + 3h_r)}{6h_r}, \quad \gamma_2 = -\frac{h_{r+1}^3}{6h_r(h_{r+1} + h_r)} \tag{4.62}$$

with local truncation error dominated by

$$T_{r+1} = -\tfrac{1}{72}h_{r+1}^3(h_{r+1} + 2h_r)y_r^{iv}. \tag{4.63}$$

When using methods with more steps we have to do a fair amount of calculation to produce formulae like (4.61), (4.62) and (4.63), but for a system of non-linear first-order equations this has to be done once only and therefore represents only a small proportion of the total amount of work.

4.9 ADDITIONAL NOTES

*Section 4.3* It is not too difficult to prove our intuition that any one-step consistent method is convergent. The application of (4.18) to (4.10) with $k = 1$ gives the recurrence relation

$$(1 - h\lambda\beta_0)Y_{r+1} + (\alpha_1 - h\lambda\beta_1)Y_r = 0, \tag{4.64}$$

and this has the single solution

$$Y_r = q^r y_0, \quad q = -(\alpha_1 - h\lambda\beta_1)/(1 - h\lambda\beta_0). \tag{4.65}$$

Now

$$y_r = e^{rh\lambda} y_0, \tag{4.66}$$

$$q^r = (-\alpha_1)^r(1 - h\lambda\beta_1/\alpha_1)^r/(1 - h\lambda\beta_0)^r. \tag{4.67}$$

As $r \to \infty$, with $x_r = rh$ fixed, we have

$$q^r = (-\alpha_1)^r e^{-\lambda\beta_1 x_r/\alpha_1} e^{\lambda\beta_0 x_r}, \tag{4.68}$$

and $Y_r \to y$, if $\alpha_1 = -1$, $\beta_0 + \beta_1 = 1$, which are the consistency conditions (4.14) and (4.15) with $k = 1$.

*Section 4.5* In the predictor–corrector process we have assumed that the correction is performed enough times to ensure that the corrector formula is essentially completely satisfied. In some contexts results are accepted which are the result of one prediction followed by a small number of corrections, perhaps only one, and this can affect the partial stability criterion.

Consider, for example, the one-step trapezoidal rule corrector and the forward Euler predictor which we used in the last chapter. For the test example $y' = \lambda y$ the predictor gives

$$Y_{r+1}^{(0)} = Y_r + h\lambda Y_r, \tag{4.69}$$

and a single correction then produces the result

$$Y_{r+1} = Y_r + \tfrac{1}{2}h\{\lambda Y_r + \lambda(Y_r + h\lambda Y_r)\}. \tag{4.70}$$

These equations give

$$Y_{r+1} = pY_r, \quad p = 1 + h\lambda + \tfrac{1}{2}h^2\lambda^2, \tag{4.71}$$

and for negative $\lambda$ partial stability requires that $|p| < 1$ which is satisfied only if $h|\lambda| < 2$. The corrector properly solved gives

$$Y_{r+1} = \left(\frac{1 + \tfrac{1}{2}h\lambda}{1 - \tfrac{1}{2}h\lambda}\right) Y_r, \tag{4.72}$$

which as we know has unrestricted stability for negative $\lambda$.

In general the incomplete use of the corrector in this way will reduce the size of the interval needed for partial stability. For example, we saw in Table 4.3 that formula (4.6) has partial stability for $h \leqslant 6/|\lambda|$, but when (4.6) is used just once with (4.39) as predictor it turns out that the stable $h$ is decreased to approximately $h \leqslant 2.4/|\lambda|$.

The literature has a special notation for these methods. If $P$ means the use of the predictor to obtain an estimate of $Y_{r+1}$, $E$ means the evaluation of $f(x_{r+1}, Y_{r+1})$, and $C$ means the use of the corrector to improve the previous estimate of $Y_{r+1}$, then particular modes of application could be $P(EC)^m$ or $P(EC)^m E$, where $m$ is specified. Our two previous examples had $m = 1$ and ignored the final $E$, though the latter is really required so that the accepted derivative of the final $Y_{r+1}$ does satisfy the differential equation, and this also has a beneficial effect on the stability.

Finally, we stated that the predictor–corrector iteration is not satisfactory for a non-linear stiff system of first-order equations, and that the corresponding Newton iteration involves much computation. In fact efficient methods of implementing the Newton technique are still the subject of much current research. In typical problems the order of the Jacobian matrix might be about 50, but many of its elements are zero and it is often quite possible to take advantage of the sparse nature of the matrix to improve the efficiency of the computation. Also it is common practice to update the Jacobian matrix infrequently with consequent savings in function evaluations and subsequent matrix computations. Even then the labour involved may appear to be somewhat restrictive, but the unrestricted stability of the trapezoidal rule means that a large range may be covered with a reasonably small number of steps.

---

## EXERCISES

1. If formula (4.13) is used to solve the problem of equation (4.20), show, without performing any step-by-step computation, that for very small values of $h$ errors will ultimately increase, in successive steps, by a factor of about $-18.5$.

2. Show that the two-step explicit formula

$$Y_{r+1} - Y_r = h(\beta_1 Y'_r + \beta_2 Y'_{r-1})$$

   has smallest local truncation error $\frac{5}{12} h^3 y'''_r$ with $\beta_1 = \frac{3}{2}, \beta_2 = -\frac{1}{2}$. Show also, with respect to (4.22) for negative $\lambda$, that the method is partially stable only if $h < 1/|\lambda|$.

3. Show that the formula

$$11Y_{r+1} + 27Y_r - 27Y_{r-1} - 11Y_{r-2} = 3h(Y'_{r+1} + 9Y'_r + 9Y'_{r-1} + Y'_{r-2})$$

   is the most accurate three-step implicit formula. Show also that it is strongly unstable, first by examining the appropriate recurrence relation, and second by finding the order of its local truncation error and using the theorem quoted near the end of Section 4.3.

4. Show that the formula corresponding to (4.6) and of next higher order is

$$Y_{r+1} - Y_r = \tfrac{1}{24}h(9Y'_{r+1} + 19Y'_r - 5Y'_{r-1} + Y'_{r-2}),$$

with dominant local truncation error $-\tfrac{19}{720}h^5 y_r^v$. By finding the largest negative root of the equation corresponding to (4.26) with $h = 6/|\lambda|$ for $\lambda < 0$, verify that the interval for partial stability is here somewhat smaller with an increase in order of the formula.

5. For the equations

$$y' = -12.5y + 6.3z - (1 + x)^{-2} + 18.8(1 + x)^{-1}$$
$$z' = 5y - 2.4z + (1 + x)^{-2} - 7.4(1 + x)^{-1}$$

show that the use of formulae relevant to (4.6) will produce partial instability if $h > 0.4$. Verify this by performing some step-by-step computations for $h$ just less than and just greater than 0.4, with the initial conditions $y(0) = 2$, $z(0) = 1$. (Compare with similar results in Exercises 5 and 6 of Chapter 3.)

6. Show that the four-step explicit formula

$$Y_{r+1} - Y_{r-3} = \tfrac{4}{3}h(2Y'_r - Y'_{r-1} + 2Y'_{r-2})$$

has local truncation error dominated by $\tfrac{14}{45}h^5 y_r^v$. Is the formula convergent? Show also that for the equation $y' = \lambda y$, with $\lambda < 0$, the equation governing partial stability for this formula has a negative root of magnitude exceeding unity for all $h > 0$. Is this also true for the Simpson rule formula (4.11)?

7. The formula of Exercise 6 is used as a predictor for the Simpson rule corrector. With these formulae repeat the analysis and computation of the first part of Section 4.5, showing that the corrector will converge if $\tfrac{1}{3}h|\partial f/\partial Y| < 1$, and that the available estimate for this number obtained by the computation is about 0.040 while its correct value is about 0.039 at $x = 0.6$.

8. With the predictor and corrector of Exercise 7 show, with the notation of equation (4.55), that

$$y_{r+1} - Y_{r+1}^{(1)} = -\tfrac{1}{29}(Y_{r+1}^{(1)} - Y_{r+1}^{(0)}),$$

and that the estimated error is about 0.0000018 while the true error is about 0.0000012.

9. The implicit formula (4.6) and the explicit formula (4.27) have respective limits $h \leqslant 6/|\lambda|$ and $h \leqslant 1/|\lambda|$ for partial stability. If the corrector is used just once, following the predictor, show by constructing a relevant table corresponding to Table 4.3 that the new limit for partial stability lies between these two numbers.

**10.** The deferred approach to the limit can be used with diagrams like Fig. 4.2 or Fig. 4.3 if, for example, more accurate results were obtained by halving each interval. The dominant error at common mesh points is then $O(h^3)$ for the formulae quoted in Section 4.8, and this can be eliminated by a suitable combination of the two solutions. Try this for the problem of equation (4.20) for Fig. 4.3, finding approximations first at the points $x = 0.0, 0.2, 0.4, 0.5, 0.6$, then at the points $x = 0.0, 0.1, 0.2, 0.3, 0.4, 0.45, 0.50, 0.55, 0.60$ and correcting the first set of results by the deferred approach.

# 5

# Initial-value methods for boundary-value problems

## 5.1 INTRODUCTION

As we saw in Chapter 1, a boundary-value problem is one in which conditions associated with the differential equations are specified at more than one point. Here we shall concentrate on the existence of just two boundary points, which is the most usual case. We may be interested in a single differential equation of $n$th order, or a set of lower-order equations equivalent to this, a special case of which is a simultaneous set of $n$ first-order equations. In all cases there will be $n$ associated conditions, the *boundary conditions*, either separated or unseparated. In the separated case there will be $p$ conditions at one boundary point and $q$ at the other, where $p + q = n$, and in the unseparated case at least some of the $n$ conditions will involve combinations of the values of the functions or their derivatives at both boundary points.

The simplest case is the second-order equation

$$y'' = f(x, y, y'), \tag{5.1}$$

for which separated conditions are

$$g_1(y, y') = 0 \quad \text{at} \quad x = a, \qquad g_2(y, y') = 0 \quad \text{at} \quad x = b, \tag{5.2}$$

and relevant unseparated conditions are typified by

$$\left.\begin{aligned} g_1\{y(a), y'(a), y(b), y'(b)\} &= 0 \\ g_2\{y(a), y'(a), y(b), y'(b)\} &= 0 \end{aligned}\right\}. \tag{5.3}$$

For a set of first-order equations such as

$$\left.\begin{aligned} y_1' &= f_1(x, y_1, y_2, y_3) \\ y_2' &= f_2(x, y_1, y_2, y_3) \\ y_3' &= f_3(x, y_1, y_2, y_3) \end{aligned}\right\} \tag{5.4}$$

separated conditions may typically have the form

$$y_1(a) = \alpha, \quad y_2(a) + y_3(a) = \beta, \quad y_1(b) = \gamma, \tag{5.5}$$

where $\alpha, \beta$ and $\gamma$ are given numbers. The most general form of unseparated conditions is clearly

$$g_i\{y_1(a), y_2(a), y_3(a), y_1(b), y_2(b), y_3(b)\} = 0, \quad i = 1, 2, 3. \tag{5.6}$$

Within the field of boundary-value problems we include eigenvalue problems typified by

$$y'' + \{f(x) + \lambda\}y = 0, \quad y(a) = 0, \quad y(b) = 0, \tag{5.7}$$

whose solutions (*eigenfunctions*) exist only for certain discrete values of $\lambda$ (*the eigenvalues*), and we usually need to determine one or more *eigensolutions* (both eigenvalues and associated eigenfunctions). For example, the simplest problem of this type, given by

$$y'' + \lambda y = 0, \quad y(0) = 0, \quad y(1) = 0, \tag{5.8}$$

has eigensolutions

$$\lambda_n = n^2\pi^2, \quad y_n = A_n \sin n\pi x, \quad n = 1, 2, \ldots, \tag{5.9}$$

where the $A_n$ are arbitrary constants. There are no other solutions.

The other class of problem we consider briefly is that in which, for example, the second boundary position is not specified in advance but has to be determined, together with the solution of the differential equation and associated conditions, with the assistance of one additional boundary condition. For example, it is easy to verify that the problem

$$y'' + y = 1, \quad y(0) = 0, \quad y(b) = 1, \quad y'(b) = 2 \tag{5.10}$$

has the solution

$$y = 1 + \cot b \sin x - \cos x, \tag{5.11}$$

for those discrete values of $b$ for which

$$\sin b = \tfrac{1}{2}. \tag{5.12}$$

There are several different types of numerical methods for solving boundary-value problems, and in this chapter we concentrate essentially on methods which solve several initial-value problems for which, as we have seen, there are available a number of accurate and economic computer library routines. Other methods are considered in Chapters 6 and 7.

## 5.2 THE SHOOTING METHOD: LINEAR PROBLEMS

Our first method is called the 'shooting method' for fairly obvious reasons. Consider, for example, the simple second-order boundary-value probem

$$y'' = f(x, y, y'), \quad y(a) = \alpha, \quad y(b) = \beta. \tag{5.13}$$

If we knew the value of $y'(a)$ which satisfies this problem we could use an initial-value technique for integrating the differential equation. The shooting method effectively guesses a value of $y'(a)$, 'shoots' to $x = b$ and compares the computed value of $y(b)$ with the required $\beta$. It then uses some systematic procedure for adjusting the guess to produce the correct value of $y'(a)$. All differential problems which are solved in the process are of initial-value type.

In Section 2.6 on sensitivity analysis we suggested a method which is applicable to linear boundary-value problems without 'guesses' and 'adjustments' and we here adopt this suggestion, using as an illustrative example the most general linear problem involving one second-order equation. This is given by the differential equation

$$y'' + f(x)y' + g(x)y = k(x), \tag{5.14}$$

with general unseparated boundary conditions

$$\left. \begin{array}{l} \alpha_0 y(a) + \alpha_1 y'(a) + \alpha_2 y(b) + \alpha_3 y'(b) = \alpha_4 \\ \beta_0 y(a) + \beta_1 y'(a) + \beta_2 y(b) + \beta_3 y'(b) = \beta_4 \end{array} \right\}. \tag{5.15}$$

We have learnt that the most general solution of (5.14) is

$$y = y^{(p)} + A_1 y^{(1)} + A_2 y^{(2)}, \tag{5.16}$$

where $y^{(p)}$ is a particular integral of (5.14), $y^{(1)}$ and $y^{(2)}$ are two independent solutions of the homogeneous form of (5.14), and $A_1$ and $A_2$ are constants. For $y^{(p)}$ we could take the initial conditions

$$y^{(p)}(a) = 0, \quad y^{(p)\prime}(a) = 0, \tag{5.17}$$

or any other selected values for these quantities, and $y^{(1)}$ and $y^{(2)}$ are guaranteed to be independent with the choices

$$y^{(1)}(a) = 1, \quad y^{(1)\prime}(a) = 0, \quad y^{(2)}(a) = 0, \quad y^{(2)\prime}(a) = 1. \tag{5.18}$$

With a numerical method, for example a Runge–Kutta method, we can solve the initial-value problems for $y^{(p)}, y^{(1)}$, and $y^{(2)}$. The solution (5.16) must then be made to satisfy the boundary conditions (5.15), and this produces two linear equations for the determination of the constants $A_1$ and $A_2$. Noting from (5.16), (5.17) and (5.18) that $y(a) = A_1$, $y'(a) = A_2$, we find the linear equations

$$\left. \begin{array}{l} \{\alpha_0 + \alpha_2 y^{(1)}(b) + \alpha_3 y^{(1)\prime}(b)\} A_1 + \{\alpha_1 + \alpha_2 y^{(2)}(b) + \alpha_3 y^{(2)\prime}(b)\} A_2 \\ \quad = \alpha_4 - \alpha_2 y^{(p)}(b) - \alpha_3 y^{(p)\prime}(b) \\ \{\beta_0 + \beta_2 y^{(1)}(b) + \beta_3 y^{(1)\prime}(b)\} A_1 + \{\beta_1 + \beta_2 y^{(2)}(b) + \beta_3 y^{(2)\prime}(b)\} A_2 \\ \quad = \beta_4 - \beta_2 y^{(p)}(b) - \beta_3 y^{(p)\prime}(b) \end{array} \right\} \tag{5.19}$$

for the determination of these constants. Having computed them we retrieve the required solution from (5.16), or solve (5.14) again as an initial-value problem now with our computed values of $y(a) = A_1$, $y'(a) = A_2$, the required initial conditions.

In this technique we have solved three initial-value problems and possibly a fourth, though the last is more of a check than a necessity. More favourable boundary conditions will reduce this number by one. Suppose, for example, that the first of (5.15) is replaced by

$$y(a) = \alpha_4 \quad \text{(with } \alpha_0 = 1, \quad \alpha_1 = \alpha_2 = \alpha_3 = 0 \text{ in (5.15))}, \tag{5.20}$$

so that only $y'(a)$ is needed to reduce our problem to initial-value form. We notice that

$$\bar{y}^{(p)} = y^{(p)} + A_1 y^{(1)} = y^{(p)} + \alpha_4 y^{(1)} \tag{5.21}$$

automatically satisfies the full differential equation (5.14) and the initial conditions $\bar{y}^{(p)}(a) = \alpha_4$, $\bar{y}^{(p)\prime}(a) = 0$. The first of (5.19) is automatically satisfied, and the second of (5.19) produces the required $A_2 = y'(a)$ from the equation

$$\{\beta_1 + \beta_2 y^{(2)\prime}(b) + \beta_3 y^{(2)\prime\prime}(b)\}A_2 = \beta_4 - \beta_0 \alpha_4 - \beta_2 \bar{y}^{(p)}(b) - \beta_3 \bar{y}^{(p)\prime}(b)\}. \tag{5.22}$$

The solution of only two initial-value problems is now strictly necessary, one for $\bar{y}^{(p)}$ and one for $y^{(2)}$.

## 5.3 THE SHOOTING METHOD. NON-LINEAR PROBLEMS

It is in the treatment of non-linear problems that the guessing and systematic adjustment becomes necessary. As we have seen, there is here no concept of 'particular' and 'complementary' solutions, but we can in fact use something similar to the variational equations also described in Chapter 2, together with an iterative scheme such as that of Newton.

Consider, for example, the non-linear 'boundary-layer' problem

$$y''' + yy'' + \beta(y'^2 - 1) = 0, \quad y(0) = y'(0) = 0, \quad y'(b) = 1, \tag{5.23}$$

for given values of $\beta$ and $b$. The obvious parameter, the knowledge of which turns (5.23) into an initial-value problem, is clearly

$$y''(0) = t. \tag{5.24}$$

We can shoot from $x = 0$ with any value of $t$, and by some systematic process compute a value of $t$ for which

$$y'(b) = 1. \tag{5.25}$$

The notation gets a bit complicated, since we shall need to differentiate with respect to $t$ as well as with respect to $x$. It seems most convenient to retain the 'dash notation' of (5.23), (5.24) and (5.25) for differentiation with respect to $x$, and the notation $y_t$ or $\partial y/\partial t$, whichever is contextually convenient, for differentiation with respect to $t$. The latter is used only to produce the variational equation with respect to $t$, and this notation was essentially the one used in the first introduction of the variational equation near the end of Section 2.3.

Our aim is the iterative solution for $t$ of equation (5.25), and a useful method for this is the Newton iteration, defined by

$$t^{(n+1)} = t^{(n)} - \frac{y^{(n)\prime}(b) - 1}{\dfrac{\partial}{\partial t}\{y^{(n)\prime}(b) - 1\}} = t^{(n)} - \frac{y^{(n)\prime}(b) - 1}{y_t^{(n)\prime}(b)}. \qquad (5.26)$$

Here $t^{(0)}$ is the first specified guess for $t$, and subsequent improvements are given by (5.26) in which $t = t^{(n)}$ is used in the computation of $y^{(n)}(x)$ and $y_t^{(n)}(x)$ and their first derivatives at $x = b$. The function $y^{(n)}(x)$ comes from the initial-value problem (5.23), with its last condition replaced by (5.24) with $t = t^{(n)}$. The function $y_t^{(n)}(x)$ then comes from the variational equation of all this with respect to $t$, given by

$$y_t''' + yy_t'' + y_t y'' + 2\beta y_t' y' = 0, \qquad y_t(0) = y_t'(0) = 0, \qquad y_t''(0) = 1, \quad (5.27)$$

in which $y = y^{(n)}$ and has just been computed, and the resulting $y_t$ is $y_t^{(n)}$.

As we noted in Section 2.3, this variational equation is linear, which is normally an advantage, but its form is different from that of the original equation in (5.23) and therefore needs a separate computer program. Moreover, its coefficients depend on the values of $y$ already computed, which adds a little further inconvenience. In these circumstances, as we noted in Section 2.7, it may be more convenient to calculate an approximation to $y_t$ by computing $y$ from the original equation (5.23) with two slightly different values of $t$. The convergence of the iteration in (5.26) is not then quite so fast, but in some cases this is economically tolerable.

In this example the boundary conditions were separated, and only one parameter was involved. We now consider a case of unseparated boundary conditions which needs the determination of two parameters, thereby demonstrating the generalized Newton technique. Consider the first-order equations

$$y' = y^2 + z^2, \qquad z' = \sin(yz), \qquad (5.28)$$

with boundary conditions

$$\left. \begin{aligned} g_1 &= y^2(a) + z(a)y(b) + y(a)\sin y(b) - 1.5 = 0 \\ g_2 &= y(a)z(b) - e^{y(a)}\sin y(b) + z(a) = 0 \end{aligned} \right\}. \qquad (5.29)$$

To produce an initial-value problem we need both $y(a)$ and $z(a)$ (or $y(b)$ and $z(b)$), and we start with the parameters

$$y(a) = p, \quad z(a) = q, \qquad (5.30)$$

observing that we then need to satisfy (5.29) in the form of the simultaneous non-linear equations

$$g_1(p, q) = 0, \qquad g_2(p, q) = 0. \tag{5.31}$$

In this case the Newton iteration equations are given by

$$\left. \begin{aligned} &\frac{\partial g_i}{\partial p^{(n)}} \delta p^{(n)} + \frac{\partial g_i}{\partial q^{(n)}} \delta q^{(n)} + g_i(p^{(n)}, q^{(n)}) = 0, \quad i = 1, 2 \\ &p^{(n+1)} = p^{(n)} + \delta p^{(n)}, \quad q^{(n+1)} = q^{(n)} + \delta q^{(n)} \end{aligned} \right\}. \tag{5.32}$$

Clearly

$$g_1 = p^2 + qy(b) + p \sin y(b) - 1.5, \quad g_2 = pz(b) - e^p \sin y(b) + q \tag{5.33}$$

and the required partial derivatives in (5.32) are obtained by successive differentiation with respect to $p$ and $q$ of equations (5.28), (5.30) and (5.33). We find

$$y_p' = 2yy_p + 2zz_p, \quad z_p' = (zy_p + yz_p) \cos(yz), \quad y_p(a) = 1, \quad z_p(a) = 0, \quad (5.34)$$

and

$$y_q' = 2yy_q + 2zz_q, \quad z_q' = (zy_q + yz_q) \cos(yz), \quad y_q(a) = 0, \quad z_q(a) = 1, \tag{5.35}$$

where $y_p = \partial y / \partial p$, $y_p' = d/dx(\partial y / \partial p)$, etc., and (5.34) and (5.35) are initial-value problems for the determination of $y_p, z_p, y_q, z_q$. From these we can compute

$$\left. \begin{aligned} &\frac{\partial g_1}{\partial p} = 2p + qy_p(b) + \sin y(b) + py_p(b) \cos y(b) \\ &\frac{\partial g_1}{\partial q} = y(b) + qy_q(b) + py_q(b) \cos y(b) \\ &\frac{\partial g_2}{\partial p} = z(b) + pz_p(b) - e^p \sin y(b) - e^p y_p(b) \cos y(b) \\ &\frac{\partial g_2}{\partial q} = pz_q(b) - e^p y_q(b) \cos y(b) + 1 \end{aligned} \right\}, \tag{5.36}$$

and with everything evaluated at the current stage we now complete one further step of the iteration by solving the linear equations (5.32) for $\delta p^{(n)}$ and $\delta q^{(n)}$.

These examples are typical, and should be sufficient to illustrate how to cope with any kind and distribution of boundary conditions. The numerical techniques with which the relevant initial-value problems are solved, such as Runge–Kutta methods, single or multi-step methods, are determined, as in all cases, by the nature of the relevant differential equations.

## 5.4 THE SHOOTING METHOD. EIGENVALUE PROBLEMS

We now adapt our methods of the previous section to the numerical solution of eigenvalue problems typified by equations (5.7). Consider, for example, the particular problem (5.8). This already has the required initial conditions to turn (5.8) into an initial-value problem. For $y(0)$ is already specified to be zero, and clearly if $y(x)$ is a solution of the problem then $ky(x)$ is also a solution for any value of $k$. Taking $y'(0) = 1$ then effectively fixed the value of $k$, removes the ambiguity, and gives the required second initial condition.

Note that this idea is relevant to any other type of separated condition, such as

$$y(0) + 5y'(0) = 0 \qquad (5.37)$$

at one end of the range. All eigenvalue problems of the type we are considering have homogeneous differential equations and homogeneous boundary conditions, so that the zero on the right of (5.37) is necessary. But clearly we cannot now have $y(0) = 0$, since then $y'(0) = 0$ from (5.37) and $y''$ and all higher derivatives are also all zero at $x = 0$ to satisfy the differential equation and the equations formed by successive differentiation thereof. This gives $y(x) = 0$ everywhere, an accurate but trivial solution. We therefore take $y(0) = 1$, and obtain $y'(0) = -\frac{1}{5}$ directly from (5.37), giving the correct number of required initial conditions.

What is lacking here to turn the problem into one of initial-value type is the value of the currently unknown $\lambda$, and this is our parameter. For any value of $\lambda$ we can solve the initial-value problem and compute $y(x)$, its values depending on $\lambda$. For problem (5.8) we want to choose $\lambda$ so that

$$y(1) = 0, \qquad (5.38)$$

and for any other separated condition at $x = 1$ we have a similar non-linear equation to solve for $\lambda$. The Newton iteration for (5.38) is very similar to that for (5.25) and is given by

$$\lambda^{(n+1)} = \lambda^{(n)} - \frac{y^{(n)}(1)}{y_\lambda^{(n)}(1)}, \qquad (5.39)$$

where $y^{(n)}$ satisfies the differential equation in (5.8) with $y^{(n)}(0) = 0$, $y^{(n)\prime}(0) = 1$ and $\lambda = \lambda^{(n)}$, and $y_\lambda^{(n)}$ satisfies the variational equation

$$y_\lambda^{(n)\prime\prime} + \lambda^{(n)} y_\lambda^{(n)} + y^{(n)} = 0, \qquad y_\lambda^{(n)}(0) = y_\lambda^{(n)\prime}(0) = 0, \qquad (5.40)$$

with $y^{(n)}$ and $\lambda^{(n)}$ already known.

With unseparated boundary conditions like

$$y(1) = -\tfrac{1}{2}y(0), \quad y'(1) = 2y'(0) \qquad (5.41)$$

we cannot avoid a second parameter. We can take $y(0) = 1$, which is just a normalizing factor, but $y'(0)$ is not then obtainable, and we must take

$$y'(0) = q \qquad (5.42)$$

as our second parameter. We then have to solve the equations

$$y(1) + \tfrac{1}{2} = 0, \quad y'(1) - 2q = 0, \qquad (5.43)$$

where $y(1)$ and $y'(1)$ depend on both $\lambda$ and $q$, and the Newton iteration is

$$\left.\begin{array}{l} y^{(n)}_\lambda(1)\delta\lambda^{(n)} + y^{(n)}_q(1)\delta^{(n)}_q + y^{(n)}(1) + \tfrac{1}{2} = 0 \\ y^{(n)\prime}_\lambda(1)\delta\lambda^{(n)} + \{y^{(n)\prime}_q(1) - 2\}\delta q^{(n)} + y^{(n)\prime}(1) - 2q^{(n)} = 0 \end{array}\right\} \quad (5.44)$$

for additive corrections $\delta\lambda^{(n)}$, $\delta q^{(n)}$ to current estimates $\lambda^{(n)}, q(n)$. The function $y^{(n)}(x)$ satisfies the differential equation in (5.8) with $y^{(n)}(0) = 1$, $y'^{(n)}(0) = q^{(n)}$, and $y^{(n)}_\lambda(x)$ and $y^{(n)}_q(x)$ satisfy the relevant variational equations, given by

$$\left.\begin{array}{ll} y^{(n)\prime\prime\prime}_\lambda + \lambda^{(n)}y^{(n)}_\lambda + y^{(n)} = 0, & y^{(n)}_\lambda(0) = 0, \quad y^{(n)\prime}_\lambda(0) = 0 \\ y^{(n)\prime\prime\prime}_q + \lambda^{(n)}y^{(n)}_q = 0, & y^{(n)}_q(0) = 0, \quad y^{(n)\prime}_q(0) = 1 \end{array}\right\}. \quad (5.45)$$

The various solutions give the wanted terms in the linear equations (5.44) for $\delta\lambda^{(n)}$ and $\delta q^{(n)}$.

For the differential equation in (5.8) with conditions (5.41) one solution is

$$y = -\tfrac{1}{2}\sin\frac{\pi}{2}x + \cos\frac{\pi}{2}x, \quad \lambda = \pi^2/4 \sim 2.4674, \quad q = y'(0) = -\frac{\pi}{4} \sim -0.7854.$$

$$(5.46)$$

With just one step of Newton iteration we can solve analytically, and without great difficulty, all the relevant initial-value problems, and a reasonably accurate start and one step of the iteration give the results $\lambda^{(0)} = 2.56$, $q^{(0)} = -0.8$, $\lambda^{(1)} = 2.4640$, $q^{(1)} = -0.7859$, which again confirm the second-order Newton rate of convergence.

Our simple examples should give enough information to permit the treatment of virtually any eigenvalue problem in which the solution is smooth and the range of integration is finite, but there is one important point that merits further attention. Our eigenvalue problems have an infinite number of solutions, and a natural question is whether our iterative technique converges and to what solution it converges in the affirmative case.

With regard to convergence of the iterative process we merely note that convergence to any solution is not guaranteed from any starting approximation, but the nearer the latter is to a solution the more likely is convergence to that solution. For example, for problem (5.8) with the first approximation $\lambda^{(0)} = 0$ we find convergence to the smallest eigenvalue $\lambda = \pi^2$, and with the first approximation $\lambda^{(0)} = 100$ we converge to the third eigenvalue $\lambda = 9\pi^2 \sim 88.83$. The question now arises as to whether, without having the advantage of an analytical solution, we are confident that there is no other solution between $\lambda = \pi^2$ and $\lambda = 9\pi^2$.

For some problems we can settle our doubts by inspection of the computed eigenfunctions. A very common class of eigenvalue problems is typified by the equation

$$\frac{\mathrm{d}}{\mathrm{d}x}\left\{p(x)\frac{\mathrm{d}y}{\mathrm{d}x}\right\} + \{\lambda r(x) + q(x)\}y = 0, \quad (5.47)$$

with $p(x)$ and $r(x)$ positive for $a \leqslant x \leqslant b$, and with separated boundary conditions

$$\alpha y(a) + \alpha' y'(a) = 0, \qquad \beta y(b) + \beta' y'(b) = 0. \qquad (5.48)$$

For such problems it is known that there is an infinite set of eigenvalues such that

$$\lambda_0 < \lambda_1 < \cdots < \lambda_n < \cdots, \qquad (5.49)$$

and that the eigenfunction $y_n$ corresponding to the eigenvalue $\lambda_n$ has exactly $n$ zeros in the range $a \leqslant x \leqslant b$ (that is not counting a possible zero at either boundary point). This theorem gives all the required information. In our current example the eigenfunction corresponding to $\lambda = \pi^2$ has no internal zero, and that corresponding to $\lambda = 9\pi^2$ has two internal zeros. We are therefore confident that there is in fact one eigenvalue between these two numbers, whose eigenfunction will have one internal zero.

A good starting value for the iteration designed to find this result is easily obtained by trial and error, choosing a $\lambda^{(0)}$ which, without necessarily satisfying the second boundary condition, gives the correct number of zeros in the first computation of $y(x, \lambda^{(0)})$. For example, with the guess $\lambda^{(0)} = 36$ we have the solution $y = \frac{1}{6}\sin 6x$ which satisfies $y(0) = 0$, $y'(0) = 1$. It does not satisfy $y(1) = 0$ but it has zeros at $x = \pi/6, \pi/3, \pi/2, \ldots$, of which only the first is in the range $0 < x < 1$. This is therefore a satisfactory first approximation.

Nothing quite so simple is available for general problems. For example, the $\lambda$ in (5.46) is the smallest eigenvalue of its particular problem, but the eigenfunction has one internal zero in the range.

## 5.5 THE SHOOTING METHOD. PROBLEMS WITH UNKNOWN BOUNDARIES

We turn finally to the problem of type (5.10) in which an extra boundary condition is imposed which enables us to determine the position of one boundary. Again it is probably more informative if we use the problem (5.10) as a particular example.

There are obviously two parameters to compute, one the value of $y'(0)$ which turns our boundary-value problem into an initial-value problem, and the other the value of $b$, there being the two conditions

$$y(b) - 1 = 0, \quad y'(b) - 2 = 0 \qquad (5.50)$$

for the determination of these parameters. The technique becomes obvious when we demonstrate the use of Newton iteration with analytical solution, which is here possible, of the relevant differential equations.

It is easy to show that the solution of the initial-value problem

$$y'' + y = 1, \quad y(0) = 0, \quad y'(0) = q, \qquad (5.51)$$

is

$$y(x) = 1 - \cos x + q \sin x, \quad y'(x) = \sin x + q \cos x. \tag{5.52}$$

Then the Newton iteration, for corrections $\delta q^{(0)}$ and $\delta b^{(0)}$ to respective starting approximations $q^{(0)}$ and $b^{(0)}$ solves for this purpose the linear equations

$$\left. \begin{array}{l} y_q(b)\delta q^{(0)} + y_b(b)\delta b^{(0)} + y(b^{(0)}) - 1 = 0 \\ y_q'(b)\delta q^{(0)} + y_b'(b)\delta b^{(0)} + y'(b^{(0)}) - 2 = 0 \end{array} \right\}, \tag{5.53}$$

the coefficients being evaluated at $q = q^{(0)}$, $b = b^{(0)}$.

Now from (5.52) we have

$$y(b) = 1 - \cos b + q \sin b, \quad y'(b) = \sin b + q \cos b, \tag{5.54}$$

and

$$y_q(b) = \sin b, \quad y_q'(b) = \cos b. \tag{5.55}$$

Note, as in all previous cases, that these results can be obtained from the variational problem for $q$, which is clearly

$$y_q'' + y_q = 0, \quad y_q(0) = 0, \quad y_q'(0) = 1. \tag{5.56}$$

But it is important to observe, and to note that this phenomenon did not appear in our previous examples, that $y(x)$ in (5.52) is not an explicit function of $b$, and we cannot find $y_b(b)$ and $y_b'(b)$ from any variational problem. Their computation is in fact a much easier operation. We can differentiate with respect to $b$ the expressions for $y(b)$ and $y'(b)$ given in (5.54), obtaining the required numbers

$$y_b(b) = \sin b + q \cos b, \quad y_b'(b) = \cos b - q \sin b. \tag{5.57}$$

Even more simply, we observe that

$$y_b(b) = y'(x)_{x=b}, \tag{5.58}$$

which is already known from (5.52), and

$$y_b'(b) = \frac{\partial}{\partial b} \{y'(b)_{x=b}\} = y''(x)_{x=b} = 1 - y(b) \tag{5.59}$$

for the particular differential equation in (5.10), and this is also known from (5.52).

Our method produces satisfactory numerical results. In (5.12) we observed that $b = \pi/6 \sim 0.5236$ is a solution, and it is easy to see that with this solution

$$q = \cot b = \sqrt{3} \sim 1.732. \tag{5.60}$$

Starting with $q^{(0)} = 1.7$, $b^{(0)} = 0.5$, we find from one step of Newton iteration the better values $q^{(1)} = 1.731$, $b^{(1)} = 0.5242$, confirming the second-order rate of convergence of the iteration.

In practice, of course, we shall not have analytic solutions like (5.52), and (5.58) will be computed by numerical differentiation, preferably using the central difference formula of equation (3.113), the values of $y(x)$ obtained at mesh points in the numerical solution of (5.51) being extended as far as necessary for this purpose. This problem also has a new feature, in that the range of integration varies in each step of the iteration which complicates somewhat the repeated calculation of $y'(x)$ at successive values of $b$. If we use the same interval $h$ throughout the whole computation we have to use special formulae, at least with finite-difference methods and at least near the end of the range. This is usefully avoided by dividing each computed range into the same number $n$ of equal intervals. Approximate values of $y(x)$ are then computed at different points in successive iterative steps, but when the process has converged, measured by the behaviour of successive approximations $q^{(r)}$ and $b^{(r)}$, the final function values are obtained at equal intervals.

## 5.6 INDUCED INSTABILITIES OF SHOOTING METHODS

The essence of the shooting methods is the replacement of boundary-value problems by initial-value problems. So far we have assumed that all the latter are soluble analytically and exactly. This, of course, is most unusual, and in practice we usually replace 'analytical solution' by 'numerical solution', doing the best we can with the numerical methods of Chapters 3 and 4 according to the accuracy and economy required in each particular context. But it may happen that with some problems the shooting methods produce an unsatisfactory amount of induced instability, even in well-conditioned problems. In the following two subsections we concentrate on linear recurrence relations, which make most of the important points, with a few remarks on linear and non-linear differential equations in subsection (iii).

### (i) A simple linear difference equation

Near the end of Section 2.4 we showed that the recurrence-relation boundary-value problem

$$y_{r+1} - 10.1y_r + y_{r-1} = -1.35r, \quad y_0 = 0, \quad y_N = \tfrac{1}{6}N \qquad (5.61)$$

is extremely well conditioned with respect to small uncertainties in the boundary values, but in the further note in Section 2.7 referring to Section 2.6 we indicated that the shooting method may exhibit an unsatisfactory amount of induced instability. In analogy with the material in the vicinity of equations (5.20)–(5.22), the shooting method solves for $\bar{y}_r^{(p)}$ from the system

$$\bar{y}_{r+1}^{(p)} - 10.1\bar{y}_r^{(p)} + \bar{y}_{r-1}^{(p)} = -1.35r, \quad \bar{y}_0^{(p)} = 0, \quad \bar{y}_1^{(p)} = 0, \qquad (5.62)$$

and for $y_r^{(2)}$ from the system

$$y_{r+1}^{(2)} - 10.1y_r^{(2)} + y_{r-1}^{(2)} = 0, \quad y_0^{(2)} = 0, \quad y_1^{(2)} = 1, \qquad (5.63)$$

**Table 5.1**

| $r$ | $y_r$ | $y_r(N=6)$ | $y_r(N=12)$ | $y_r(N=18)$ |
|---|---|---|---|---|
| 0 | 0.000000 | 0.000000 | 0.000000 | 0.000000 |
| 1 | 0.166667 | 0.166667 | 0.166667 | 0.166667 |
| 2 | 0.333333 | 0.333334 | 0.333333 | 0.333333 |
| 3 | 0.500000 | 0.500002 | 0.499998 | 0.499994 |
| 4 | 0.666667 | 0.666687 | 0.666641 | 0.666611 |
| 5 | 0.833333 | 0.833496 | 0.833008 | 0.832642 |
| 6 | 1.000000 | 1.000000 | 0.996094 | 0.992188 |
| 7 | 1.166667 | | 1.140625 | 1.109375 |
| 8 | 1.333333 | | 0.875000 | 0.625000 |
| 9 | 1.500000 | | $-2.000000$ | $-4.000000$ |
| 10 | 1.666667 | | $-32.000000$ | $-64.000000$ |
| 11 | 1.833333 | | $-256.000000$ | $-512.000000$ |
| 12 | 2.000000 | | 0.000000 | $-3072.000000$ |
| 13 | 2.166667 | | | $-16384.000000$ |
| 14 | 2.333333 | | | $-262144.000000$ |
| 15 | 2.500000 | | | $-2097152.000000$ |
| 16 | 2.666667 | | | $-16777216.000000$ |
| 17 | 2.833333 | | | $-134217728.000000$ |
| 18 | 3.000000 | | | 0.000000 |

and then computes $q$ from the equation

$$\bar{y}_N^{(p)} + q y_N^{(2)} = \tfrac{1}{6}N \tag{5.64}$$

to satisfy the second boundary condition. The required solution is then

$$y_r = \bar{y}_r^{(p)} + q y_r^{(2)}. \tag{5.65}$$

The exact solution is $y_r = \tfrac{1}{6}r$, with $q = \tfrac{1}{6}$, and in Table 5.1 we give six-decimal values of this quantity in comparison with the $y_r$ obtained by computation for three cases, with $N$ having the respective values 6, 12 and 18. In each case, of course, the computed $q$ is also the computed $y_1$.

For small $N$ we see that the results are reasonably satisfactory, but as $N$ increases they become poorer for the larger values of $r$, and as $N$ increases still further the error creeps back a little to affect the results for smaller values of $r$. We also observe that for large $N$ we do not even reproduce accurately the specified value of $y_N$ from (5.65), even though $q$ was deliberately chosen for this purpose from (5.64). To explain these phenomena we need to examine briefly the details of the arithmetic.

Suppose, for example, that in the computation we can do all the arithmetic exactly but need to round the computed $\bar{y}_r^{(p)}$ for storage in a four-decimal machine. Then $\bar{y}_0^{(p)} = 0$, $\bar{y}_1^{(p)} = 0$, $\bar{y}_2^{(p)} = -1.35$ are exact, but

$$\bar{y}_3^{(p)} = 10.1(-1.35) - 2(1.35) = -16.335, \tag{5.66}$$

which has to be rounded to $-16.34 = -0.1634 \times 10^2$. Next

$$\bar{y}_4^{(p)} = 10.1(-16.34) - (-1.35) - 3(1.35) = -167.734, \qquad (5.67)$$

which the machine rounds to $-167.7 = -0.1677 \times 10^3$. Similarly for $\bar{y}_5^{(p)}$ we round the computed $-1682.83$ to $-1683 = -0.1683 \times 10^4$, and finally, for our current purpose,

$$\bar{y}_6^{(p)} = 10.1(-1683) + 167.7 - 5(1.35) = -16837.35, \qquad (5.68)$$

which has to be rounded to $-0.1684 \times 10^5$, with a local error of as much as 2.65, some 40% of the local value of $1.35r$. It follows, even at this early stage of the total computation, that the computed and stored values of $\bar{y}_r^{(p)}$ are not satisfying the recurrence exactly. In fact, with $\bar{y}_4^{(p)} = -167.7$ and $\bar{y}_5^{(p)} = -1683$, we should with our computer get *the same* $\bar{y}_6^{(p)} = -0.1684 \times 10^5$ if we replaced the specified $-1.35r = -6.75$ at this point by any number in the approximate range $-4.4$ to $-14.4$. And of course with this example this phenomenon rapidly gets more pronounced as $r$ increases.

Consider now the computation of $q$ when $N = 6$. Equation (5.64) gives

$$q = (1 - \bar{y}_6^{(p)})/y_6^{(2)}, \qquad (5.69)$$

and with a $\bar{y}_6^{(p)}$ of $-0.1684 \times 10^5$ this extra unit in the numerator makes no contribution at all in our four-decimal machine. Indeed, any specified $y_6$ in the range $-5$ to $+5$ would give exactly the same four-digit numerator in (5.69) and the same value of $q$.

The quantities $y_r^{(2)}$ obtained by the appropriate recurrence will also have absolute errors increasing with $r$, and without going into excessive detail one can show that the finally computed $y_r$, instead of satisfying (5.61), will be the exact solution of the 'perturbed' problem

$$\left.\begin{array}{c} y_{r+1} - 10.1y_r + y_{r-1} = -1.35r + \varepsilon_r \\ y_0 = 0, \quad y_N = \tfrac{1}{6}N + \varepsilon_N \end{array}\right\}, \qquad (5.70)$$

where $\varepsilon_r$ increases rapidly with $r$ and $\varepsilon_N$ increases significantly with $N$.

All this, of course, is related to the fact that for this particular example the initial-value problems for $\bar{y}_r^{(p)}$ and $y_r^{(2)}$ are inherently absolutely ill conditioned, this being the reason for the ultimately large $\bar{y}_r^{(p)}$ and $y_r^{(2)}$ and the unsatisfactory resulting induced instability. The initial-value problem was discussed in Chapter 2, and equation (2.41) reveals that $q$ (the $\beta$ of that equation) must be accurate to many figures for us to have a chance of getting good step-by-step results for large $r$.

It is perhaps rather surprising to find such good results for small $r$, but in fact our boundary-value problem is extremely well conditioned. The perturbations $\varepsilon_r$ in (5.70) are small for small $r$, and later perturbations affect much earlier values only slightly. In particular, the perturbation $\varepsilon_N$ in (5.70) affects $y_r$ by an amount very close to $10^{r-N}\varepsilon_N$, and we need an $\varepsilon_N$ of as much as $5 \times 10^4$ to affect $y_1$ by as much as five units in the first neglected decimal in Table 5.1.

## (ii) Simultaneous linear difference equations

A slightly different kind of instability, but also associated with ill-conditioning of a corresponding initial-value problem, is manifest in the solving by shooting methods of particular sets of simultaneous difference equations. Consider, for example, the simple set of difference equations given by

$$y_{r+1} = a_1 y_r + b_1 z_r + c_r, \qquad z_{r+1} = a_2 y_r + b_2 z_r + d_r, \tag{5.71}$$

where $a_1, b_1, a_2$ and $b_2$ are constants and $c_r$ and $d_r$ are known functions of $r$, together with the unseparated boundary conditions

$$\alpha_1 y_0 + \beta_1 z_0 + \gamma_1 y_N + \delta_1 z_N = \varepsilon_1, \qquad \alpha_2 y_0 + \beta_2 z_0 + \gamma_2 y_N + \delta_2 z_N = \varepsilon_2. \tag{5.72}$$

This situation is analogous to that of the first part of Section 5.2, with two parameters to compute. The general solution of the recurrence relation is

$$y_r = y_r^{(p)} + p y_r^{(1)} + q y_r^{(2)}, \qquad z_r = z_r^{(p)} + p z_r^{(1)} + q z_r^{(2)}, \tag{5.73}$$

where

$$y_{r+1}^{(p)} = a_1 y_r^{(p)} + b_1 z_r^{(p)} + c_r, \qquad z_{r+1}^{(p)} = a_2 y_r^{(p)} + b_2 z_r^{(p)} + d_r, \quad y_0^{(p)} = 0, \quad z_0^{(p)} = 0, \tag{5.74}$$

$$y_{r+1}^{(1)} = a_1 y_r^{(1)} + b_1 z_r^{(1)}, \qquad z_{r+1}^{(1)} = a_2 y_r^{(1)} + b_2 z_r^{(1)}, \qquad y_0^{(1)} = 1, \quad z_0^{(1)} = 0, \tag{5.75}$$

$$y_{r+1}^{(2)} = a_1 y_r^{(2)} + b_1 z_r^{(2)}, \qquad z_{r+1}^{(2)} = a_2 y_r^{(2)} + b_2 z_r^{(2)}, \qquad y_0^{(2)} = 0, \quad z_0^{(2)} = 1, \tag{5.76}$$

the initial conditions in (5.75) and (5.76) being chosen so that the relevant solutions are independent. The required parameters $p$ and $q$ are then obtained by substituting (5.73)–(5.76) in (5.72), and obtaining for $p$ and $q$ the two linear equations corresponding to (5.19) given by

$$\left. \begin{aligned} (\alpha_1 + \gamma_1 y_N^{(1)} + \delta_1 z_N^{(1)})p + (\beta_1 + \gamma_1 y_N^{(2)} + \delta_1 z_N^{(2)})q = \varepsilon_1 - \gamma_1 y_N^{(p)} - \delta_1 z_N^{(p)} \\ (\alpha_2 + \gamma_2 y_N^{(1)} + \delta_2 z_N^{(1)})p + (\beta_2 + \gamma_2 y_N^{(2)} + \delta_2 z_N^{(2)})q = \varepsilon_2 - \gamma_2 y_N^{(p)} - \delta_2 z_N^{(p)} \end{aligned} \right\} . \tag{5.77}$$

The question is whether or not $p$ and $q$ can be computed sufficiently accurately from (5.77). The danger here is the existence of one quite dominating eigenvalue of the matrix

$$A_1 = \begin{bmatrix} a_1 & b_1 \\ a_2 & b_2 \end{bmatrix} \tag{5.78}$$

of equations (5.71), and the point is best made with a numerical example. For $a_1 = 6$, $b_1 = 2$, $a_2 = 10$ and $b_2 = 5$, the eigenvalues are 10 and 1, the respective eigenvectors are normalizing multiples of $\begin{bmatrix} 1 \\ 2 \end{bmatrix}$ and $\begin{bmatrix} 2 \\ -5 \end{bmatrix}$, and it can be shown, and easily verified from (5.75) and (5.76), that

$$\left. \begin{aligned} y_r^{(1)} &= \tfrac{1}{9}(5 \cdot 10^r + 4 \cdot 1^r), & z_r^{(1)} &= \tfrac{1}{9}(10 \cdot 10^r - 10 \cdot 1^r) \\ y_r^{(2)} &= \tfrac{1}{9}(2 \cdot 10^r - 2 \cdot 1^r), & z_r^{(2)} &= \tfrac{1}{9}(4 \cdot 10^r + 5 \cdot 1^r) \end{aligned} \right\}. \tag{5.79}$$

The computed values will be affected by rounding errors discussed in the previous subsection, but we note the additional fact that for large $t = N$, and for normal-sized coefficients in the boundary conditions (5.72), the matrix for $p$ and $q$ in (5.77) is dominated by

$$B \sim \begin{bmatrix} \tfrac{5}{9}\gamma_1 \cdot 10^N + \tfrac{10}{9}\delta_1 \cdot 10^N & \tfrac{2}{9}\gamma_1 \cdot 10^N + \tfrac{4}{9}\delta_1 \cdot 10^N \\ \tfrac{5}{9}\gamma_2 \cdot 10^N + \tfrac{10}{9}\delta_2 \cdot 10^N & \tfrac{2}{9}\gamma_2 \cdot 10^N + \tfrac{4}{9}\delta_2 \cdot 10^N \end{bmatrix}, \tag{5.80}$$

and this matrix is singular. We would hardly expect to get good results from a matrix which differs from this only by rounding errors, and again the problem arises through the ill-conditioning of problem (5.71) with initial conditions at $r = 0$. The solutions with suffixes (1) and (2) in (5.75) and (5.76) rapidly lose their linear independence as $r$ increases.

Consider, for example, the problem

$$y_{r+1} = 6y_r + 2z_r + r - 6, \qquad z_{r+1} = 10y_r + 5z_r + 2r - 17, \tag{5.81}$$

with boundary conditions

$$y_0 + z_0 + 3y_{12} + z_{12} = 6, \qquad 4y_0 + 6z_0 + 2y_{12} + z_{12} = 1, \tag{5.82}$$

which has the solution

$$y_r = 1 + r, \qquad z_r = 1 - 3r. \tag{5.83}$$

With our seven-decimal machine equations (5.77) look like

$$\left. \begin{aligned} 2.777777 \times 10^{12}p + 1.111111 \times 10^{12}q &= 3.888889 \times 10^{12} \\ 2.222222 \times 10^{12}p + 0.888889 \times 10^{12}q &= 3.11111 \times 10^{12} \end{aligned} \right\}. \tag{5.84}$$

The matrix is very nearly singular, but with the numbers stored in the machine the results appear with $p = 1.00000$, $q = 2.00000$, of which the second is far from the truth, and at $r = 12$ equation (5.73) gives $y_r = 0.222222 \times 10^{12}$, $z_r = 0.444444 \times 10^{12}$, which are catastrophically in error.

### (iii) Linear and non-linear differential equations

We should expect somewhat similar results in the shooting-method solution of linear differential equations of similar form. It is perhaps worth noting that the set of first-order differential equations corresponding to the set of difference equations in (5.71) is given by

$$y' = A(x)y + b(x), \tag{5.85}$$

and if the matrix $A(x)$ is independent of $x$ the complementary solution of (5.85) contains multiples of $e^{\lambda x}\xi$, where $\lambda$ is an eigenvalue of $A$ and $\xi$ the corresponding eigenvector.

We should also expect similar problems with non-linear differential equations which are ill conditioned in the initial-value sense. Either the original equation, or the variational equations, or both, may produce rapidly increasing numbers in the step-by-step solutions, with various components of the differential equation or boundary conditions making no significant contributions in the solving process. When more than one parameter is involved, as in problems of type (5.28) and (5.29), we may find ill-conditioning in the variational equations of type (5.34) and (5.35), with the linear dependence imposed by the conditions in (5.34) and (5.35) disappearing as the step-by-step method proceeds and with the possibility of near-singularity in the matrix like that of (5.32) for the determination of new parametric values.

## 5.7 AVOIDING INDUCED INSTABILITIES

### (i) Shooting the other way

In our previous discussion we have performed the shooting method from one boundary point and integrated or carried out the recurrence as far as the second boundary point. But there are, or course, other possibilities which avoid or to some extent reduce the dangers of instability which, as we saw, depend on a possible ill-conditioning of the corresponding initial-value problem.

One possibility is to shoot in the opposite direction in cases in which there is little ill-conditioning with initial conditions specified at the second boundary point. Consider, for example, the problem similar to (5.61) and with the same solution given by

$$y_{r+1} - 7y_r - 8y_{r-1} = -\tfrac{7}{3}r + \tfrac{3}{2}, \quad y_0 = 0, \quad y_N = \tfrac{1}{6}N. \tag{5.86}$$

Shooting from $r = 0$ we expect failure for large $N$ due to the fact that the complementary solutions of (5.86) are multiples of $8^r$ and $(-1)^r$, of which the first increases very rapidly. In fact, with $N = 16$ we produce, by the method of Section 5.6(i), the poor results shown in Table 5.2. In the reverse direction, however, the complementary solutions are multiples of $8^{r-N}$ and $(-1)^{r-N}$, neither of which increases rapidly as $r$ increases, or even at all, and shooting backwards from $r = N$ gives the excellent results shown in Table 5.3.

**Table 5.2**

| $r$ | 0 | 1 | 2 | 4 | 8 | 12 | 16 |
|-----|---|---|---|---|---|----|----|
| $y_r$ | 0 | 0.16667 | 0.33333 | 0.66666 | 1.34375 | 0.00000 | 0.00000 |

**Table 5.3**

| $r$ | 16 | 15 | 14 | 12 | 8 | 4 | 0 |
|---|---|---|---|---|---|---|---|
| $y_r$ | 2.66667 | 2.50000 | 2.33333 | 2.00000 | 1.33333 | 0.66667 | 0 |

In the analysis leading to the equation corresponding to (5.70) both the $\varepsilon_r$ and the $\varepsilon_N$ are now very small, and the computed solution satisfies an only very slight perturbation of the original problem. Since the latter is very well conditioned the computed solution is very good.

For similar reasons the problem of Section 5.6(ii) is soluble accurately by reversing the shooting direction. The relevant eigenvalues are now $10^{-1}$ and 1. The contribution from the first of these will soon disappear in the backward shooting, but again this introduces only very small rounding errors. Moreover, although this contribution has disappeared for large $N$ from the relevant coefficients in (5.77) the other coefficients $\alpha_1, \beta_1, \alpha_2$ and $\beta_2$ are now contributing their full weight, and the matrix of (5.77) is not singular (at least from this cause). In fact backward shooting for this problem produces, with our seven-decimal machine with $N = 16$, results with a maximum error in $y_r$ and $z_r$ of 0.00005.

### (ii) Matching in the middle

Note, however, that the problem of (5.61) cannot be solved in this way. The complementary solutions in forward shooting are $10^r$ and $(1/10)^r$, and in backward shooting are $10^{r-N}$ and $(1/10)^{r-N}$, with similar effect. A possible solution in this case is to prevent the occurrence of very large numbers by shooting in both directions and 'matching' at some intermediate point. In connection with problem (5.61), for example, we would then proceed as follows, the superscripts f and b in the sequel representing 'forward' and 'backward' recurrence. The change of notation from (5.62) and (5.63) to (5.87) has no significance, designed solely to make eligible the introduction of the subscripts.

1. Calculate (forwards) the values of $u_r^{(f)}$ and $v_r^{(f)}$ for $r = 2, 3, \ldots, m-1, m$ from

$$\left.\begin{array}{ll} u_{r+1}^{(f)} - 10.1u_r^{(f)} + u_{r-1}^{(f)} = -1.35r, & u_0^{(f)} = 0, \quad u_1^{(f)} = 0 \\ v_{r+1}^{(f)} - 10.1v_r^{(f)} + v_{r-1}^{(f)} = 0, & v_0^{(f)} = 0, \quad v_1^{(f)} = 1 \end{array}\right\}. \tag{5.87}$$

2. Calculate (backwards) the values of $u_r^{(b)}$ and $v_r^{(b)}$ for $r = N-2, N-3, \ldots, m, m-1$ from

$$\left.\begin{array}{ll} u_{r-1}^{(b)} - 10.1u_r^{(b)} + u_{r+1}^{(b)} = -1.35r, & u_N^{(b)} = \tfrac{1}{6}N, \quad u_{N-1}^{(b)} = 0 \\ v_{r-1}^{(b)} - 10.1v_r^{(b)} + v_{r+1}^{(b)} = 0, & v_N^{(b)} = 0, \quad v_{N-1}^{(b)} = 1 \end{array}\right\}. \tag{5.88}$$

3. Then $y_r^{(f)} = u_r^{(f)} + q_f v_r^{(f)}$ satisfies the recurrence and the first boundary condition, and $y_r^{(b)} = u_r^{(b)} + q_b v_r^{(b)}$ satisfies the recurrence and the second boundary

condition, and they match at the two points $r = m - 1$ and $r = m$ if

$$
\left.
\begin{aligned}
u^{(f)}_{m-1} + q_f v^{(f)}_{m-1} &= u^{(b)}_{m-1} + q_b v^{(b)}_{m-1} \\
u^{(f)}_m + q_f v^{(f)}_m &= u^{(b)}_m + q_b v^{(b)}_m
\end{aligned}
\right\}.
\tag{5.89}
$$

These two linear equations suffice for the determination of $q_f$ and $q_b$ and hence for the complete solution, and of course matching of $y^{(f)}$ and $y^{(b)}$ is necessary *at two points* to guarantee that $y^{(f)}$ and $y^{(b)}$, solutions of a second-order recurrence relation, are identical at all other points.

The idea is that by covering only part of the range in each shooting no relevant numbers get too large, and the effect of this device on problem (5.61) is revealed with a comparison of Table 5.1 and the current Table 5.4 whose results are obtained by matching at the points 3, 4 for $N = 6$, then 5, 6 for $N = 12$, and 8, 9 for $N = 18$.

As we might expect from the earlier discussion of one-way shooting for this problem, and particularly from the remarks at the end of Section 5.6 (i), the values of $q_f$ and $q_b$, which are those of $y_1$ and $y_{N-1}$, are in fact quite accurate. The results are obviously poorest near the matching points, but they are clearly better than those of Table 5.1. A similar technique applied to problems of the type of Section 5.6 (ii) will obviously reduce the possibility of near singularity of the relevant matrix in (5.77).

**Table 5.4**

| $r$ | $y_r$ | $y_r(N = 6)$ | $y_r(N = 12)$ | $y_r(N = 18)$ |
|---|---|---|---|---|
| 0 | 0.000000 | 0.000000 | 0.000000 | 0.000000 |
| 1 | 0.166667 | 0.166667 | 0.166667 | 0.166667 |
| 2 | 0.333333 | 0.333333 | 0.333333 | 0.333333 |
| 3 | 0.500000 | 0.500000 | 0.499996 | 0.499992 |
| 4 | 0.666667 | 0.666666 | 0.666626 | 0.666580 |
| 5 | 0.833333 | 0.833333 | 0.832886 | 0.832397 |
| 6 | 1.000000 | 1.000000 | 1.000000 | 0.990234 |
| 7 | 1.166667 |  | 1.167969 | 1.078125 |
| 8 | 1.333333 |  | 1.333496 | 0.375000 |
| 9 | 1.500000 |  | 1.500015 | 0.000000 |
| 10 | 1.666667 |  | 1.666668 | 2.000000 |
| 11 | 1.833333 |  | 1.833333 | 1.750000 |
| 12 | 2.000000 |  | 2.000000 | 2.000000 |
| 13 | 2.166667 |  |  | 2.167969 |
| 14 | 2.333333 |  |  | 2.333496 |
| 15 | 2.500000 |  |  | 2.500031 |
| 16 | 2.666667 |  |  | 2.666672 |
| 17 | 2.833333 |  |  | 2.833334 |
| 18 | 3.000000 |  |  | 3.000000 |

### (iii) Multiple shooting

With the intention of making still smaller the total range of each partial solution we have the final possibility of *multiple shooting*, in which separate solutions over successive small ranges, obtained by forward or backward solution, or both, are matched not only at the extreme boundary points but also at the junction points of the ranges. A general discussion of the method involves too much notation, and a simple application to the non-linear problem (5.13) should indicate how it would be applied to other problems. Moreover, our discussion here will effectively extend the methods of the previous subsection from the treatment of linear difference equations to that of the other extreme of non-linear differential equations.

To simplify things even further we consider multiple shooting in one direction only, generally called *parallel* shooting, and with reference to problem (5.13) we divide the complete region $a \leqslant x \leqslant b$ into a number of sub-regions, which for convenience we here take to be three. The arrangement is shown in Fig. 5.1.

For $y_1(x)$ we know its initial value $y_1(a) = \alpha$ but not its initial slope, so that in the first sub-region we use the parameter $y_1'(a) = q_1$. In sub-region II we know neither the starting value nor its derivative, so we use the two parameters $y_2(x_1) = p_2$, $y_2'(x_1) = q_2$, and in sub-region III we have the same problem with starting parameters $y_3(x_2) = p_3$, $y_3'(x_2) = q_3$. The initial-value problems we choose to solve are then given by

$$\left.\begin{aligned}
y_1'' &= f(x, y_1, y_1'), & y_1(a) = \alpha, & \quad y_1'(a) = q_1, & a \leqslant x \leqslant x_1 \\
y_2'' &= f(x, y_2, y_2'), & y_2(x_1) = p_2, & \quad y_2'(x_1) = q_2, & x_1 \leqslant x \leqslant x_2 \\
y_3'' &= f(x, y_3, y_3'), & y_3(x_2) = p_3, & \quad y_3'(x_2) = q_3, & x_2 \leqslant x \leqslant b
\end{aligned}\right\}, \quad (5.90)$$

and the continuity conditions at $x_1$ and $x_2$ and the boundary condition at $x = b$ give for solution the equations

$$\left.\begin{aligned}
y_1(x_1, q_1) - p_2 &= 0 \\
y_1'(x_1, q_1) - q_2 &= 0 \\
y_2(x_2, p_2, q_2) - p_3 &= 0 \\
y_2'(x_2, p_2, q_2) - q_3 &= 0 \\
y_3(b, p_3, q_3) - \beta &= 0
\end{aligned}\right\}, \quad (5.91)$$

where the notation indicates the respective parameters on which each partial solution depends.

The solution of (5.91) is effected by the Newton iterative method. After first

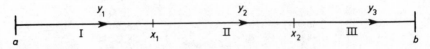

**Figure 5.1**

solving (5.90) with estimated parameters $q_1^{(0)}$, $p_2^{(0)}$, $q_2^{(0)}$, $p_3^{(0)}$ and $q_3^{(0)}$ we obtain corrections $\delta q_1^{(0)}$, etc., to produce the next approximations $q_1^{(1)} = q_1^{(0)} + \delta q_1^{(0)}$, etc., from the linear equations

$$
\left.
\begin{aligned}
&\frac{\partial y_1}{\partial q_1}(x_1)\delta q_1 \quad - \delta p_2 && = p_2 - y_1(x_1) \\[2mm]
&\frac{\partial y_1'}{\partial q_1}(x_1)\delta q_1 \qquad - \delta q_2 && = q_2 - y_1'(x_1) \\[2mm]
&\frac{\partial y_2}{\partial p_2}(x_2)\delta p_2 + \frac{\partial y_2}{\partial q_2}(x_2)\delta q_2 - \delta p_3 && = p_3 - y_2(x_2) \\[2mm]
&\frac{\partial y_2'}{\partial p_2}(x_2)\delta p_2 + \frac{\partial y_2'}{\partial q_2}(x_2)\delta q_2 \qquad - \delta q_3 && = q_3 - y_2'(x_2) \\[2mm]
&\qquad\qquad \frac{\partial y_3}{\partial p_3}(b)\delta p_3 + \frac{\partial y_3}{\partial q_3}(b)\delta q_3 = \beta - y_3(b)
\end{aligned}
\right\}, \quad (5.92)
$$

where all the values of functions and derivatives on the right of (5.92) are obtained from the solutions of (5.90) just computed. The partial derivatives on the left of (5.92) are obtained in the usual way from the variational equations typified by

$$
\left(\frac{\partial y_2}{\partial p_2}\right)'' = \frac{\partial f}{\partial y_2}\left(\frac{\partial y_2}{\partial p_2}\right) + \frac{\partial f}{\partial y_2'}\frac{\partial y_2'}{\partial p_2}, \quad \frac{\partial y_2}{\partial p_2}(x_1) = 1, \quad \frac{\partial y_2'}{\partial p_2}(x_1) = 0,
$$
$$(5.93)$$

or more approximately from (5.90) solved for two slightly different values of $p_2$. The details should now be obvious.

This type of method is now incorporated in very respectable libraries of numerical software, and decisions like the length of each sub-interval are decided automatically, being mainly dependent on the rate of growth in the size of the partial solutions. Whether or not one shoots partially from the other end, or shoots partially from both ends and matches somewhere in the interior of the total range, are matters which are usually decided by the computer user, guided by some of the considerations of this chapter.

## 5.8 INVARIANT EMBEDDING FOR LINEAR PROBLEMS

Finally we mention a somewhat different method in which a boundary-value problem is solved using one or more initial-value techniques. We consider only the linear case, since both the method and its analysis are very complicated for the non-linear case, and insist that the boundary conditions should be of separable kind. Consider first the single second-order equation

$$
y'' + f(x)y' + g(x)y = k(x), \quad (5.94)
$$

with the boundary conditions

$$y'(a) + \alpha y(a) = \gamma, \quad F\{y(b), y'(b)\} = 0, \tag{5.95}$$

the second of which can be non-linear. Now, with the first of (5.95) in mind, we make the deliberate substitution

$$y'(x) + s(x)y(x) = v(x), \tag{5.96}$$

and try to find first-order equations satisfied by $s$ and $v$. Differentiating (5.96), substituting in (5.94) and eliminating $y'$ by using (5.96), we find the equation

$$\{v' + v(f - s) - k\} + y\{g - s(f - s) - s'\} = 0, \tag{5.97}$$

which is clearly satisfied if $s$ and $v$ satisfy the respective equations

$$\left.\begin{array}{l} s' + s(f - s) = g \\ v' + v(f - s) = k \end{array}\right\}. \tag{5.98}$$

Comparison of (5.96) with the first boundary condition shows that the respective initial conditions for (5.98) are given by

$$s(a) = \alpha, \quad v(a) = \gamma, \tag{5.99}$$

with which we can integrate (5.98) from $x = a$ to $x = b$.

The knowledge of $s(b)$ and $v(b)$ then gives from (5.96) the relation

$$y'(b) = v(b) - s(b)y(b), \tag{5.100}$$

and the value $y(b)$ is then obtained by 'solving' the second boundary condition written in the form

$$F\{y(b), v(b) - s(b)y(b)\} = 0. \tag{5.101}$$

With the value of $y(b)$ calculated from (5.101) we can then obtain the required $y(x)$ by integrating the first-order equation (5.96) in the reverse direction from $x = b$ to $x = a$, with the initial value $y(b)$ now known and $s(x)$ and $v(x)$ already computed.

The method needs a little modification if the first boundary condition does not contain the term $y'(a)$, since we cannot then relate (5.96) to this condition. More generally, let us replace the first boundary condition in (5.95) with

$$y(a) + \beta y'(a) = \gamma, \tag{5.102}$$

where we can allow $\beta$ to be zero. The previous analysis can be modified, but it is useful to consider an alternative in which the second-order equation is replaced by the pair of first-order equations

$$\left.\begin{array}{l} y' = z \\ z' = -gy - fz + k \end{array}\right\}, \tag{5.103}$$

with boundary conditions

$$y(a) + \beta z(a) = \gamma, \qquad F\{y(b), z(b)\} = 0. \qquad (5.104)$$

Corresponding to (5.96) we now make the substitution

$$y(x) + t(x)y'(x) = w(x) \Rightarrow y + tz = w. \qquad (5.105)$$

Then

$$y' = w' - t'z - tz' = w' - t'z + t(gy + fz - k), \qquad (5.106)$$

and on substituting for $y$ from (5.105) and using the first of (5.103) we find

$$(w' + gtw - kt) - z(t' + 1 + gt^2 - ft) = 0. \qquad (5.107)$$

This is satisfied if we make both bracketed terms vanish, producing for $w$ and $t$ the first-order equations

$$\left.\begin{array}{c} w' + gtw - kt = 0 \\ t' - ft + gt^2 + 1 = 0 \end{array}\right\}, \qquad (5.108)$$

and by comparing (5.105) with the first boundary condition of (5.104) we deduce the required and respective initial conditions

$$w(a) = \gamma, \quad t(a) = \beta. \qquad (5.109)$$

We therefore integrate (5.108) with (5.109) by initial-value techniques from $x = a$ to $x = b$, and as in our previous method we can now compute $z(b)$ from the second boundary condition in the form

$$F\{w(b) - t(b)z(b), \quad z(b)\} = 0. \qquad (5.110)$$

The function $z(x)$ is then obtained by integrating backwards, from $x = b$ to $x = a$, the single first-order equation obtained by eliminating $y$ from (5.105) and the second of (5.103), in the form

$$z' = (gt - f)z + k - gw. \qquad (5.111)$$

Then $y$ is recovered easily from (5.105).

The method of invariant embedding has two main advantages, at least for linear problems. First, it shares with the shooting method the advantage of solving the boundary-value problem with initial-value techniques. But, second, it tends to avoid the sort of induced instability which the shooting method reveals when the corresponding initial-value problem has a rapidly increasing solution. Consider, for example, the differential equation

$$y'' = p^2 y + \phi(x), \qquad (5.112)$$

with boundary conditions

$$y(a) = \alpha, \quad F\{y(b), \quad y'(b)\} = 0. \qquad (5.113)$$

The transformations just outlined produce the first-order systems

$$t' - p^2t^2 + 1 = 0, \quad t(a) = 0 \left. \right\}.$$
$$w' - p^2tw - t\phi(x) = 0, \quad w(a) = \alpha \left. \right\} \tag{5.114}$$

The first of (5.114) can be solved exactly, and we find

$$t = -p^{-1}\tanh p(x - a), \tag{5.115}$$

a solution which is perfectly stable in the sense that $|t|$ does not increase exponentially as $x$ increases beyond $x = a$.

Moreover, $t$ is everywhere negative, so that in the second of (5.114) the factor $p^2t$ is everywhere negative and the complementary solution is of decreasing exponential type, another stable effect for forward integration from $x = a$. Finally, the computation of $z$ from (5.111) entails the solution of

$$z' + p^2tz = \phi(x) + p^2w. \tag{5.116}$$

Here $p^2t$ is everywhere negative, but we are now integrating backwards with $x$ decreasing from $b$ to $a$, so that the complementary solution is again of decreasing exponential form and we have stability at every stage.

It is also worth noticing that the invariant embedding method can be used to solve certain classes of problems in which one of the boundary positions is not specified in advance, and indeed without even computing the solution of the differential equation. Consider, for example, problem (5.10) with solutions (5.11) and (5.12). Our recent analysis produces the embedding equations

$$t' + t^2 + 1 = 0, \quad t(0) = 0 \left. \right\},$$
$$w' + tw - t = 0, \quad w(0) = 0 \left. \right\}, \tag{5.117}$$

which here we can solve analytically to find

$$t = -\tan x, \quad w = 1 - \sec x. \tag{5.118}$$

The conditions at $x = b$ say that

$$y(b) = 1, \quad z(b) = 2, \tag{5.119}$$

and the transformation (5.105) gives immediately

$$1 + 2t = w \quad \text{at } x = b, \tag{5.120}$$

that is

$$1 - 2\tan b = 1 - \sec b, \tag{5.121}$$

producing the correct solution (5.12) for the required equation for $b$. The solution $y(x)$, if required, is obtained by integrating backwards from $x = b$ the equation (5.111), which here is given by

$$z' + z\tan x = \sec x, \quad z(b) = 2, \tag{5.122}$$

followed by the computation from (5.105) of

$$y = 1 - \sec x + z \tan x, \qquad (5.123)$$

and it is easy to confirm that the known solution (5.11) satisfies (5.122) and (5.123).


## 5.9 ADDITIONAL NOTES

*Section 5.1* With boundary conditions of type (5.5) for a system of first-order equations it is not necessary that all the variables should appear in some form at least once. For equation (5.1), for example, with the conditions $y(a) = \alpha$, $y(b) = \beta$, the corresponding first-order system is given by

$$y' = z, \quad z' = f(x, y, z), \qquad \text{with } y(a) = \alpha, \quad y(b) = \beta, \qquad (5.124)$$

and $z$ does not appear at all in the boundary conditions.

*Section 5.2* There is, of course, some restriction on the coefficients $\alpha_r$ and $\beta_r$ in the unseparated boundary conditions (5.15), in that the two equations must be independent and we cannot have $\beta_r = k\alpha_r$ for $r = 0, 1, 2, 3$.

*Section 5.3* With problem (5.23) we could, of course, choose as parameters the values of $y(b)$ and $y''(b)$, giving an initial-value problem for integrating in the reverse direction. This clearly involves more work, but may be necessary in some contexts to avoid or mitigate the effects of induced instability mentioned in Section 5.6.

With the unseparated boundary conditions (5.29) we clearly need to determine two parameters $y$ and $z$ at either $a$ or $b$, but this would not be necessary with what one might call 'partially separated' conditions, with for example the first of (5.29) replaced by the condition $y(a) = \alpha$.

The problem (5.28) and (5.29) is sufficiently complicated to make it very likely more economic to find approximations to the partial derivatives in (5.32) by evaluating $g_1$ and $g_2$ for slightly different values first of $p$ and then of $q$. This avoids not only the differentiation needed to produce the variational equations but, in addition, that involved in (5.36).

*Section 5.4.* For the problem of the differential equation in (5.8) and the unseparated boundary conditions (5.41) we said that we can take $y(0) = 1$ but this, of course, would not produce any solution for which $y(0) = 0$. Such a solution, however, cannot exist for this particular problem, since if we multiply the differential equation in (5.8) by $2y'$ and integrate, we produce the equation

$$y'^2 + \lambda y^2 = \text{constant}. \qquad (5.125)$$

Then the quantity on the left has the same value at both ends of the range, and if $y(0) = 0$ so that $y(1) = 0$ from the first of (5.41), then $y'(0) = y'(1)$ and the second of (5.41) is then not satisfied for any $\lambda$ unless $y'(0) = y'(1) = 0$, giving a trivial solution.

In addition to the eigenvalue problems mentioned here, in which both the differential equation and the boundary conditions are homogeneous, there is another class, conveniently but not quite accurately referred to as 'eigenvalue problems', in which neither the differential equation nor the boundary conditions need be homogeneous. Here, however, the differential equation does contain a parameter $\lambda$ whose value is determined by the fact that the solution of the differential equation satisfies a third associated condition, rather like the corresponding situation in the problem of determining an unknown boundary point. A typical problem involves the computation of one or more values of $\lambda$, and of the corresponding solution $y(\lambda, x)$, which satisfy the system

$$y'' + \{g(x) + \lambda\}y = k(x), \quad y(a) = y(b) = 0, \quad \int_a^b y^2 \, dx = 1. \quad (5.126)$$

The procedure for the shooting method should be obvious, with parameters $\lambda$ and $y'(a) = q$ chosen to satisfy the last two conditions in (5.126) by a Newton-type iteration associated with the solution of the relevant variational equations.

*Section 5.5* The case of unseparated boundary conditions may need an additional parameter in the problem of an unknown boundary point, but there is no new principle even if all the relevant equations are non-linear. Consider, for example, the problem

$$y'' = f(x, y, y'), \quad g_i\{y(a), y(b), y'(a), y'(b)\} = 0, \quad i = 1, 2, 3, \quad (5.127)$$

where $a$ is known and we need to determine $b$ and $y(x)$. With parameters $y(a) = p$, $y'(a) = q$, and the second boundary point $b$, we can find $\partial y/\partial p = y_p$ and $\partial y/\partial q = y_q$, and of course their derivatives with respect to $x$, from the variational equations

$$\left.\begin{array}{l} y_p'' = \dfrac{\partial f}{\partial y} y_p + \dfrac{\partial f}{\partial y'} y_p', \quad y_p(a) = 1, \quad y_p'(a) = 0 \\[3mm] y_q'' = \dfrac{\partial f}{\partial y} y_q + \dfrac{\partial f}{\partial y'} y_q', \quad y_q(a) = 0, \quad y_q'(a) = 1 \end{array}\right\}, \quad (5.128)$$

from which we obtain $y_p(b)$, $y_q(b)$ and their derivatives for use in the Newton equations. The other required quantities, the partial derivatives of the $g_i$ with respect to $b$, are just

$$\frac{\partial g_i}{\partial b} = \frac{\partial g_i}{\partial y(b)} \frac{\partial y(b)}{\partial b} + \frac{\partial g_i}{\partial y'(b)} \frac{\partial y'(b)}{\partial b}$$

$$= \frac{\partial g_i}{\partial y(b)} y'(b) + \frac{\partial g_i}{\partial y'(b)} f\{x, y(b), y'(b)\}, \quad (5.129)$$

according to the relevant parts of equation (5.59). The Newton equations for corrections $\delta p, \delta q$ and $\delta b$ to previous estimates can then be expressed in terms of easily computed quantities, and in fact they are given by

$$\left\{ \frac{\partial g_i}{\partial p} + \frac{\partial g_i}{\partial y(b)} y_p(b) + \frac{\partial g_i}{\partial y'(b)} y'_p(b) \right\} \delta p + \left\{ \frac{\partial g_i}{\partial q} + \frac{\partial g_i}{\partial y(b)} y_q(b) + \frac{\partial g_i}{\partial y'(b)} y'_q(b) \right\} \delta q$$

$$+ \left\{ \frac{\partial g_i}{\partial y(b)} y'(b) + \frac{\partial g_i}{\partial y'(b)} f\{x, y(b), y'(b)\} \right\} \delta b + g_i\{p, y(b), q, y'(b)\} = 0,$$

$$i = 1, 2, 3. \quad (5.130)$$

*Section 5.6* If the second-order equation (5.1) has the specified conditions

$$y(a) = \alpha, \quad y(b) = \beta \qquad (5.131)$$

its solution in terms of the corresponding pair of first-order equations is best found using the parameter $y'(a) = q$, and the same remark is true if (5.1) is solved directly using the Taylor series method. In the direct solution of (5.1) by a finite-difference method, in which (5.1) is approximated in the first place by a three-term recurrence relation, a more useful parameter is

$$y(a + h) = y_1 = q. \qquad (5.132)$$

This, of course, is effectively the parameter in the solution of second-order recurrence relations in the technique of equations (5.62)–(5.65).

Whenever a second-order equation or its boundary conditions involve a first derivative, it is in some ways easier to treat the problem as a pair of first-order equations, since no special formulae have to be used for any such first derivative. In particular, a Runge–Kutta method would have no difficulty with ranges of different lengths in problems in which one boundary point has to be determined, since equal-interval step-lengths have no importance in such methods.

*Section 5.7 (iii)* The multiple parallel shooting method has its limiting case when each separate shoot is performed over just one step.

In Fig. 5.2 we propose to shoot from $r = 0, 1$ to $r = 2$, from $r = 1, 2$ to $r = 3$, from $r = 2, 3$ to $r = 4$, and from $r = 3, 4$ to $r = 5$, which is the second boundary. For the linear recurrence relation

$$y_{r+1} + a_r y_r + b_r y_{r-1} = c_r, \text{ with } y_0 = \alpha, \quad y_5 = \beta, \qquad (5.133)$$

we clearly have

$$y_2^{(\mathrm{I})} = c_1 - a_1 y_1^{(\mathrm{I})} - b_1 y_0^{(\mathrm{I})}, \dots, y_5^{(\mathrm{IV})} = c_4 - a_4 y_4^{(\mathrm{IV})} - b_4 y_3^{(\mathrm{IV})}. \qquad (5.134)$$

For the continuity conditions we have $y_1^{(\mathrm{I})} = y_1^{(\mathrm{II})}, \dots, y_4^{(\mathrm{III})} = y_4^{(\mathrm{IV})}$, and to satisfy the boundary conditions we must take $y_0^{(\mathrm{I})} = y_0 = \alpha, y_5^{(\mathrm{IV})} = y_5 = \beta$. The

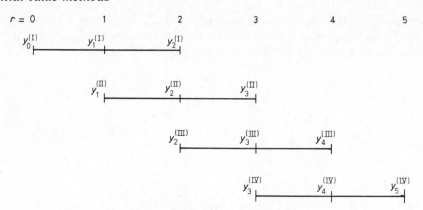

**Figure 5.2**

equations can then be recast immediately in the form

$$
\left.
\begin{aligned}
a_1 y_1 + y_2 && = c_1 - b_1 \alpha \\
b_2 y_1 + a_2 y_2 + y_3 && = c_2 \\
b_3 y_2 + a_3 y_3 + y_4 && = c_3 \\
+ b_4 y_3 + a_4 y_4 && = c_4 - \beta
\end{aligned}
\right\}, \tag{5.135}
$$

which is a set of four linear algebraic equations for the unknowns $y_1, y_2, y_3$ and $y_4$. These equations could, of course, have been written down at the start without any thought of shooting, and it turns out that the solution of these equations, by standard methods, is a very good way of solving the boundary-value recurrence problem. We discuss this in the next chapter, together with methods of a comparable kind for solving the corresponding boundary-value differential equation problem.

*Section 5.8* The analysis of the invariant embedding method explains why 'we make the deliberate substitution' given in (5.96). One explanation comes from the attempt to 'split' the second-order equation (5.94) into two first-order equations, that is write

$$
\left\{ \frac{d}{dx} + r(x) \right\} \left\{ \frac{d}{dx} + s(x) \right\} y = y'' + f(x)y' + g(x)y = k(x). \tag{5.136}
$$

Carrying out the differentiation on the left and equating corresponding terms in (5.136), we find

$$
r + s = f, \quad s' + rs = g, \tag{5.137}
$$

with the immediate result

$$
s' + s(f - s) = g, \tag{5.138}
$$

which is the first of (5.98). Then if in the left of (5.136) we write

$$y' + sy = v, \tag{5.139}$$

which is the substitution (5.96), we have directly from (5.136) and the first of (5.137) the equation

$$v' + v(f - s) = k, \tag{5.140}$$

which is the second of (5.98). The remaining analysis follows as before.

---

## EXERCISES

1. Rather more complicated than (5.10) is the problem

$$y'' - y = 1, \quad y(0) = 0, \quad y(b) = e^b, \quad y'(b) = -1,$$

   in which the unknown $b$ also gives an unknown boundary value in addition to an unknown boundary point. This particular problem can be solved analytically, and in this way show that the only real solution is

$$b = \log(\sqrt{2} - 1), \quad y = \tfrac{1}{2}e^x + \tfrac{1}{2}e^{-x} - 1.$$

2. For the problem of Exercise 1 repeat the analysis of equations (5.51)–(5.59) to find, starting with $q^{(0)} = 0.1$, $b^{(0)} = -0.9$, the first Newton correction which gives $q = -0.002$, $b = -0.878$, the solution of the problem being $q = 0$, $b = -0.882$.

3. Also solve the problem of Exercise 1 by numerical solution of the differential equations, using: (a) the method suggested at the end of Section 5.5, and (b) a Runge–Kutta method.

4. For the problem (5.23) carry out two steps of Newton iteration using the variational equations for the case $\beta = 0.1$, $b = 5$, using the standard fourth-order Runge–Kutta method for the solution of the relevant first-order equations, and starting with the approximation $y''(0) = 0.3$.

5. Repeat Exercise 4 without using the variational equation but using the method suggested in the paragraph following equation (5.27), and compare the rates of convergence of the two methods.

6. Repeat Exercise 4 by using appropriate finite-difference equations for the original third-order differential equation.

7. For problem (5.28) and (5.29) use a fourth-order Runge–Kutta method and enough Newton iterations to guarantee convergence with $a = 0$, $b = 1$. It is not obvious what is a good choice of starting values for $p$ and $q$, so try various choices and see what happens. The choice $p = -1$, $q = -0.5$ should give convergence.

8. Using a suitable finite-difference method for the solution of the relevant differential equations, find to a few figures an $O(h^2)$ approximation to the first eigensolution of the problem

$$y'' - \lambda y/x = 0, \quad y(0) = y(4) = 0.$$

9. For problem (5.61) suppose that we compute $\bar{y}_r^{(p)}$ from (5.62) with its initial conditions changed to $\bar{y}_0^{(p)} = 0$, $\bar{y}_1^{(p)} = \frac{1}{6}$, and $y_r^{(2)}$ from (5.63) without change. With exact arithmetic (5.64) should give $q = 0$. Would you expect this choice of $\bar{y}_r^{(p)}$ to give better results with computer arithmetic than the choice $\bar{y}_1^{(p)} = 0$? Check your view by computing Table 5.1 on your machine with both choices.

10. Prove that the general solution of equations (5.81) is

$$y_r = A_1 + B_1 10^r + 1 + r, \qquad z_r = -2.5A_1 + 2B_1 10^r + 1 - 3r,$$

that for the given boundary conditions (5.82) both $A_1$ and $B_1$ are zero, and that for boundary conditions perturbed by amounts $\varepsilon_1$ and $\varepsilon_2$ on the right of (5.82) the values of $A_1$ and $B_1$ are respectively and approximately $0.08\varepsilon_1 - 0.09\varepsilon_2$ and $10^{-12}(0.2\varepsilon_1 - 0.02\varepsilon_2)$. Hence deduce that the poor results following equation (5.84), due to the near singularity of the matrix in (5.77), are caused by the induced instability of the method and not by the ill-conditioning of the problem.

11. In some circumstances the matrix in (5.77) can be nearly singular, again producing poor results, due to the ill-conditioning of the problem. Consider, for example, the problem

$$y_{r+1} = 0.1y_r + z_r + 3.9r + 0.9,$$
$$z_{r+1} = 0.04y_r + 0.1z_r - 2.74r - 2.14,$$

with boundary conditions

$$y_0 + z_0 + 3y_{12} + z_{12} = 6,$$
$$y_0 + z_0 - 2y_{12} - z_{12} = 11.$$

Prove that the general solution of the recurrence relations is

$$y_r = A_1(0.3)^r + B_1(-0.1)^r + 1 + r$$
$$z_r = 0.2A_1(0.3)^r - 0.2B_1(-0.1)^r + 1 - 3r,$$

and $A_1 = B_1 = 0$ for the given boundary conditions. If the latter are perturbed by respective amounts $\varepsilon_1$ and $\varepsilon_2$ on the right-hand sides, show that for this problem we have approximately

$$A_1 = 0.35 \times 10^6(\varepsilon_1 - \varepsilon_2),$$
$$B_1 = 0.70 \times 10^6(\varepsilon_1 - \varepsilon_2),$$

demonstrating the ill-conditioning of the problem. Record the matrix in

(5.77), and verify that it is nearly singular. What is the main numerical difference between the problems of Exercises 10 and 11?

12. Use a Runge–Kutta technique to solve, by a method analogous to that of Section 5.6(ii), the equations

$$y' = 10y - 8z + 14 + 10x - 10x^2, \quad z' = 8y - 10z + 23 + 10x - 8x^2$$

with $y(0) = 1$, $z(12) = 15$. How good is your answer?

13. Try to improve the results of Exercise 12 by a method analogous to that of Section 5.7(ii), 'matching' at $x = 1.5$.

14. Write down the equations relevant to a pair of first-order linear differential equations of boundary-value type using the parallel shooting method of Section 5.7(iii). Use this method for the problem of Exercise 12, using three sub-regions of equal length.

15. Find the analytical solution of Exercise 1 by using an invariant embedding method. Also find numerically the required value of $b$, using the method discussed at the end of Section 5.8, solving the relevant first-order differential equations by a Runge–Kutta method and noting that the value of $b$ is here obtained as the value of $x$ for which $e^x - t(x) = w(x)$. By tabulating this quantity at successive mesh points compute the result by inverse interpolation.

# 6

# Global (finite-difference) methods for boundary-value problems

## 6.1 INTRODUCTION

In the initial-value methods of the last chapter at least some of the boundary conditions were not directly involved in the numerical solution of the derived initial-value problems. In global methods all the boundary conditions play equal parts from the beginning and are involved in all parts of the computation. There are essentially two main types of method. The first, considered in this chapter, is the class of finite-difference methods, in which the finite-difference equations and boundary-condition equations are essentially solved simultaneously rather than in step-by-step succession. The second, considered in Chapter 7, is the class of *expansion* methods, in which the solution is represented in the approximate form

$$y = \sum_{r=0}^{N} a_r \phi_r(x),$$  \hfill (6.1)

where the $\phi_r(x)$ are a preselected set of functions of which the most obvious is the sequence of polynomials

$$\phi_r(x) = x^r, \quad r = 0, 1, 2, \ldots.$$  \hfill (6.2)

The constants $a_r$ are computed so that the differential equation and boundary conditions are satisfied to some degree of approximation.

One of the main differences between initial-value methods and global methods is the following. For the former each contributing differential equation solution can be made as accurate as we please by the use of techniques in which the combined selection of interval size and order of method for a required accuracy is more or less automatic in modern computer routines. In global methods these parameters must be specified from the start, and only at the end can we say anything about the resulting accuracy. There is therefore more need for correcting devices such as the 'deferred approach to the limit' or 'deferred correction', at least with finite-difference methods, than in the corresponding initial-value problems of Chapters 3 and 4.

We mentioned earlier that in global methods relevant equations are solved simultaneously rather than in a step-by-step manner, and we start by considering a standard technique for the solution of sets of linear simultaneous algebraic equations.

## 6.2 SOLVING LINEAR ALGEBRAIC EQUATIONS

Consider the system of three simultaneous algebraic equations given by

$$\left.\begin{array}{l} a_{11}x_1 + a_{12}x_2 + a_{13}x_3 = b_1 \\ a_{21}x_1 + a_{22}x_2 + a_{23}x_3 = b_2 \\ a_{31}x_1 + a_{32}x_2 + a_{33}x_3 = b_3 \end{array}\right\}. \tag{6.3}$$

In the *elimination* process we reduce the equations to a simpler form as follows. If $a_{11} \neq 0$ we add the multiple $m_{21} = -a_{21}/a_{11}$ of the first equation to the second to produce a new second equation, and the multiple $m_{31} = -a_{31}/a_{11}$ of the first equation to the third to produce a new third equation, the complete and equivalent set now looking like

$$\left.\begin{array}{l} a_{11}x_1 + a_{12}x_2 + a_{13}x_3 = b_1 \\ a_{22}^{(2)}x_2 + a_{23}^{(2)}x_3 = b_2^{(2)} \\ a_{32}^{(2)}x_2 + a_{33}^{(2)}x_3 = b_3^{(2)} \end{array}\right\}. \tag{6.4}$$

If $a_{22}^{(2)} \neq 0$ we now add the multiple $m_{32} = -a_{32}^{(2)}/a_{22}^{(2)}$ of the second of (6.4) to the third of (6.4) to produce a new third equation, and the final simpler set looks like

$$\left.\begin{array}{l} a_{11}x_1 + a_{12}x_2 + a_{13}x_3 = b_1 \\ a_{22}^{(2)}x_2 + a_{23}^{(2)}x_3 = b_2^{(2)} \\ a_{33}^{(3)}x_3 = b_3^{(3)} \end{array}\right\}. \tag{6.5}$$

The matrix on the left of (6.5) is *upper triangular*, with zeros below the main diagonal, and this is the required simpler form since we can now solve (6.5) easily by working backwards to find in order $x_3, x_2$ and $x_1$. This process is in fact called *back-substitution*.

The technique for three equations extends obviously to any number $n$ of equations with $n$ unknowns. Notice also that if we wanted to solve several sets with the same matrix but different right-hand sides, the elimination part, on the left of (6.5), is the same for every set. It can also be shown that the number of numerical operations, counting multiplication as the most important and assuming $n$ large so that its highest power is the dominating term, is $\frac{1}{3}n^3$ for the elimination on the left and $\frac{1}{2}n^2$ for each right-hand side belonging to the upper triangular set of equations, and $\frac{1}{2}n^2$ for the back-substitution for each right-hand side.

We must now consider (i) the numerical stability of the method, and (ii) any economization produced by the particular types of equations which come from ordinary boundary-value problems.

## (i) Stability

The process outlined must be modified if any of the *pivotal* elements, the numbers $a_{11}$ and $a_{22}^{(2)}$ in the $(3 \times 3)$ example, are zero. If the final pivot, here $a_{33}^{(3)}$, is zero, then we cannot perform the back-substitution and indeed the matrix is singular. But there is a significant danger of *induced instability* if any earlier pivotal element is small compared with the other elements beneath it. Consider, for example, the equations

$$\left. \begin{array}{r} -0.001014x_1 + 1.403x_2 = 1.501 \\ 4.964x_1 + 0.6892x_2 = 0.4963 \end{array} \right\}, \tag{6.6}$$

which we shall solve with four-figure floating-point decimal arithmetic. The multiplier for the single elimination involved is $(4.964)/(0.001014)$, which is rounded to 4895 in this arithmetic. The new elements in the second equation are then

$$\left. \begin{array}{l} 4895(1.403) + 0.6892 = 6868 + 0.6892 = 6869 \\ 4895(1.501) + 0.4963 = 7347 + 0.4963 = 7347 \end{array} \right\} \tag{6.7}$$

in this arithmetic, and we have to back-substitute in the equations

$$\left. \begin{array}{r} -0.001014x_1 + 1.403x_2 = 1.501 \\ 6869x_2 = 7347 \end{array} \right\}. \tag{6.8}$$

We obtain $x_2 = 1.070$, $x_1 = 0$, the latter number coming from the fact that in this arithmetic $(1.403)(1.070) = 1.501$ and cancels out the right-hand side of the first of (6.8). The correct answer to four significant figures is

$$x_1 = -0.04855, \quad x_2 = 1.070, \tag{6.9}$$

so that our computed $x_2$ is as good as we could expect, whereas the computed $x_1$ is significantly inaccurate.

This result is not surprising. Even if the computed $x_2$ is correct to four figures it might have an error of as much as $5 \times 10^{-4}$ which, from the first of (6.8), may produce from this source alone an error in $x_1$ of as much as 0.7. And if $x_1$ is not large there must be some cancellation between 1.501 and $1.403x_2$ to give fewer than four significant figures in this number.

But this is not the only instability caused by a small pivot. In equations (6.7) the original coefficients 0.6892 and 0.4963 make little or no contribution to the numbers in the second of (6.8), and the information they contain is essentially lost. Indeed, the results would have been the same if the 0.6892 had been any number in the range 0.5001 to 1.499, and if the 0.4963 had been any number in the range $-0.4999$ to $+0.4999$. In fact it can be verified that (6.8) would have been produced by exact arithmetic, with no rounding errors, if the original equations were the 'perturbed' set

$$\left.\begin{array}{r} -0.001014x_1 + 1.403x_2 = 1.501 \\ 4.96353x_1 + 1.315x_2 = -0.395 \end{array}\right\}, \qquad (6.10)$$

and at this stage we are trying to solve equations (6.10), which are quite a bit different from (6.6) in the second equation.

This effect did not show up forcefully in this example, but it produces catastrophic results in the solution of the three equations

$$\left.\begin{array}{r} -0.001014x_1 + 1.403\ x_2 - 0.9456x_3 = 1.501 \\ 4.964x_1 + 0.6892x_2 + 0.3296x_3 = 0.4963 \\ -1.499x_1 - 2.684\ x_2 + 0.1001x_3 = -0.2896 \end{array}\right\}. \qquad (6.11)$$

The first elimination with respective multipliers 4895 and $-1478$ gives the new second and third equations

$$\left.\begin{array}{r} 6869x_2 - 4629x_3 = 7347 \\ -2077x_2 + 1398x_3 = -2218 \end{array}\right\}, \qquad (6.12)$$

and with multiplier 0.3024 the new third equation after another elimination is

$$-2x_3 = 4, \qquad (6.13)$$

with only one digit available on both sides of (6.13). The pivots $-2$ in (6.13) and 6869 in the first of (6.12) are not 'small', but the resulting solutions

$$x_3 = -2, \quad x_2 = -0.2782, \quad x_1 = 0 \qquad (6.14)$$

are all incorrect, compared with the true four-figure results

$$x_3 = -1.708, \quad x_2 = -0.08129, \quad x_1 = 0.2247. \qquad (6.15)$$

To avoid this induced instability we perform elimination with *interchanges*, sometimes called *pivoting*, the equations being rearranged at each stage so that the element on the diagonal of the relevant column is the largest in absolute value of the possible pivotal rows, and then no multiplier exceeds unity in absolute value. With the set (6.11), for example, we take the largest element in the first column as pivot, and interchange the first and second equations. The elimination produces the equations

$$\left.\begin{array}{r} 4.964x_1 + 0.6892x_2 + 0.3296x_3 = 0.4963 \\ 1.403\ x_2 - 0.9455x_3 = 1.501 \\ -2.476\ x_2 + 0.1996x_3 = -0.1397 \end{array}\right\}. \qquad (6.16)$$

We have now finished with the first equation, but before eliminating in the remaining $(2 \times 2)$ set we must interchange these two equations, to put $-2.476$ on the diagonal, and the final upper triangular form is

$$\left. \begin{array}{r} 4.964x_1 + 0.6892x_2 + 0.3296x_3 = 0.4963 \\ -2.476x_2 + 0.1996x_3 = -0.1397 \\ -0.8324x_3 = 1.422 \end{array} \right\}, \tag{6.17}$$

with no small diagonal terms and no large multipliers. Back-substitution gives

$$x_3 = -1.708, \quad x_2 = -0.08126, \quad x_1 = 0.2246, \tag{6.18}$$

very good results compared with the correct values in (6.15).

### (ii) Economy

The algebraic equations which arise in our current context are of a very special structured form which involves less arithmetic than we quoted for a general set of equations, which is at least $\frac{1}{3}n^3$ multiplications for a single set of $n$ equations. Consider, for example, the difference equation problem of boundary-value type given by

$$y_{r+1} - 10.1y_r + y_{r-1} = -1.35r, \quad y_0 = 0, \quad y_N = \tfrac{1}{6}N. \tag{6.19}$$

We observed in the last chapter that the shooting method exhibited a great deal of induced instability but that this could be avoided by recording the recurrence equation in (6.19) for $r = 1, 2, \ldots, N-1$, taking the known $y_0$ and $y_N$ to the right-hand side, and solving the simultaneous algebraic equations given for $N = 5$ by

$$\left. \begin{array}{rl} -10.1y_1 + y_2 & = -1.35 - 0 = -1.35 \\ y_1 - 10.1y_2 + y_3 & = -2(1.35) = -2.70 \\ y_2 - 10.1y_3 + y_4 & = -3(1.35) = -4.05 \\ y_3 - 10.1y_4 & = -4(1.35) - \tfrac{1}{6}(5) = -6.233\ldots \end{array} \right\}. \tag{6.20}$$

If we perform the elimination process we find that no interchanges are necessary here, in each column there is only one elimination to perform, and in each new row there is only one new number on the left and of course the one on the right. For example, the first elimination produces to six decimal digits the new second equation

$$-10.0010y_2 + y_3 = -2.83366, \tag{6.21}$$

and nothing else is altered. This has needed effectively one division and one multiplication on the left, so that for $n$ equations with $n-1$ such eliminations there is a total of $n-1$ multiplications and $n-1$ divisions for the production of the final upper triangular matrix. For one right-hand side there are $n-1$ multiplications to produce the final right-hand column.

The total is very much smaller than that for a general $(n \times n)$ system, and this economy is preserved in the back-substitution. The upper triangular equations have the form typified for $n = 4$ by

$$\begin{bmatrix} \times & \times & 0 & 0 \\ & \times & \times & 0 \\ & & \times & \times \\ & & & \times \end{bmatrix} \begin{bmatrix} \times \\ \times \\ \times \\ \times \end{bmatrix}, \tag{6.22}$$

and again each back-substitution involves one division for each unknown and one multiplication for all but the last unknown, a total of $n$ divisions and $n - 1$ multiplications.

The reason for this is that the original matrix, of so-called *triple-diagonal* form, has a lot of zeros in such ideal places that in the elimination process these zeros are never changed to non-zero numbers. There is no *fill-in*, in the modern jargon. With interchanges, however, there may be fill-in with the triple-diagonal form. For example, if the first two rows are exchanged in (6.20) (and only two *successive* rows are involved at any stage), the respective pictures before and after the first elimination look on the left like

$$\begin{bmatrix} \times & \times & \times & \\ \times & \times & & \\ & \times & \times & \times \\ & & \times & \times \end{bmatrix}, \begin{bmatrix} \times & \times & \times & \\ & \times & \times & \\ & \times & \times & \times \\ & & \times & \times \end{bmatrix}. \tag{6.23}$$

The arithmetic has one more multiplication than before in this elimination, but for the next elimination, not involving the very first equation, the relevant $(n - 1, n - 1)$ matrix is again of triple-diagonal form. If an interchange is necessary at every step there is an extra multiplication in each, and the upper triangular form then has an extra non-zero element in all but the last two rows.

More fill-in can be expected with non-separated boundary conditions. Consider, for example, the recurrence in (6.19) with the boundary conditions

$$y_0 + y_N = \tfrac{1}{6}N, \quad y_1 + y_{N-1} = \tfrac{1}{6}N, \tag{6.24}$$

a problem which has the same solution $y_r = \tfrac{1}{6}r, r = 0, 1, 2, \dots, N$, as (6.19). We now have six unknowns for $N = 5$, and the matrix has the form

$$\begin{bmatrix} \times & & & & & \times \\ \times & \times & \times & & & \\ & \times & \times & \times & & \\ & & \times & \times & \times & \\ & & & \times & \times & \times \\ & \times & & & & \times \end{bmatrix}, \tag{6.25}$$

with the boundary conditions (6.24) written as the first and last equations in

(6.25). Elimination without interchanges produces at some stage below the diagonal and at all stages above the diagonal (which with the diagonal is the final upper triangular form), the fill-in shown by the solid circles in the array

$$\begin{bmatrix} \times & & & & & \times \\ \times & \times & \times & & & \bullet \\ & \times & \times & \times & & \bullet \\ & & \times & \times & \times & \bullet \\ & & & \times & \times & \times \\ & & \times & \bullet & \bullet & \times & \bullet \end{bmatrix}. \tag{6.26}$$

Not all our structured sets of equations, however, will be as simple to deal with as (6.25), and the fill-in could be quite substantial with separated boundary conditions, more so with unseparated boundary conditions, and even more so if we have to perform interchanges to preserve stability. Consider, for example, the simultaneous pair of second-order difference equations given by

$$\left.\begin{array}{l} a_r y_{r+1} + b_r y_r + c_r y_{r-1} + d_r z_{r+1} + e_r z_r + f_r z_{r-1} = g_r \\ A_r y_{r+1} + B_r y_r + C_r y_{r-1} + D_r z_{r+1} + E_r z_r + F_r z_{r-1} = G_r \end{array}\right\}, \tag{6.27}$$

with boundary conditions specifying $y_0, y_N$ and $z_0, z_N$. There are various orders in which we can write the equations, that is all of (6.27) for $r = 1, 2, \ldots, N-1$, incorporating the boundary values, and in which we can arrange the unknowns, that is all of the $y_r$ and $z_r$ for $r = 1, 2, \ldots, N-1$.

If we arrange the unknowns in order $y_1, \ldots, y_{N-1}, z_1, \ldots, z_{N-1}$ and write down all of the first of (6.27) followed by all of the second of (27), the matrix for $N = 5$ looks like

$$\begin{array}{cccccccc} y_1 & y_2 & y_3 & y_4 & z_1 & z_2 & z_3 & z_4 \end{array}$$

$$\left[\begin{array}{cccc|cccc} \times & \times & & & \times & \times & & \\ \times & \times & \times & & \times & \times & \times & \\ & \times & \times & \times & & \times & \times & \times \\ & & \times & \times & & & \times & \times \\ \hline \times & \times & & & \times & \times & & \\ \times & \times & \times & & \times & \times & \times & \\ & \times & \times & \times & & \times & \times & \times \\ & & \times & \times & & & \times & \times \end{array}\right]. \tag{6.28}$$

With non-separated conditions there will be a few more equations, and some extra elements in the top right and bottom left of the matrix.

Another possible arrangement, with unknowns in order $y_1, z_1, y_2, z_2, \ldots,$ and one of the first of equations (6.27) followed immediately by its partner in the second of (6.27), looks like

$$
\begin{array}{cccccccc}
y_1 & z_1 & y_2 & z_2 & y_3 & z_3 & y_4 & z_4
\end{array}
$$

$$
\left[
\begin{array}{cc|cc|cc|cc}
\times & \times & \times & \times & & & & \\
\times & \times & \times & \times & & & & \\
\hline
\times & \times & \times & \times & \times & \times & & \\
\times & \times & \times & \times & \times & \times & & \\
\hline
 & & \times & \times & \times & \times & \times & \times \\
 & & \times & \times & \times & \times & \times & \times \\
\hline
 & & & & \times & \times & \times & \times \\
 & & & & \times & \times & \times & \times \\
\end{array}
\right], \qquad (6.29)
$$

and again with more equations and more terms in the top right and bottom left for unseparated boundary conditions. Form (6.29) is obviously more compact, with no fill-in if there are no interchanges in the elimination. Form (6.28) obviously does produce fill-in even without interchanges, and both obviously have more fill-in with unseparated boundary conditions and also with interchanges.

For space-saving reasons we have not been able to make very obvious the sparsity of matrices like (6.28) and (6.29), but normally there will be many more rows and columns with a maximum of six elements in each row, and such a matrix is sparse and looks sparse. We shall not pursue the general question of how to deal with such large sparse sets in order to achieve economy with stability and therefore accuracy. This is a matter of considerable current research, but there are routines which, among other things, check on the amount of fill-in which stems from a particular row interchange, and also measures the size of the multipliers assuming no interchange and no significant fill-in. If the multipliers are not too large then the small inaccuracy is tolerated in favour of economy, and there is some case history of what can be tolerated while still producing sufficiently accurate results.

## 6.3 LINEAR DIFFERENTIAL EQUATIONS OF ORDERS TWO AND FOUR

### (i) Second-order equations

We turn now to our main preoccupation, the numerical solution of ordinary differential equations with associated boundary conditions. The most obvious and easiest such problem is the linear second-order equation

$$
y'' + f(x)y' + g(x)y = k(x), \quad y(a) = \alpha, \quad y(b) = \beta. \qquad (6.30)
$$

Our plan, as with the initial-value problems, is to compute approximations at discrete points, which for the present we take to be at equal intervals $h$, so that

$$nh = b - a, \tag{6.31}$$

and the unknowns we have to compute are approximations $Y_r$ to $y_r = y(a + rh)$ for $r = 1, 2, \ldots, n - 1$.

We do this by attempting to satisfy the differential equation at each mesh point using central-difference approximations for the derivatives in (6.30). At the point $x_r = a + rh$ we have

$$\left. \begin{array}{l} y_r' = \dfrac{y_{r+1} - y_{r-1}}{2h} + O(h^2) \\[2mm] y_r'' = \dfrac{y_{r+1} - 2y_r + y_{r-1}}{h^2} + O(h^2) \end{array} \right\}, \tag{6.32}$$

so at this point we replace the differential equation by the algebraic equation

$$(1 + \tfrac{1}{2}hf_r)y_{r+1} - (2 - h^2 g_r)y_r + (1 - \tfrac{1}{2}hf_r)y_{r-1} = h^2 k_r + c(y_r), \tag{6.33}$$

where $c(y_r)$ is the local truncation error, of order $O(h^4)$, expressible in terms of derivatives or central differences as in Chapter 3.

In the first place we neglect the $c(y_r)$, replace $y_r$ by its approximation $Y_r$, and compute the latter from equations like (6.33) recorded at $r = 1, 2, \ldots, n - 1$ and with $Y_0 = y(a)$ and $Y_n = y(b)$ given their specified values. The resulting equations are

$$\left. \begin{array}{ll} -(2 - h^2 g_1)Y_1 \quad + (1 + \tfrac{1}{2}hf_1)Y_2 & = h^2 k_1 - (1 - \tfrac{1}{2}hf_1)\alpha \\[1mm] (1 - \tfrac{1}{2}hf_2)Y_1 \quad - (2 - h^2 g_2)Y_2 + (1 + \tfrac{1}{2}hf_2)Y_3 & = h^2 k_2 \\[1mm] \qquad\qquad \vdots & \\[1mm] (1 - \tfrac{1}{2}hf_{n-2})Y_{n-3} - (2 - h^2 g_{n-2})Y_{n-2} + (1 + \tfrac{1}{2}hf_{n-2})Y_{n-1} = h^2 k_{n-2} \\[1mm] (1 - \tfrac{1}{2}hf_{n-1})Y_{n-2} - (2 - h^2 g_{n-1})Y_{n-1} = h^2 k_{n-1} - (1 + \tfrac{1}{2}hf_{n-1})\beta \end{array} \right\} \tag{6.34}$$

with a nice triple-diagonal matrix which we have already discussed.

The boundary conditions in (6.30) are separated, but a rather more complicated linear separated condition would have the form

$$py(a) + qy'(a) = r. \tag{6.35}$$

Here $y_0 = y(a)$ is not known, and the first of equations (6.34) has to be replaced by three other equations. The first of these is a finite-difference representation of the boundary condition (6.35), and with the use of our favoured central-difference formulae, here the first of (6.32) for a first derivative, we have the equation

$$-qy_{-1} + 2hpy_0 + qy_1 = 2hr + c'(y_0), \tag{6.36}$$

where $c'(y_0)$ is a local truncation error of order $O(h^3)$, again expressible in terms of central differences or derivatives at $x = a$. Two other equations are needed because we have introduced two extra unknowns, $y_{-1}$ and $y_0$. The first is (6.33) applied at the boundary point $x = a$, and the second is (6.33) applied at the first internal point $x = a + h$, this being effectively the first of (6.34) with the term $-(1 - \frac{1}{2}hf_1)\alpha$ on the right now being the unknown $-(1 - \frac{1}{2}hf_1)Y_0$ and therefore moved back to the left-hand side. The first three equations of the new set corresponding to (6.34) are then

$$\left.\begin{array}{llll} -qY_{-1} & +2hpY_0 & +qY_1 & = 2hr \\ (1 - \frac{1}{2}hf_0)Y_{-1} - (2 - h^2g_0)Y_0 + (1 + \frac{1}{2}hf_0)Y_1 & & & = h^2k_0 \\ (1 - \frac{1}{2}hf_1)Y_0 - (2 - h^2g_1)Y_1 + (1 + \frac{1}{2}hf_1)Y_2 = h^2k_1 \end{array}\right\}. \quad (6.37)$$

The corresponding matrix is not quite triple-diagonal, but it is so after the very first elimination, equivalent to the replacement of the first two of (6.37) by the single equation

$$\{(1 - \frac{1}{2}hf_0)2hpq^{-1} - (2 - h^2g_0)\}Y_0 + 2Y_1 = (1 - \frac{1}{2}hf_0)2hrq^{-1} + h^2k_0. \quad (6.38)$$

For non-separated boundary conditions a quite general linear set has the form

$$\left.\begin{array}{l} p_1 y(a) + q_1 y'(a) + r_1 y(b) = s_1 \\ p_2 y(b) + q_2 y'(b) + r_2 y(a) = s_2 \end{array}\right\}. \quad (6.39)$$

If $q_1 \neq 0$ we would, with aims similar to those of the previous treatment, approximate to the first of (6.39) by the equation

$$-q_1 Y_{-1} + 2hp_1 Y_0 + q_1 Y_1 + 2hr_1 Y_n = 2hs_1. \quad (6.40)$$

There would be a similar treatment at the point $x_n = b$ for the second of (6.39), and we note the extra terms in the top right and bottom left positions of the matrix.

If $q_2 = 0$, but $q_1 \neq 0$, because otherwise (6.39) reduces to the conditions in (6.30), then the two equations at the bottom end are just the reordered last of (6.34) and the adjusted second of (6.39), given by

$$\left.\begin{array}{ll} (1 - \frac{1}{2}hf_{n-1})Y_{n-2} - (2 - h^2g_{n-1})Y_{n-1} + (1 + \frac{1}{2}hf_{n-1})Y_n = h^2k_{n-1} \\ r_2 Y_0 \qquad\qquad\qquad\qquad + p_2 Y_n \qquad\qquad = s_2 \end{array}\right\}, \quad (6.41)$$

and there is no need to try to satisfy the differential equation at the second boundary point $x_n = b$.

## (ii) Fourth-order equations

Many practical problems, as for example the bending of an elastic beam supported in some way at its ends, involve the solution of fourth-order equations, the most general linear case having the form

$$y^{iv} + f(x)y''' + g(x)y'' + k(x)y' + l(x)y = m(x). \quad (6.42)$$

A fourth-order equation needs four associated conditions to provide a unique solution, and in a boundary-value problem at least one condition must be specified at each end of the range, $x = a$ and $x = b$. The 'clamped' beam, for example, has conditions

$$y(a) = 0, \quad y'(a) = 0, \quad y(b) = 0, \quad y'(b) = 0; \tag{6.43}$$

with a 'simply supported' beam we have

$$y(a) = 0, \quad y''(a) = 0, \quad y(b) = 0, \quad y''(b) = 0; \tag{6.44}$$

and the most general form of any linear separated boundary condition would be

$$py''' + qy'' + ry' + sy = t \tag{6.45}$$

at either boundary.

To obtain the approximating algebraic equations from (6.42) we use the suitably truncated central-difference formulae for the first four derivatives given in the notation of Chapter 3 by

$$\left.\begin{aligned}
h^4 y_r^{iv} &= \delta^4 y_r - \tfrac{1}{6}\delta^6 y_r + \cdots = \delta^4 y_r + O(h^6) \\
h^3 y_r''' &= \mu\delta^3 y_r - \tfrac{1}{4}\mu\delta^5 y_r + \cdots = \mu\delta^3 y_r + O(h^5) \\
h^2 y_r'' &= \delta^2 y_r - \tfrac{1}{12}\delta^4 y_r + \tfrac{1}{90}\delta^6 y_r - \cdots = \delta^2 y_r - \tfrac{1}{12}\delta^4 y_r + O(h^6) \\
h y_r' &= \mu\delta y_r - \tfrac{1}{6}\mu\delta^3 y_r + \tfrac{1}{30}\mu\delta^5 y_r - \cdots = \mu\delta y_r - \tfrac{1}{6}\mu\delta^3 y_r + O(h^5)
\end{aligned}\right\} \tag{6.46}$$

If we use the terms given explicitly on the right of (6.46), and express them in terms of mesh values, we find that at the general point $x_r$ the equation involves the five adjacent unknowns $Y_{r-2}, Y_{r-1}, Y_r, Y_{r+1}, Y_{r+2}$, and this is clearly the most compact form possible for a fourth-order equation. The dominant term in the truncation error is $O(h^6)$.

It is clearly desirable to use central differences for the derivatives in the boundary condition (6.45), and in fact to use the last three of (6.46) to maintain satisfactory accuracy. With reference to Fig. 6.1, two conditions like (6.45) at the boundary point $x_0 = a$ would introduce two equations involving the external values at points $-1$ and $-2$. With similar arrangements at the second boundary we easily see that if we satisfy the differential equation with the relevant equations

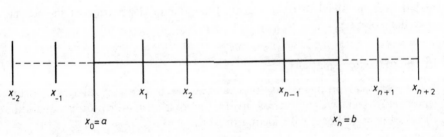

**Figure 6.1**

(6.46) at all internal and boundary points, we have just enough algebraic equations for all the $n + 5$ unknowns $Y_{-2}, Y_{-1}, Y_0, \ldots, Y_n, Y_{n+1}, Y_{n+2}$.

The structure of the resulting matrix has the early form

$$
\begin{array}{ccccc}
Y_{-2} & Y_{-1} & Y_0 & Y_1 & Y_2 \quad \cdots
\end{array}
$$

$$
\begin{bmatrix}
\times & \times & \times & \times & \times & & \\
\times & \times & \times & \times & \times & & \\
\times & \times & \times & \times & \times & & \\
 & \times & \times & \times & \times & \times & \\
 & & \times & \times & \times & \times & \times \\
 & & & & \cdots
\end{bmatrix}, \tag{6.47}
$$

the first two rows representing the boundary conditions.

With the more special conditions (6.43) and (6.44) one of the first two rows reduces to the single statement that $Y_0$ is specified, so that with the removal of $Y_0$ the early part of the resulting matrix in each case looks like

$$
\begin{array}{ccccccc}
Y_{-2} & Y_{-1} & Y_1 & Y_2 & Y_3 & Y_4 & Y_5
\end{array}
$$

$$
\begin{bmatrix}
\times & \times & \times & \times & & & \\
\times & \times & \times & \times & & & \\
 & \times & \times & \times & \times & & \\
 & & \times & \times & \times & \times & \\
 & & \times & \times & \times & \times & \times \\
 & & & \cdots
\end{bmatrix} \tag{6.48}
$$

And of course we may have three conditions at one end and one condition at the other end, for example

$$
y(a), y'(a), y''(a), y'''(b) \text{ all specified.} \tag{6.49}
$$

In this case the same ideas as we have used so far produce equations whose matrix structure for the range $Y_1$ to $Y_5$, with external values $Y_{-2}, Y_{-1}, Y_6$ and $Y_7$, looks like

$$
\begin{array}{ccccccccc}
Y_{-2} & Y_{-1} & Y_1 & Y_2 & Y_3 & Y_4 & Y_5 & Y_6 & Y_7
\end{array}
$$

$$
\begin{bmatrix}
\times & \times & \times & \times & & & & & \\
\times & \times & \times & \times & & & & & \\
\times & \times & \times & \times & & & & & \\
 & \times & \times & \times & \times & & & & \\
 & & \times & \times & \times & \times & & & \\
 & & \times & \times & \times & \times & \times & & \\
 & & & \times & \times & \times & \times & \times & \\
 & & & & \times & \times & \times & \times & \times \\
 & & & & \times & \times & \times & \times & \times
\end{bmatrix} \tag{6.50}
$$

with $Y_0$ known and not therefore included in the array.

It is also clear that the use of similar equations will enable us to deal with any unseparated boundary conditions with extra elements in the top right and bottom left regions of the matrix.

### (iii) Eigenvalue problems

In the discussion in Chapter 5 of shooting methods for linear boundary-value problems we mentioned the treatment of eigenvalue problems. The corresponding *boundary-value method* effectively approximates to the solution of the differential problem by finding an eigenvalue and eigenvector of the type of matrix we have produced in previous sections.

Consider first the solution of the second-order problem

$$\frac{d}{dx}\left\{p(x)\frac{dy}{dx}\right\} + \{q(x) - \lambda\}y = 0, \quad y(a) = 0, \quad y(b) = 0. \tag{6.51}$$

From the first of (6.32) we can record, at the point $x_r = a + rh$ and with $x_{r+\frac{1}{2}} = a + (r + \frac{1}{2})h$, the approximation

$$h^2 \frac{d}{dx}\left\{p(x)\frac{dy}{dx}\right\} \sim h(p_{r+\frac{1}{2}}y'_{r+\frac{1}{2}} - p_{r-\frac{1}{2}}y'_{r-\frac{1}{2}})$$

$$\sim p_{r+\frac{1}{2}}(y_{r+1} - y_r) - p_{r-\frac{1}{2}}(y_r - y_{r-1})$$

$$= p_{r+\frac{1}{2}}y_{r+1} - (p_{r+\frac{1}{2}} + p_{r-\frac{1}{2}})y_r + p_{r-\frac{1}{2}}y_{r-1}. \tag{6.52}$$

Using this approximation at points $r = 1, 2, 3, \ldots, n-1$, and inserting the boundary conditions, we produce algebraic equations which can be written in matrix–vector form as

$$(A - \mu I)Y = 0, \quad \mu = h^2\bar{\lambda}, \tag{6.53}$$

where $\bar{\lambda}$ is an approximation to $\lambda$, the vector $Y$ has components $Y_1, Y_2, \ldots, Y_{n-1}$, the approximations to $y_1, y_2, \ldots, y_{n-1}$, and $A$ is the matrix

$$A = \begin{bmatrix} h^2q_1 - (p_{\frac{1}{2}} + p_{\frac{3}{2}}) & p_{\frac{3}{2}} & & & \\ p_{\frac{3}{2}} & h^2q_2 - (p_{\frac{3}{2}} + p_{\frac{5}{2}}) & p_{\frac{5}{2}} & & \\ & & \vdots & & \\ & & p_{n-\frac{5}{2}} & h^2q_{n-2} - (p_{n-\frac{5}{2}} + p_{n-\frac{3}{2}}) & p_{n-\frac{3}{2}} \\ & & & p_{n-\frac{3}{2}} & h^2q_{n-1} - (p_{n-\frac{3}{2}} + p_{n-\frac{1}{2}}) \end{bmatrix}. \tag{6.54}$$

For any matrix $A$ the algebraic eigenvalue problem (6.53) has had much attention, and all computer-program libraries have good routines for solving for both eigenvalues and eigenvectors. The details of the methods are too space-consuming to be given here, but they can be obtained from most numerical analysis textbooks or occasionally from the relevant annotation of the computer-program library.

One fact, however, is worthy of note. The eigenvalues and vectors of a symmetric matrix are much easier to calculate than those of an unsymmetric matrix, and in addition they have better inherent stability properties. The form (6.51) therefore represents a very convenient problem, the matrix in (6.54) being symmetric. The perhaps more obviously general form

$$y'' + f(x)y' + \{g(x) - \lambda\}y = 0, \quad y(a) = 0, \quad y(b) = 0, \tag{6.55}$$

when treated by the methods of the previous sections, will still be approximated by an equation of type (6.53) but with a matrix which is unsymmetric.

If the first boundary condition in (6.55) is replaced by (6.35) with $r = 0$, then we have to use a procedure analogous to that which produced equations (6.37). In the relevant equations stemming from (6.55) and this new boundary condition, all but the first, the boundary-condition equation, includes a coefficient $-h^2\bar{\lambda}$ on the diagonal, and to produce the form (6.53) we again eliminate the unknown $Y_{-1}$ by a process analogous to that which produced equation (6.38).

Similar things can be done with non-separated conditions analogous to (6.39), and an example is given in Exercise 11 at the end of the chapter. We may also have eigenvalue problems in connection with differential equations of fourth order, for example in the vibration of beams, but there is no major problem, with almost any relevant boundary conditions, in setting up for solution an approximating algebraic eigenvalue problem of type (6.53) with five adjacent elements in a general row of the matrix.

## 6.4 SIMULTANEOUS LINEAR DIFFERENTIAL EQUATIONS OF FIRST ORDER

Since a single first-order equation needs only one associated condition it can hardly be of boundary-value type, but simultaneous sets may appear in their own right or may be derived from a boundary-value problem of higher order. Of what we have so far considered, equation (6.42) with conditions typified by (6.45) is the most complicated example, and the substitutions

$$y' = u, \quad y'' = v, \quad y''' = w \tag{6.56}$$

give rise to the four simultaneous first-order equations

$$y' = u, \quad u' = v, \quad v' = w, \quad w' + f(x)w + g(x)v + k(x)u + l(x)y = m(x). \tag{6.57}$$

The most general linear separated boundary condition, from (6.45), is then

$$pw + qv + ru + sy = t, \quad \text{at } x = a \quad \text{or} \quad x = b, \tag{6.58}$$

and there will be four of these with at least one at each boundary point.

We used the trapezoidal rule formula for the solution of initial-value problems and this is still useful and popular for boundary-value problems. With this formula the first of (6.57), for example, is replaced in the interval $x_r$ to $x_{r+1}$ by the approximation

$$Y_{r+1} - Y_r = \tfrac{1}{2}h(U_{r+1} + U_r).$$ (6.59)

For the $4(n+1)$ unknowns $Y_r, U_r, V_r, W_r$ for $r = 0, 1, \ldots, n$, we have $4n$ equations of type (6.59) for the $n$ available intervals, and four other equations for the four boundary conditions of type (6.58).

As a particular case suppose that we want to solve the differential equations (6.57) with boundary conditions given by the general (6.58) at $x = a$ and three particular forms of (6.58), namely the separate specification of $Y, U$ and $V$ at $x = b$. With $n = 4$ the resulting algebraic equations have a matrix of the form

| $Y_0$ | $U_0$ | $V_0$ | $W_0$ | $Y_1$ | $U_1$ | $V_1$ | $W_1$ | $Y_2$ | $U_2$ | $V_2$ | $W_2$ | $Y_3$ | $U_3$ | $V_3$ | $W_3$ | $W_4$ |
|---|---|---|---|---|---|---|---|---|---|---|---|---|---|---|---|---|
| × | × | × | × | | | | | | | | | | | | | |
| × | × | | | × | × | | | | | | | | | | | |
| | × | × | | | × | × | | | | | | | | | | |
| | | × | × | | | × | × | | | | | | | | | |
| × | × | × | × | × | × | × | × | | | | | | | | | |
| | | | | × | × | | | × | × | | | | | | | |
| | | | | | × | × | | | × | × | | | | | | |
| | | | | | | × | × | | | × | × | | | | | |
| | | | | × | × | × | × | × | × | × | × | | | | | |
| | | | | | | | | × | × | | | × | × | | | |
| | | | | | | | | | × | × | | | × | × | | |
| | | | | | | | | | | × | × | | | × | × | |
| | | | | | | | | × | × | × | × | × | × | × | × | |
| | | | | | | | | | | | | × | × | | | |
| | | | | | | | | | | | | | × | × | | |
| | | | | | | | | | | | | | | × | × | × |
| | | | | | | | | | | | | × | × | × | × | × |

(6.60)

The ordering of the unknowns is shown in (6.60), and in the ordering of the equations the first is the single boundary condition at the first point, the next four blocks are the successive equations of type (6.59) belonging to the differential equations (6.57), and three boundary conditions at the final point are satisfied by specifying $Y_4, U_4, V_4$ and therefore removing them from the unknowns. Any other arrangement of boundary conditions can obviously be treated successfully on its merits, always obtaining the correct number of equations for all the unknowns. Non-separated conditions, as usual, give extra elements in the top right and bottom left regions of the matrix.

6.5 CONVENIENCE AND ACCURACY OF METHODS

The qualities of convenience and accuracy are to some extent correlated. For equations of second and fourth order treated directly we use only the simplest finite-difference equations, in which the differential equation of order $m$ is approximated essentially by a difference equation of order $m$. For our proposed method for simultaneous first-order equations each such equation is also approximated by a difference equation of first order. In all cases, then, we have made the matrix of our resulting algebraic equations as sparse and compact as possible, with obvious economic advantages in their solution. But this treatment also has an advantage with respect to stability. As in our step-by-step methods, here also we must be concerned with spurious parasitic solutions, and clearly our treatment introduces none of these. We therefore expect zero-stability, that is convergence to the correct solution as $h \to 0$.

For second- and fourth-order equations treated directly we have the convenience that the number of unknowns is equal only to the number of mesh points, whereas when treated as simultaneous first-order equations these numbers are multiplied respectively by two and four. The extra arithmetic involved in the solution of the linear algebraic equations is then also increased quite considerably. The simultaneous equations have a significant advantage, however, in that all the boundary conditions are satisfied exactly, whereas in the direct treatment we have two sources of local error, one from the 'discretization' of the differential equation and one from that of any boundary condition involving any derivative. The use of the trapezoidal rule in this context, moreover, means that at this stage at least we need not concentrate on equal intervals, whereas in the direct solution of the higher-order equations unequal intervals require the construction of special formulae.

The treatment of the simultaneous first-order equations also has a significant arithmetic advantage over the direct treatment of higher-order equations, in that with the latter method machine rounding errors introduce errors in the computed solutions which increase quite rapidly as the interval is reduced. The reasons for this are not difficult to find. With the second-order equation (6.30) the right-hand sides of the resulting finite-difference equations (6.34), which ignore the difference correction, include the power $h^2$. When the interval is halved these numbers are divided by four, and since the matrix is only doubled in size the elements of the inverse must increase roughly by a factor of two at each step. For a fourth-order equation this last number becomes eight, a very significant increase. Moreover, if, for example, in equation (6.34) the number $g_2 = 0.09675$, then for $h = 0.05$ the value of $2 - h^2 g_2 = 1.999758125$, and the inability to store this exactly, say with our seven-decimal machine, gives an additional significant source of error. Similarly, for the fourth-order equation (6.65) below, the main diagonal term of the finite-difference matrix is $6 - 4h^2 + h^4$, which is $5.75390625$ when $h = 0.25$. This is a rather large value of $h$, but again our small machine cannot store all these numbers, and the neglect even of the last two could give very large errors when the elements of the inverse are quite large. In the latter case at least one late pivot in the elimination with interchanges will be quite small for small $h$, and at least one unknown will be computed as the division of two small numbers with few accurate significant figures in each.

The same arguments suggest that the treatment of simultaneous first-order equations is likely to give much better accuracy. The dangerous factors $h^2$ and $h^4$ in second- and fourth-order equations are now just $h$, with no general increase in the elements of the inverse matrix as $h$ is reduced, and the ability to store accurately the elements in the finite-difference equations is also much greater.

With all the numerical illustrations which appear in the rest of this chapter we should bear these facts in mind, but in each case the arithmetic is here performed with sufficient accuracy to guarantee that any inaccuracy in the accepted solutions is not caused by machine storage or computer rounding errors. This is obviously necessary for our current purpose, which is to show how well our methods and correcting devices make successful allowance for local truncation errors, since such illustrations are invalid if they are contaminated by other sources of error.

The first relevant question of this kind is: what is the nature of the global error of the approximate solution $Y_r$ which is obtained by approximating to any derivative in a differential equation or boundary condition by the simplest central-difference formula, or with a slight variation of this for fourth-order equations in the extreme right of the last two of (6.46)? It can be shown that in the absence of singularities in the problem, the global error, for sufficiently small $h$, is given at the point $x_r$ by the formula

$$y(x_r) - Y_r = h^2 w_2(x_r) + h^4 w_4(x_r) + \cdots, \tag{6.61}$$

where the functions $w_2(x), w_4(x), \ldots$, are independent of $h$.

This is effectively identical with the corresponding formula recorded in Chapter 3 in connection with the trapezoidal rule method for the step-by-step solution of initial-value problems. For all the problems and all the methods mentioned so far in this chapter it follows that the global error is the rather large $O(h^2)$ and, in contrast to the methods of Chapter 4 for initial-value problems, here in general we shall need methods for improving the accuracy of all our first approximations.

## 6.6 IMPROVEMENT OF ACCURACY

One obvious technique for improving the accuracy is the method of the deferred approach to the limit, already discussed in Chapter 3 for initial-value problems. A second technique is the method of deferred correction, discussed in Section 3.9, but which is now probably the major technique for boundary-value problems. Compared with the deferred approach it appears to be more economic, and by using the same interval throughout it avoids the sort of numerical error mentioned a little earlier which could be caused by successive halvings of the interval. We discuss the techniques separately in the next two subsections.

### (i) The deferred approach to the limit

In virtually all our illustrations we are using formulae of type (6.61), so that our successive improved estimates are obtained from formulae like those of (3.71) and (3.72). In each numerical example we have used the machine's full word length in all the computations, unlike some of the corresponding results of Chapter 3, so that these results are not affected by rounding errors exceeding half a unit in the last figure. We start with the simple example

$$y'' + \left(\frac{2}{1+x}\right)y' - \left(\frac{3+x}{1+x}\right)y = -\frac{3+x}{(1+x)^2}, \quad y'(0) = 0, \quad y(1) = e + \tfrac{1}{2}, \qquad (6.62)$$

whose solution is

$$y = e^x + (1+x)^{-1}, \qquad (6.63)$$

the latter being used in a different context in Chapter 3.

Treating it directly as a second-order equation we use equations of type (6.34) modified by (6.37), and for $h = 0.2, 0.1$ and $0.05$ we obtain the results of which a small sample is given in Table 6.1 for the computed values at $x = 0$ and $x = 0.6$. It is worth remarking, incidentally, that with our smaller machine the computed values for $h = 0.05$ have significant errors in the last figure, for reasons discussed in Section 6.5. The table also shows the results of the two extrapolations corresponding to those of Table 3.7, the respective correct values to six decimals being 2 and 2.447119.

Similar results are shown in Table 6.2 for the solution of (6.62) treated in terms of the simultaneous first-order equations

$$\left. \begin{array}{c} y' = z, \quad z' = \left(\frac{3+x}{1+x}\right)y - \left(\frac{2}{1+x}\right)z - \frac{3+x}{(1+x)^2} \\ z(0) = 0, \quad y(1) = e + \tfrac{1}{2} \end{array} \right\}, \qquad (6.64)$$

**Table 6.1**

| $h$ | $Y(x=0.0)$ | | | $Y(x=0.6)$ | | |
|-----|-----------|--------|----------|-----------|----------|----------|
| 0.2 | 1.992018 | | | 2.447601 | | |
| | | 1.999929 | | | 2.447112 | |
| 0.1 | 1.997951 | | 1.999999 | 2.447234 | | 2.447118 |
| | | 1.999995 | | | 2.447118 | |
| 0.05 | 1.999484 | | | 2.447147 | | |
| Error $O(h^2)$ | $O(h^4)$ | $O(h^6)$ | | $O(h^2)$ | $O(h^4)$ | $O(h^6)$ |

**Table 6.2**

| h | Y(x = 0.2) | | | Z(x = 0.4) | | |
|---|---|---|---|---|---|---|
| 0.2 | 2.048016 | | | 0.988125 | | |
| | | 2.054767 | | | 0.981568 | |
| 0.1 | 2.053079 | | 2.054736 | 0.983207 | | 0.981621 |
| | | 2.054738 | | | 0.981618 | |
| 0.05 | 2.054323 | | | 0.982015 | | |
| Error $O(h^2)$ | | $O(h^4)$ | $O(h^6)$ | $O(h^2)$ | $O(h^4)$ | $O(h^6)$ |

with equations of type (6.61) holding for both $y(x_r)$ and $z(x_r)$. The extrapolation is performed for $Y(0.2)$ and $Z(0.4)$, where the errors of the unextrapolated results are largest. The final extrapolated values agree respectively with $y(0.2)$ and $z(0.4)$ to all the figures quoted, again confirming the applicability of formula (6.61).

It is worth noting that it would hardly be possible to obtain an equally good result for the first derivative by numerical differentiation of the results of the method of Table 6.1. For, in the first place, even with $h = 0.2$ the first differences of the correct results have at least one significant figure less than the seven figures of $Y$. And, in the second place, at this interval, which is the only one for which we can extrapolate $Y$ to agree with $y$ to full seven-figure accuracy, the difference table contains too few central differences (there is only one fifth difference) to provide accurate numerical differentiation, especially near the ends of the range.

As an example of a fourth-order problem we take the differential equation

$$y^{iv} + 2y''' + 2y'' + 2y' + y = 5 + 4x + x^2, \tag{6.65}$$

with boundary conditions

$$y(0) = 0, \quad y'(0) = 1, \quad y''(1) = -1, \quad y'''(1) = 0.5. \tag{6.66}$$

The equation differs from a general linear equation only in the not very important numerical point that the coefficients in (6.65) are independent of $x$ so that the main rows of the finite-difference matrix have the same coefficients, moved one to the right in successive rows. We have seen how to set up the approximating algebraic equations for (6.65) and (6.66), and the first half of Table 6.3 shows the computed values at $x = 1$ (where the error is greatest for all of $h = 0.2, 0.1$ and 0.05) and the two possible extrapolation steps which give a final result with an error of only one unit in the sixth decimal place.

We can also solve (6.65) and (6.66) in terms of four simultaneous first-order

**Table 6.3**

| h | Y(x = 1) | | | Y'''(x = 0) | | |
|---|---|---|---|---|---|---|
| 0.2 | 1.495325 | | | −11.522081 | | |
| | | 1.509697 | | | −11.632602 | |
| 0.1 | 1.506104 | | 1.509360 | −11.604972 | | −11.632907 |
| | | 1.509381 | | | −11.632888 | |
| 0.05 | 1.508562 | | | −11.625909 | | |
| Error $O(h^2)$ | $O(h^4)$ | $O(h^6)$ | | $O(h^2)$ | $O(h^4)$ | $O(h^6)$ |

equations and boundary conditions given by

$$\left. \begin{aligned} y' = u, \quad u' = v, \quad v' = w, \quad w' + 2w + 2v + 2u + y = 5 + 4x + x^2 \\ y(0) = 0, \quad u(0) = 1, \quad v(1) = -1, \quad w(1) = 0.5 \end{aligned} \right\}, \quad (6.67)$$

and we have also seen how to set up the approximating algebraic equations in this case. The error in the computed $Y''' = W$ is greatest with each interval at $x = 0$, and the second part of Table 6.3 shows the values of this quantity at $h = 0.2, 0.1$ and $0.05$ and the two extrapolations which give a final result accurate to within one unit in the fifth decimal place, that is the seventh significant figure.

For reasons already mentioned, we have here even less chance than in the second-order problem of using numerical differentiation of the accurate $y$ values, obtained from (6.65) and (6.66) and improved by extrapolation, to obtain similar accurate results for $y'$ and especially for $y''$ and $y'''$. If these derivatives are really required it is certainly preferable to solve the simultaneous first-order equations (6.67). It is also worth remarking that with our smaller computer the effect of rounding errors in the computation of $Y$ in the first half of Table 6.3 is considerably greater, as we expected in Section 6.5, than in the corresponding computation for second-order equations. With this computer the values corresponding to those for $Y$ in Table 6.3 have respective errors of 5, 350 and 3884 in the sixth decimal place. By comparison this small computer gave results corresponding to the $Y'''$ values in Table 6.3 with errors of only a few units in the sixth decimal place.

Finally we treat the fairly simple eigenvalue problem

$$y'' - (\lambda + 2q \cos 2x)y = 0, \quad y(0) = y(\pi) = 0, \quad (6.68)$$

of which in the literature the differential equation is called Mathieu's equation. Here we take the value $q = 1$. At interval $h$ the general approximate algebraic equation is

$$Y_{r+1} - (2 + h^2 \bar{\lambda} + 2h^2 \cos 2x_r) Y_r + Y_{r-1} = 0, \quad r = 1, 2, \ldots, n - 1, \quad (6.69)$$

with $Y_0$ and $Y_n$ having zero values and with $nh = \pi$. The algebraic eigenvalue problem then has the form of (6.53) where the matrix is given by

$$A = \begin{bmatrix} -(2 + 2h^2 \cos 2x_1) & 1 & & & \\ 1 & -(2 + 2h^2 \cos 2x_2) & 1 & & \\ & \vdots & & & \\ & 1 & -(2 + 2h^2 \cos 2x_{n-2}) & 1 & \\ & & 1 & -(2 + 2h^2 \cos 2x_{n-1}) \end{bmatrix}.$$

$$(6.70)$$

We here normalize the computed eigenvector by simple scaling so that for some particular value of $x$ the approximation $Y(x)$ has the same value, say unity, for all relevant $h$. Here it is not difficult to see that there is a solution which is symmetric about the point $x = \pi/2$, so that we choose this point to have a fixed value of unity. For such a normalization it can be shown that (6.61) holds for the eigenvector, but it is also true that the computed eigenvalue $\bar{\lambda}$ satisfies the equation

$$\lambda - \bar{\lambda} = Ah^2 + Bh^4 + \cdots, \qquad (6.71)$$

where $A$ and $B$ are constants independent of $h$.

Table 6.4 gives some results for $\bar{\lambda}$ at intervals $h = \pi/8, \pi/16, \pi/32$ and for $Y$ at $x = 3\pi/8$, where the error is largest for all $h$. In the final extrapolated results the number for the eigenvalue has an error of one unit in the last figure, and that for the eigenvector is in error by three units in the sixth decimal place. A further interval reduction produces six accurate figures in both cases.

An interesting question with eigenvalue problems is to what particular solution have we found approximations. The differential system (equation and boundary conditions) usually has an infinity of solutions whereas the approximating problem has only a finite number. It is reasonable to suppose that we shall get approximations only to those solutions in which the eigenfunctions can be approximated reasonably with the given number of mesh points.

Consider, for example, a case in which $y$ vanishes at the two end points 0 and 1, and we obtain successive approximations at intervals $\frac{1}{2}, \frac{1}{3}$ and $\frac{1}{4}$.

**Table 6.4**

| $h$ | $Y(x = 3\pi/8)$ | | | $\bar{\lambda}$ | | |
|---|---|---|---|---|---|---|
| $\pi/8$ | 0.856308 | | | 0.136440 | | |
| | | 0.865328 | | | 0.110035 | |
| $\pi/16$ | 0.863073 | | 0.865051 | 0.116636 | | 0.110250 |
| | | 0.865068 | | | 0.110236 | |
| $\pi/32$ | 0.864570 | | | 0.111836 | | |
| Error $O(h^2)$ | $O(h^4)$ | $O(h^6)$ | | $O(h^2)$ | $O(h^4)$ | $O(h^6)$ |

The number of possible independent distributions of signs in the eigenvectors is just

$$h = \tfrac{1}{2} \quad 0 \qquad\qquad + \qquad\qquad 0$$

$$h = \tfrac{1}{3} \quad \begin{array}{ccccc} 0 & & + & + & 0 \\ 0 & & - & + & 0 \end{array}$$

$$h = \tfrac{1}{4} \quad \begin{array}{cccccc} 0 & + & + & + & 0 \\ 0 & - & + & + & 0 \\ 0 & - & - & + & 0 \\ 0 & + & - & + & 0 \end{array}$$

With $h = \tfrac{1}{2}$ our matrix is a single element and we can approximate at all reasonably only to an eigenfunction with no internal zero. With $h = \tfrac{1}{3}$ we get two solutions, one of which will be an improvement on the previous 'no-zero' solution and the other can be a rather poor approximation to a 'one-zero' solution. With $h = \tfrac{1}{4}$ we have three solutions, and we obviously have the likelihood both of improving the two previous solutions and of producing rather a poor approximation to a 'two-zero' solution.

We remarked in Section 5.4 that for quite general second-order equations with general separated boundary conditions (equations (5.47) and (5.48)), the smallest eigenvalue gives a 'no-zero' eigenfunction, the next highest a 'one-zero' eigenfunction, and so on. The corresponding matrix, for example that of order $n - 1$ in (6.70), can and probably will then give approximations to the first $n - 1$ eigenvalues in increasing order of magnitude and to the corresponding eigenfunctions, the accuracy decreasing steadily from 'very good' for the first to 'very bad' for the last. In our example we have, of course, recorded only the 'no-zero' solution with the certainty that we have found the smallest eigenvalue.

We also observed in Chapter 5 that unseparated boundary conditions for second-order equations may not permit this one-to-one correspondence between the position of a particular eigenvalue in the spectrum of eigenvalues and the number of internal zeros of its eigenfunction. With equations of higher order the useful property may not exist even with separated boundary conditions, and in such circumstances we may need some additional 'physical' knowledge to supplement the information provided by our computation.

### (ii) The deferred correction

The method of deferred correction is performed with virtually the same techniques as we discussed in Section 3.9 for initial-value problems. For the linear second-order problem (6.30), for example, the first approximation $Y_r^{(1)}$ is obtained at internal mesh points by solving the linear algebraic equations (6.34).

External values are then calculated beyond both ends of the range by using (6.33) with appropriate $r$ and with the difference-correction neglected, in a quite obvious step-by-step manner. From (6.46) we easily find that the difference correction in (6.33) is given by

$$c(Y_r^{(1)}) = \{hf_r(\tfrac{1}{6}\mu\delta^3 - \tfrac{1}{30}\mu\delta^5 + \tfrac{1}{140}\mu\delta^7 - \cdots) + (\tfrac{1}{12}\delta^4 - \tfrac{1}{90}\delta^6 + \tfrac{1}{560}\delta^8 - \cdots)\} Y_r^{(1)},$$
(6.72)

and this is then computed at all interior mesh points and a few external points. The interior mesh-point values of $c(Y_r^{(1)})$ are inserted on the right of equations (6.34), which are then solved to give $Y_r^{(2)}$ at these points, and these are again extended a few points beyond the boundaries using (6.33) with the insertion on the right-hand side of the $c(Y_r^{(1)})$ at boundary points and external points. The general iterative process is an obvious extension of all this, but a few other remarks are relevant.

First, there is an important practical point regarding the solution of the successive sets of linear algebraic equations. The matrix is the same each time, so that the production by elimination of the relevant upper triangular matrix has to be performed once only. Only the new right-hand sides in the upper triangular equations have to be computed in the elimination process, followed by the back-substitution. It follows that the method of deferred correction is quite economical in computer time and storage.

Second, one can, as in equations (3.123), use fewer orders of differences in the deferred correction in the early stages of the iteration. More specifically, it has been proved in recent years that if in relation to (6.72) we take

$$\left.\begin{aligned}
c^{(1)}(Y_r^{(1)}) &= (hf_r \cdot \tfrac{1}{6}\mu\delta^3 + \tfrac{1}{12}\delta^4) Y_r^{(1)} \\
c^{(2)}(Y_r^{(2)}) &= \{hf_r(\tfrac{1}{6}\mu\delta^3 - \tfrac{1}{30}\mu\delta^5) + (\tfrac{1}{12}\delta^4 - \tfrac{1}{90}\delta^6)\} Y_r^{(2)} \\
c^{(3)}(Y_r^{(3)}) &= \{hf_r(\tfrac{1}{6}\mu\delta^3 - \tfrac{1}{30}\mu\delta^5 + \tfrac{1}{140}\mu\delta^7) + (\tfrac{1}{12}\delta^4 - \tfrac{1}{90}\delta^6 + \tfrac{1}{560}\delta^8)\} Y_r^{(3)}
\end{aligned}\right\},$$
(6.73)

and so on, the error in $Y_r^{(1)}$ is $O(h^2)$, as already stated, while the errors in $Y_r^{(2)}, Y_r^{(3)}, Y_r^{(4)}, \ldots$ are $O(h^4), O(h^6), O(h^8), \ldots$. The technique adds two extra orders of differences at each stage, and deliberately stops when the next orders of differences have no effect to the required accuracy. The comments following equation (3.123) are then relevant here also.

For simultaneous first-order equations similar results have been proved. For equations (6.57), for example, the most complicated formula is that for the last of (6.57), and is given by

$$w_{r+1} - w_r = \tfrac{1}{2}h(w'_{r+1} + w'_r) + c(w_r),$$
(6.74)

where

$$c(w_r) = (-\tfrac{1}{12}\delta^3 + \tfrac{1}{120}\delta^5 - \tfrac{1}{840}\delta^7 + \cdots)w_{r+\frac{1}{2}},$$
(6.75)

and the derivatives of $w_{r+1}$ and $w_r$ on the right of (6.74) are replaced from the fourth of (6.57) by a linear combination of the unknown $u, v$ and $w$ and other known terms at $r$ and $r+1$. Equations of type (6.74) will be part of the simultaneous set for mesh points in the range of interest, and used in step-by-step form for external values. Again, the first approximation has errors $O(h^2)$, and these reduce to $O(h^4), O(h^6)$ and $O(h^8)$ in successive approximations if we use in successive difference corrections just the terms involving $-\frac{1}{12}\delta^3, -\frac{1}{12}\delta^3 + \frac{1}{120}\delta^5, -\frac{1}{12}\delta^3 + \frac{1}{120}\delta^5 - \frac{1}{840}\delta^7$, and so on.

Some other interesting points are illustrated in the following examples. Consider first the problem of type (6.62) in which one of the boundary conditions involves a derivative as in (6.35). With regard to the first approximation we showed how to deal with this in equations (6.36)–(6.38), but we observed that the derivative condition involves another difference correction in equation (6.36), and the form of this is given by

$$c'(y_0) = 2q(\tfrac{1}{6}\mu\delta^3 - \tfrac{1}{30}\mu\delta^5 + \tfrac{1}{140}\mu\delta^7 - \cdots)y_0. \tag{6.76}$$

One can use the method just outlined, with obvious modifications, but note that (6.76) contains a third difference not multiplied by $h$, as in (6.73), so that the difference correction $c'(y_0)$ of the boundary condition is likely to be considerably larger than the general $c(y_r)$ of the differential equation, and this could be quite a disadvantage, again in agreement with a remark at the end of Section 3.9, referring to Section 3.7, in connection with the propinquity of a singularity in the solution. In the problem of equation (6.62), with interval 0.1, the maximum errors in the computed $Y_r^{(1)}, Y_r^{(2)}, Y_r^{(3)}$ and $Y_r^{(4)}$ are respectively $10^{-6}(2049, -90, 13, -3)$ and higher orders of accuracy cannot be obtained at this interval since we are too near the singularity.

This disadvantage could be avoided, of course, by using the corresponding pair of first-order equations and boundary conditions in (6.64), the derivative condition now merely specifying a boundary value of one of the functions. Even the most general condition (6.35) merely gives a known relation between $y(a)$ and $z(a)$, and a similar sentiment is still true, with obvious modification, for the more general non-separated conditions (6.39). For this particular example, however, the function $z(x) = y'(x)$ has in its solution the term $(1+x)^{-2}$, and this has considerably more dangerous behaviour near $x = -1$ than the function $y(x)$. The standard difference correction of type (6.75) for $z(x)$ might then affect adversely the attainable accuracy for both functions. For problem (6.62), solved in form (6.64), the maximum errors at interval 0.1 in the computed $Y_r$ and $Z_r$ sequences are respectively $10^{-6}(1658, -52, 5, -1)$ and $10^{-6}(-1640, 116, -11, 2)$. Despite the previous remark these are still very good indeed, and in particular we should hardly expect to get such an accurate $Z_r$ by differentiation of the results for $Y_r$ obtained by the previous method. At interval $h = 0.05$ the maximum errors without difference correction are $413 \times 10^{-6}$ in $Y_r$ and $-407 \times 10^{-6}$ in $Z_r$. One application of the difference correction reduces these respectively to $-3 \times 10^{-6}$ and $7 \times 10^{-6}$, and a second gives at least six-decimal accuracy in both $Y_r$ and $Z_r$. To this precision the seventh differences are making no contribution to the required solution.

Similarly, though the direct treatment of fourth-order equations is obviously quite possible with the deferred correction, all four boundary conditions may involve difference corrections and the whole process becomes quite complicated. In many cases it is preferable to solve the corresponding simultaneous first-order equations, in which the general difference correction is of quite simple form and involves no function of $x$, and in which the boundary conditions involve no difference correction at all.

In Table 6.5 we show some results of this method to the problem of equations (6.67), the first-order equivalents of (6.65) and (6.66), for which some results using the deferred approach to the limit appear on the right of Table 6.3. Here we use the interval $h = 0.1$, and give in Table 6.5 the errors (with obvious notation) of the first two approximations for both $Y_r$ and $W_r (= Y_r''')$, and the difference corrections of the two first approximations which use third differences only. The next two difference corrections, using respectively third and fifth differences and third, fifth and seventh differences, in accord with equation (6.75) and the following comment, differ by only a few units in the sixth decimal place, but we see from the numbers in Table 6.5 that the function value corrections are larger than the difference correction. In particular, the error in $W_r^{(1)}$ is at least ten times the size of the difference correction $c(W_r^{(1)})$ for most of the range, so that at least seven decimal places in the latter are needed, and of course are used automatically by the computer, to get six-decimal accuracy in the required quantities. The maximum errors in $Y_r^{(3)}$ and $W_r^{(3)}$ are $4 \times 10^{-6}$ and $11 \times 10^{-6}$, and $Y_r^{(4)}$ and $W_r^{(4)}$, as well as $U_r^{(4)}$ and $V_r^{(4)}$, are correct to six decimal places. We have, incidentally, omitted from Table 6.5 the computed external values and their difference corrections.

**Table 6.5**

| $x$ | $10^6 e(Y_r^{(1)})$ | $10^6 e(Y_r^{(2)})$ | $10^6 e(W_r^{(1)})$ | $10^6 e(W_r^{(2)})$ | $10^6 c(Y_r^{(1)})$ | $10^6 c(W_r^{(1)})$ |
|-----|------|------|---------|------|------|--------|
| 0.0 | 0     | 0    | $-27945$ | $-56$ | 895  | $-1687$ |
| 0.1 | 827   | $-3$ | $-24690$ | $-45$ | 751  | $-1453$ |
| 0.2 | 1380  | $-5$ | $-21454$ | $-36$ | 619  | $-1246$ |
| 0.3 | 1678  | $-7$ | $-18274$ | $-29$ | 501  | $-1066$ |
| 0.4 | 1737  | $-8$ | $-15184$ | $-22$ | 393  | $-909$  |
| 0.5 | 1573  | $-9$ | $-12214$ | $-17$ | 295  | $-772$  |
| 0.6 | 1204  | $-10$ | $-9392$ | $-12$ | 207  | $-655$  |
| 0.7 | 644   | $-11$ | $-6740$ | $-8$  | 127  | $-555$  |
| 0.8 | $-86$ | $-13$ | $-4280$ | $-4$  | 55   | $-470$  |
| 0.9 | $-971$ | $-14$ | $-2027$ | $-2$  | $-9$ | $-400$  |
| 1.0 | $-1991$ | $-15$ | 0      | 0     | $-67$ | $-342$  |

Finally, we use a difference correction method for the eigenvalue problem (6.68). Not all difference correction methods suggested in the literature are generally applicable, but the method suggested here seems to work for all suggested problems. In the approximating algebraic equations (6.69) there are $n$ unknowns, the $n-1$ internal values of $Y_r$ and the single value $\bar{\lambda}$, whereas there are only $n-1$ equations. We intend to use a Newton iterative method, so that one more equation is needed. An obvious equation is the normalization condition which we used earlier, so that we use the accurate equations

$$\left.\begin{array}{c} y_{r+1} - (2 + h^2\lambda + 2h^2\cos 2x_r)y_r + y_{r-1} = c(y_r), \qquad r = 1, 2, \ldots, n-1 \\ y_p = 1 \end{array}\right\}, \quad (6.77)$$

where $y_0 = y_n = 0$ and

$$c(y_r) = (\tfrac{1}{12}\delta^4 - \tfrac{1}{90}\delta^6 + \tfrac{1}{560}\delta^8 - \cdots)y_r. \qquad (6.78)$$

We therefore solve the equations

$$\left.\begin{array}{c} Y_{r+1}^{(s+1)} - (2 + h^2\lambda^{(s+1)} + 2h^2\cos 2x_r)Y_r^{(s+1)} + Y_{r-1}^{(s+1)} = c(Y_r^{(s)}), \qquad r = 1, 2, \ldots, n-1 \\ Y_p^{(s+1)} = 1 \end{array}\right\},$$

$$(6.79)$$

where $p$ has a specified value.

The difference corrections are ignored for $s = 0$, and the computation of $Y_r^{(1)}$ and $\lambda^{(1)}$ is just the standard algebraic eigenvalue problem, the normalization for $s = 0$ being performed at the end of the computation. For each $s > 0$ we extend the solutions to a few external points, compute difference corrections, taking successive segments of (6.78) for this purpose (as in (6.73) with zero $f_r$), and solve (6.79) using the Newton method and the Jacobian matrix which we have described in some detail previously. The second of (6.79) is quite easily incorporated in this process, the relevant row of the Jacobian matrix having a single unit element and zeros elsewhere.

Table 6.6 gives a summary of the computation for interval $h = \pi/8$, where because of the symmetry about $x = \pi/2$, at which point we specify $Y_r = 1$, we need give only half the full table. Here the third difference correction involves

**Table 6.6**

| $x$ | $10^6 Y_r^{(1)}$ | $10^6$(Corrections) | | | $10^6 c(Y_r^{(1)})$ | $10^6 c(Y_r^{(2)})$ | $10^6 c(Y_r^{(3)})$ |
|---|---|---|---|---|---|---|---|
| $\pi/8$ | 242898 | 11584 | 799 | 217 | $-8727$ | $-9685$ | $-9758$ |
| $\pi/4$ | 543881 | 16243 | 783 | 581 | $-11128$ | $-11630$ | $-12111$ |
| $3\pi/8$ | 856308 | 7987 | 178 | 447 | $+5127$ | $+5744$ | $+5562$ |
| $\pi/2$ | 1000000 | 0 | 0 | 0 | $+19775$ | $+20309$ | $+21251$ |
| $10^6\lambda^{(s)}$ | 136440 | $-24645$ | $-156$ | $-309$ | | | |

eighth differences, but we do not have $Y_r$ consistent to six figures and the results are hardly 'smooth'. We should perhaps decide that we do not wish to use differences of order higher than the eighth, and we should therefore reduce the interval and repeat the process. With $h = \pi/16$ we find for $\lambda^{(1)}$ and its corrections the values $10^6(116636, -6293, -87, -7)$, the final result being correct to all six figures. The $Y_r^{(4)}$ values have a maximum error of one unit in the sixth decimal. These are very good results, and a comparison with those of the deferred approach verifies the relative economy of the current method.

## 6.7 NON-LINEAR PROBLEMS

The eigenvalue problem is in a sense non-linear, since the unknown number $\lambda$ multiplies the unknown function $y$, and the technique just described for eigenvalue problems is applicable with only obvious changes to the solution of problems which are non-linear in $y$ itself, such as

$$y'' = f(x, y, y'),\tag{6.80}$$

with unseparated non-linear boundary conditions

$$g_1\{y(a), y'(a), y(b), y'(b)\} = 0, \quad i = 1, 2.\tag{6.81}$$

For the direct treatment of this problem we have to solve the equations

$$\left.\begin{array}{l} y_{r+1} - 2y_r + y_{r-1} + c_1(y_r) = h^2 f[x_r, y_r, h^{-1}\{\tfrac{1}{2}y_{r+1} - \tfrac{1}{2}y_{r-1} + c_2(y_r)\}], \quad r = 0, 1, 2, \ldots, n \\ g_i[y_0, h^{-1}\{\tfrac{1}{2}y_1 - \tfrac{1}{2}y_{-1} + c_2(y_0)\}, y_n, h^{-1}\{\tfrac{1}{2}y_{n+1} - \tfrac{1}{2}y_{n-1} + c_2(y_n)\}] = 0, \quad i = 1, 2 \end{array}\right\}$$

where

$$\left.\begin{array}{l} c_1(y_r) = (-\tfrac{1}{12}\delta^4 + \tfrac{1}{90}\delta^6 - \tfrac{1}{560}\delta^8 + \cdots)y_r \\ c_2(y_r) = (-\tfrac{1}{6}\mu\delta^3 + \tfrac{1}{30}\mu\delta^5 - \tfrac{1}{140}\mu\delta^7 + \cdots)y_r \end{array}\right\}.\tag{6.83}$$

Again we have two possible methods. First we can find approximations for several values of $h$, with difference corrections neglected, and improve these with the method of the deferred approach to the limit, equation (6.61) still holding if there are no singularities in the problem. Second, we can use the method of deferred correction, for fixed $h$, with the Newton method effectively the same as we have used for the eigenvalue problem except that there is here no different special method for the first approximation.

As an alternative we can use either of these methods for the corresponding first-order equations given by

$$y' = z, \quad z' = f(x, y, z), \quad g_i\{y(a), z(a), y(b), z(b)\} = 0, \quad i = 1, 2,\tag{6.84}$$

for which the discretized equations are

$$\left.\begin{array}{l} y_{r+1} - y_r = \tfrac{1}{2}h(z_{r+1} + z_r) + c(y_r), \quad r = 0, 1, \ldots, n \\ z_{r+1} - z_r = \tfrac{1}{2}h(f_{r+1} + f_r) + c(z_r), \quad r = 0, 1, \ldots, n \\ g_i(y_0, z_0, y_n, z_n) = 0, \quad i = 1, 2 \end{array}\right\}.\tag{6.85}$$

**Table 6.7**

| h | Y(x = 2) | | | Y(x = 1.5) | | | Z(x = 1) | | |
|---|---|---|---|---|---|---|---|---|---|
| $\frac{1}{4}$ | 0.503507 | | | 0.660297 | | | −1.010211 | | |
| | | 0.500026 | | | 0.666741 | | | −1.000010 | |
| $\frac{1}{8}$ | 0.500896 | | | 0.665130 | | 0.666666 | −1.002560 | | −1.000000 |
| | | 0.500003 | | | 0.666671 | | | −1.000001 | |
| $\frac{1}{16}$ | 0.500226 | | | 0.666286 | | | −1.000641 | | |
| Error | $O(h^2)$ | $O(h^4)$ | $O(h^6)$ | $O(h^2)$ | $O(h^4)$ | $O(h^6)$ | $O(h^2)$ | $O(h^4)$ | $O(h^6)$ |

where $c(y_r)$ and $c(z_r)$ are effectively defined in (6.75) and $f_r = f(x_r, y_r, z_r)$.

To indicate the performance of these various methods we consider the not too complicated problem given by

$$y'' = 2y^3, \quad y(1) = 1, \quad y'(2) + y^2(2) = 0, \tag{6.86}$$

whose solution is $y = x^{-1}$. Table 6.7 gives on the left some representative results for the extrapolation method for the direct solution of (6.86) from equations of type (6.82) with difference corrections neglected, at intervals $h = \frac{1}{4}, \frac{1}{8}$ and $\frac{1}{16}$. On the right of Table 6.7 we give some similar results for the corresponding pair of first-order problems, using equations of type (6.85). The errors of the final extrapolated results are nowhere greater than one unit in the last figure.

In these computations we used no special knowledge for our first guess at interval $h = \frac{1}{4}$. In the first method we started with $Y_r = 1$ at all points, and after five iterations we had guaranteed convergence to six decimal places, the fourth and fifth computed $Y_r$ having the same values at all points. For the iteration with $h = \frac{1}{8}$, however, we can obtain very good starting values by interpolation at the midpoints of the intervals. With values $Y_r, Y_{r+1}, Y_{r+2}, Y_{r+3}$ at four equidistant points, good interpolated values at the half-way points are obtained from the formulae

$$\left.\begin{array}{l} Y_{r+\frac{1}{2}} = \frac{1}{16}(5Y_r + 15Y_{r+1} - 5Y_{r+2} + Y_{r+3}) \\ Y_{r+\frac{3}{2}} = \frac{1}{16}(- Y_r + 9Y_{r+1} + 9Y_{r+2} - Y_{r+3}) \end{array}\right\}, \tag{6.87}$$

which cover all possibilities.

With the resulting good start the numbers of iterations with $h = \frac{1}{8}$ and $\frac{1}{16}$ are only three and two respectively, with the final iteration in each case merely confirming the previous results to all figures required.

Similar operations with the computation for the simultaneous first-order equations obtained identical values for both $Y$ and $Z$ after six iterations with $h = \frac{1}{4}$, reducing respectively to three and two for $h = \frac{1}{8}$ and $h = \frac{1}{16}$ after interpolations of type (6.87) to give a good start.

Table 6.8 gives some results for the solution of this problem by the deferred correction method, using the relevant forms of equations (6.82) and (6.83). At interval $h = \frac{1}{8}$ we show the first approximation and the next three corrections,

**Table 6.8**

| $x$ | $Y_r^{(1)}$ | $10^6 \delta Y_r^{(1)}$ | $10^6 \delta Y_r^{(2)}$ | $10^6 \delta Y_r^{(3)}$ | $10^6 \delta^6 y$ | $10^6 \delta^8 y$ |
|-----|-------------|-------------------------|-------------------------|-------------------------|-------------------|-------------------|
| 0.750 | 1.330314 | + 3049 | − 64 | + 69 | 31746 | 88889 |
| 0.875 | 1.141893 | + 974 | − 17 | + 8 | 9524 | 17162 |
| 1.000 | 1.000000 | 0 | 0 | 0 | 3463 | 3040 |
| 1.125 | 0.889357 | − 473 | + 7 | − 2 | 1443 | 1243 |
| 1.250 | 0.800697 | − 704 | 9 | − 2 | 666 | 444 |
| 1.375 | 0.728079 | − 815 | 11 | − 3 | 333 | 178 |
| 1.500 | 0.667522 | − 864 | 11 | − 3 | 178 | 77 |
| 1.625 | 0.616260 | − 885 | 12 | − 3 | | |
| 1.750 | 0.572312 | − 893 | 11 | − 2 | | |
| 1.875 | 0.534222 | − 898 | 11 | − 2 | | |
| 2.000 | 0.500896 | − 906 | 11 | − 1 | | |
| 2.125 | 0.471498 | − 920 | 12 | − 2 | | |
| 2.250 | 0.445375 | − 941 | 12 | − 2 | | |

by which time we have used differences up to the eighth. To achieve this we have had to compute and correct several times the approximations to external values getting quite close to the singularity at $x = 0$, and the danger of this is revealed in the tabulation of the correctly rounded sixth and eighth differences of the true solution near $x = 1$. Some eighth-difference contributions are obviously not negligible, and there is no guarantee that tenth and higher differences can be ignored. Indeed, though the computed $Y_r^{(4)}$ agrees exactly to six decimals with the correct solution at all points in the range $1 \leqslant x \leqslant 2$, the error at the adjacent external point $x = \frac{3}{4}$ is − 35 in the sixth decimal place, and in fact the correction $10^6 \delta Y_r^{(3)}$ looks rather strange at this point. Clearly we have really reached the limit of what can be done at this interval in this problem.

When we treat the second-order equation as two first-order equations, one of the solutions is $(-1/x^2)$, and in analogy with a remark in the paragraph following that which contains equation (6.76) it is also true here that this has even more difficult behaviour near $x = 0$ than the $(1/x)$ of Table 6.8. And even though the difference corrections for this method have very small coefficients there is again a limit to what we can achieve at this interval, the respective maximum errors of successive approximations being 1537, − 107, 13 and − 2 in the sixth decimal for $Y_r$, and 2560, − 254, 34 and − 5 for $Z_r$. We definitely need a smaller interval, but the differences in Table 6.8 indicate that the current interval is probably quite satisfactory for at least the second half of the range, and in Chapter 8 we indicate some modern techniques for useful and automatic choices of interval or intervals.

Just as with linear problems, so here the simultaneous first-order equations are easier to deal with, but for a second-order equation, for example, the number of unknowns is about double that stemming from the direct treatment of the second-order equation, and it now happens that the solution of the relevant non-linear algebraic equations is much more difficult and time-consuming. Here the question of convergence of the Newton iteration is very important. Not all starting approximations will guarantee convergence, and though we obviously have very good starts for $s \geqslant 1$ we have no guidance for the start for $s = 0$ without the assistance of relevant mathematical or physical knowledge. This makes the general non-linear problem far more difficult than the eigenvalue problem, in which the method for $s = 0$, the solution of the corresponding algebraic eigenvalue problem, is different from that for all other $s$ but does in fact avoid all our starting difficulties. Finding a good start is in fact one of the most difficult problems in the solution of non-linear boundary-value problems and is the item which causes most of the failures in otherwise very good computer-library routines. The next section discusses one possibility for solving this problem.

## 6.8 CONTINUATION FOR NON-LINEAR PROBLEMS

We have just observed that one of the major difficulties of solving non-linear problems, by almost any iterative method, is the possibility of lack of convergence or a very slow rate of convergence of the iteration. The Newton method will give quadratic convergence, that is the error at any stage will be proportional to the square of the error in the previous step, only when we are sufficiently close to the solution required. Otherwise the iterations may not converge at all, or converge very slowly at least for the first few steps, or converge to an unwanted solution if there is more than one solution.

In our boundary-value problems the first estimate can be quite crucial. For a two-point boundary-value problem with conditions specifying $y = \alpha$ at $x = a$ and $y = \beta$ at $x = b$, a tempting and often-used first estimate is the linear form

$$\bar{y}(x) = \{(\alpha - \beta)x + a\beta - b\alpha\}/(a - b). \tag{6.88}$$

More complicated boundary conditions give no particular difficulty in the determination of such a linear form. For example, for the non-linear problem of equation (6.86) the linear starting approximation which satisfies the boundary conditions is easily found to be

$$\bar{y}(x) = \tfrac{1}{2}(-3 + \sqrt{5})x + \tfrac{1}{2}(5 - \sqrt{5}). \tag{6.89}$$

The values of (6.89) at the points $x = 1$, 1.25, 1.50, 1.75 and 2.0 are respectively near to 1, 0.905, 0.809, 0.714 and 0.618. Those of the true solution are 1, 0.800, 0.667, 0.571 and 0.500, and despite the great analytical difference between the nature of this solution and the linear approximation the latter is numerically close enough to give convergence with most iterative methods. In fact with the obvious finite-difference equations Newton's method gives a next approximation with respective values 1, 0.809, 0.681, 0.588 and 0.516, demonstrating quite satisfactory convergence.

In some cases this simple approach may produce quite unsatisfactory results. Consider, for example, the problem

$$y'' = 1 + 15y^3 - 30y, \quad y(0) = 0, \quad y(1) = 1. \tag{6.90}$$

If we use the linear starting approximation, which in this case is just $y(x) = x$, successive approximations obtained with the simple finite-difference formula show a wildly erratic behaviour. With interval $h = 0.1$ the second step of the Newton iteration produces the successive mesh values $-4.3$, $-8.3$, $-14.7$, $-27.6$, $-46.5$, $-60.0$, $-55.8$, $-38.2$, $-18.2$, and successive iterations show no signs of convergence.

We can obtain a solution of the problem by various versions of a *continuation* method. We might, for example, write (6.90) in the slightly more general form

$$y'' = 1 + 15ty^3 - 30y, \quad y(0) = 0, \quad y(1) = 1, \tag{6.91}$$

including the parameter $t$. When $t = 0$ the problem is linear, and the solution is very simple. The original problem corresponds to $t = 1$. We may therefore begin by solving the problem for $t = 0$, and then, in successive stages, for a number of values of $t$ between 0 and 1. At each stage we solve the problem (6.91) by the standard method, using as first estimate for $y(x)$ the solution obtained for the previous value of $t$. Assuming that the solution of (6.91) is a smooth function of $t$ over the range from 0 to 1, and that we use values of $t$ which are reasonably close together, the process works well. Table 6.9 gives an example of the process, using the values $t = 0, \frac{1}{4}, \frac{1}{2}, \frac{3}{4}$ and 1.

At the intermediate stages it is not necessary to obtain a highly accurate solution; all that is necessary is to find a solution which is adequate to start the next stage. The table also shows the number of Newton iterations used at

**Table 6.9**

| $x$ | $Y(t=0)$ | $Y(t=0.25)$ | $Y(t=0.5)$ | $Y(t=0.75)$ | $Y(t=1)$ |
|---|---|---|---|---|---|
| 0.0 | 0.000 | 0.000 | 0.000 | 0.000 | 0.000000 |
| 0.1 | $-0.774$ | $-0.557$ | $-0.490$ | $-0.455$ | $-0.437328$ |
| 0.2 | $-1.305$ | $-0.943$ | $-0.832$ | $-0.774$ | $-0.746004$ |
| 0.3 | $-1.435$ | $-1.068$ | $-0.957$ | $-0.903$ | $-0.883153$ |
| 0.4 | $-1.125$ | $-0.908$ | $-0.851$ | $-0.833$ | $-0.848680$ |
| 0.5 | $-0.467$ | $-0.494$ | $-0.526$ | $-0.569$ | $-0.641294$ |
| 0.6 | 0.341 | 0.074 | $-0.044$ | $-0.145$ | $-0.271080$ |
| 0.7 | 1.057 | 0.630 | 0.461 | 0.332 | 0.187470 |
| 0.8 | 1.466 | 1.016 | 0.845 | 0.724 | 0.600767 |
| 0.9 | 1.444 | 1.467 | 1.031 | 0.951 | 0.876359 |
| 1.0 | 1.000 | 1.000 | 1.000 | 1.000 | 1.000000 |
| No. of iterations | | 3 | 3 | 4 | 3 |

each stage. In this case the first problem, with $t = 0$, is linear, so no starting approximation is necessary, and the Newton iteration gives the exact solution of the finite-difference equations in one step.

It is clear, of course, that there are very many ways of 'embedding' the given problem in a more general problem involving a parameter $t$. We require only smooth behaviour as a function of $t$, that $t = 1$ should correspond to the given problem, and that $t = 0$ should give a problem which is easily solved. An interesting result is obtained by writing (6.90) in a different general form, as

$$y'' = 1 + t(15y^3 - 30y), \quad y(0) = 0, \quad y(1) = 1. \tag{6.92}$$

Here the case $t = 0$ gives the trivial solution $y = \frac{1}{2}x(x + 1)$, and the same continuation process works just as before; the result is shown in Table 6.10. Notice that we have now found another, quite different, solution to the problem. If this were a practical problem we should now have to ask which of these two solutions is physically acceptable, or whether it may be necessary to try to find other solutions, perhaps by another continuation scheme.

Practical problems are usually more difficult than this, and a particular continuation method may break down. This happens for the very similar example

$$y'' = 1 - t(y^3 + 9y), \quad y(0) = 0, \quad y(1) = 1. \tag{6.93}$$

This is very like (6.92), but with different numerical coefficients. Beginning with $t = 0$, for which the solution is again $\frac{1}{2}x(x + 1)$, and increasing $t$ in small steps, the process works well until it breaks down at about $t = 0.65$. Beyond this point the solution ceases to be a smooth function of $t$, and the Newton method does not converge. The problem in (6.93) has a solution, but a different form of continuation is required to find it.

**Table 6.10**

| $x$ | $Y(t=0)$ | $Y(t=0.25)$ | $Y(t=0.5)$ | $Y(t=0.75)$ | $Y(t=1)$ |
|---|---|---|---|---|---|
| 0.0 | 0.000 | 0.000 | 0.000 | 0.000 | 0.000000 |
| 0.1 | 0.055 | 0.197 | 0.340 | 0.437 | 0.511301 |
| 0.2 | 0.120 | 0.390 | 0.642 | 0.794 | 0.899262 |
| 0.3 | 0.195 | 0.566 | 0.877 | 1.039 | 1.136525 |
| 0.4 | 0.280 | 0.716 | 1.042 | 1.187 | 1.263037 |
| 0.5 | 0.375 | 0.836 | 1.145 | 1.266 | 1.322868 |
| 0.6 | 0.480 | 0.926 | 1.199 | 1.299 | 1.343089 |
| 0.7 | 0.595 | 0.985 | 1.212 | 1.295 | 1.333799 |
| 0.8 | 0.720 | 1.017 | 1.187 | 1.255 | 1.290298 |
| 0.9 | 0.855 | 1.022 | 1.120 | 1.165 | 1.191935 |
| 1.0 | 1.000 | 1.000 | 1.000 | 1.000 | 1.000000 |
| No. of iterations | | 5 | 5 | 4 | 5 |

## 6.9 ADDITIONAL NOTES

*Section 6.2(i)* The method for solving linear equations described here is called *Gauss elimination*, and the corresponding method with interchanges is *Gauss elimination with interchanges*, or *with pivoting*. Other theoretically identical methods are mentioned in the literature and are used in practice, especially those based on the *triangular splitting*

$$A = LU, \tag{6.94}$$

where $L$ is a lower triangular matrix and $U$ upper triangular, with either $L$ or $U$ having prescribed diagonal elements, usually unity. For order three we can easily find $l_{rs}$ and $u_{rs}$ such that

$$
\begin{matrix} L \\ \begin{bmatrix} 1 & & \\ l_{21} & 1 & \\ l_{31} & l_{32} & 1 \end{bmatrix} \end{matrix}
\begin{matrix} U \\ \begin{bmatrix} u_{11} & u_{12} & u_{13} \\ & u_{22} & u_{23} \\ & & u_{33} \end{bmatrix} \end{matrix}
=
\begin{matrix} A \\ \begin{bmatrix} a_{11} & a_{12} & a_{13} \\ a_{21} & a_{22} & a_{23} \\ a_{31} & a_{32} & a_{33} \end{bmatrix} \end{matrix}. \tag{6.95}
$$

The first row of $L$ times successive columns of $U$ gives

$$\underline{u}_{11} = a_{11}, \quad \underline{u}_{12} = a_{12}, \quad \underline{u}_{13} = a_{13}, \tag{6.96}$$

and the second row of $L$ times successive columns of $U$ gives

$$\underline{l}_{21}u_{11} = a_{21}, \quad l_{21}u_{12} + \underline{u}_{22} = a_{22}, \quad l_{21}u_{13} + \underline{u}_{23} = a_{23}, \tag{6.97}$$

the quantities which can be computed in succession being underlined. The final set of equations, for this case, then gives $l_{31}$, $l_{32}$ and $u_{33}$ from

$$\underline{l}_{31}u_{11} = a_{31}, \quad l_{31}u_{12} + \underline{l}_{32}u_{22} = a_{32}, \quad l_{31}u_{13} + l_{32}u_{23} + \underline{u}_{33} = a_{33}. \tag{6.98}$$

After the computation of $L$ and $U$ we then merely solve the two triangular sets of equations

$$Lc = b, \quad Ux = c, \tag{6.99}$$

the first by *forward substitution*, calculating the elements $c_1, c_2, c_3$ of the vector $c$ in succession from the three equations of the first of (6.99) in normal order, and the second by *back-substitution* which we have already discussed.

There is, of course, a very close connection with Gauss elimination, and it is not difficult to show that for a matrix of order three the $L$ and $U$ of (6.95) are identical with the forms

$$
L = \begin{bmatrix} 1 & & \\ -m_{21} & 1 & \\ -m_{31} & -m_{32} & 1 \end{bmatrix}, \quad
U = \begin{bmatrix} a_{11} & a_{12} & a_{13} \\ & a_{22}^{(2)} & a_{23}^{(2)} \\ & & a_{33}^{(3)} \end{bmatrix}, \tag{6.100}
$$

respectively a matrix of the multipliers in the elimination, with different sign, and the final upper triangle of the elimination. Moreover, the vector $c$ in (6.99) is of course identical with the vector on the right-hand side of equation (6.5). Finally, if in both methods all arithmetic is performed by rounding every division, multiplication and addition to the same number of figures, then the rounding errors and final results of the two methods are identical.

Why then do we show interest in the methods of equations (6.94) and (6.99)? The main reason is that, with a little more attention to the arithmetic which is possible in some computers, this method will reduce the amount of rounding error and, less importantly, the amount of temporary storage used in the procedure. Consider, for example, the last of (6.98) for the determination of $u_{33}$, which we can also write in the notation of (6.100) as

$$- m_{31} a_{13} - m_{32} a_{23}^{(2)} + a_{33}^{(3)} = a_{33}, \qquad (6.101)$$

that is

$$a_{33}^{(2)} = a_{33} + m_{31} a_{13} + m_{32} a_{23}^{(2)} \qquad (6.102)$$

for the determination of $a_{33}^{(3)}$, with everything on the right previously known or computed and stored in single-length locations. In the Gauss elimination process we first compute $a_{33} + m_{31} a_{13}$, rounding first the multiplication and then the addition to single length, and actually store this number as $a_{33}^{(2)}$ in (6.4). The required $a_{33}^{(3)}$ is then obtained by rounding the multiplication $m_{32} a_{23}^{(2)}$ and then rounding the final addition to give $a_{33}^{(3)}$ which is then stored in single length.

Here there are no fewer than four rounding errors, and for a large matrix the total number is very large. But now suppose that we can do the arithmetic on the right of (6.102) accurately, say in double length, and then make just one rounding to produce and store the required single-length $a_{33}^{(3)}$, neither rounding nor the storage of the partial sums of (6.102) being needed. We get a different result from that of Gauss elimination, and indeed a result involving far fewer rounding errors.

The effect of this is difficult to measure with respect to the computed solution of the original equations $A\bar{x} = b$, but it can be shown that with all methods the computed result $\bar{x}$ is the exact solution of a perturbed problem

$$(A + \delta A)\bar{x} = b + \delta b, \qquad (6.103)$$

and the smaller the induced perturbations $\delta A$ and $\delta b$, the better our method. They are smaller with our special arithmetic than with the standard floating-point arithmetic. In particular, the $L$ and $U$ of the splitting, using this arithmetic, are better than those of standard Gauss elimination in the sense that $LU - A$ is smaller in the first case than in the second. In our simple ($2 \times 2$) example both $L$ and $U$ are respectively the same except for $u_{22}$, which from the first of (6.7) is 6868 with our better arithmetic, and in the (2, 2) position $LU - A$ is $-0.3742$ with the better arithmetic of the splitting and 0.6258 with the Gauss version.

The other and far more important technique for keeping small the size of the perturbations is the use of interchanges. The important quantities are the elements of the $L$ and $U$ matrices. If these are large, the perturbations are large, and with interchanges we ensure that all $|l_{rs}| \leqslant 1$, which also help to keep $|u_{rs}|$ satisfactorily small. The effect of the interchanges in the matrix splitting is to find $L$ and $U$ for which

$$LU = \bar{A}, \tag{6.104}$$

a row permutation of $A$ which is easily discovered in the process of the computation. Interchanges, combined with the better arithmetic of the splitting, then give a very satisfactory algorithm, free from almost all vestige of induced instability.

*Section 6.2(ii)* It is sometimes possible to be certain from the start that no interchanges are needed, as for example with the matrix of equation (6.20). This matrix is of a form called *diagonally dominant*, the modulus of the diagonal term in each row being at least as large as the sum of the moduli of the other elements in the row. If the term 'at least as large as' can be replaced by 'greater than' for at least one row then it can be guaranteed that no interchanges are necessary, none of the multipliers exceeding unity in absolute value in Gauss elimination or the elements of the $L$ matrix in the triangular splitting.

The matrix in (6.20) is also symmetric, and this symmetry is maintained in Gauss elimination with pivots taken down the diagonal. In equation (6.4), for example, symmetry in (6.3) would cause $a_{23}^{(2)}$ and $a_{32}^{(2)}$ to be equal, and the latter would not need separate computation. There is some similar relation in the corresponding $L$ and $U$ matrices, but here the relation is more obvious if we compute the elements of $L$, including the diagonal elements, from the 'symmetric equivalent'

$$A = LL^{\mathrm{T}} \tag{6.105}$$

of the more general (6.94). With interchanges, symmetry is maintained only if row interchanges are associated with corresponding column interchanges.

For very sparse systems the attempt to restrict to unity the size of all multipliers may produce too large an amount of unwanted fill-in, and interchanges which avoid the latter must also have some consideration. Any resulting instability may be counterbalanced by a method of improving computed results called *iterative refinement*. If we denote by $\bar{x}$ the computed solution of $Ax = b$, and perform very accurately the computation of the residual vector

$$r = b - A\bar{x}, \tag{6.106}$$

then a correction to $\bar{x}$ is obtained from the equations

$$A\delta\bar{x} = r, \tag{6.107}$$

and we can repeat this and express the technique in terms of the iterative scheme

$$x^{(r+1)} = x^{(r)} + \delta x^{(r)}, \quad A\delta x^{(r)} = b - Ax^{(r)}. \tag{6.108}$$

Note that the solution of the second of (6.108) is now effected quickly with the use of (6.104) and (6.99), the matrices $L$ and $U$ having already been computed and stored. This process will usually converge rapidly unless the problem is very ill conditioned, that is when small perturbations $\delta A$ and $\delta b$ in (6.103) make large changes to the solution. This technique has the additional advantage of providing this important information about the inherent instability of the problem.

It may happen that a large amount of fill-in is virtually unavoidable, in which case an iterative rather than a direct method of solving the linear equations may be more economic if the iteration converges rapidly. A fairly simple and useful iterative technique is the Gauss–Seidel version, which performs a different splitting of $A$ in the form

$$A = L + D + U, \tag{6.109}$$

where $L$ is the part of $A$ below the diagonal, $D$ is the diagonal part and $U$ the part above the diagonal. The technique starts with an arbitrary or guessed $x^{(0)}$ and then proceeds with the iteration

$$(L + D)x^{(r+1)} = b - Ux^{(r)}. \tag{6.110}$$

Convergence is certainly guaranteed, for example, if the matrix $A$ is diagonally dominant.

*Section 6.3(ii)* If only the first term is retained on the right of equations (6.46) the local truncation error is still only $O(h^6)$, since when (6.42) is multiplied by $h^4$ the contributions from the separate neglected terms are all of this order. The retention of third and fourth differences, however, makes the size of the local truncation error rather smaller, and indeed would permit exact results for a quartic solution of (6.42). These terms do not, moreover, increase the number of terms in the general finite-difference equations and therefore do not introduce spurious solutions and induced instability.

*Section 6.3(iii)* For an even more general form than (6.55), given by

$$\left.\begin{array}{l} y'' + f(x)y' + \{g(x) - \lambda r(x)\}y = 0 \\ p_0 y(a) + q_0 y'(a) = 0, \quad p_n y(b) + q_n y'(b) = 0 \end{array}\right\}, \tag{6.111}$$

the use of our standard finite-difference methods will give rise to a matrix equation of the form

$$(A - h^2 \bar{\lambda} D)Y = 0, \tag{6.112}$$

where $A$ is a triple-diagonal matrix and $D$ a diagonal matrix. With just a few mesh points we would, for example, have arrays like

$$A = \begin{bmatrix} \alpha_1 & \beta_2 & \\ \gamma_2 & \alpha_2 & \beta_3 \\ & \gamma_3 & \alpha_3 \end{bmatrix}, \quad D = \begin{bmatrix} d_1 & & \\ & d_2 & \\ & & d_3 \end{bmatrix}, \tag{6.113}$$

and it is always possible to produce from this a symmetric eigenvalue problem of type (6.53) if the elements $\beta_i, \gamma_i$ and $d_i$ have certain properties. The first step is to write (6.112) in the form

$$(D^{-\frac{1}{2}}AD^{-\frac{1}{2}} - h^2\bar{\lambda})D^{\frac{1}{2}}Y = 0, \qquad (6.114)$$

and a further transformation to the form

$$(T^{-1}D^{-\frac{1}{2}}AD^{-\frac{1}{2}}T - h^2\bar{\lambda})T^{-1}D^{\frac{1}{2}}Y = 0, \qquad (6.115)$$

where in our simple case $T$ is the diagonal matrix

$$T = \begin{bmatrix} 1 & & \\ & (\gamma_2/\beta_2)^{\frac{1}{2}} & \\ & & (\gamma_2\gamma_3/\beta_2\beta_3)^{\frac{1}{2}} \end{bmatrix}, \qquad (6.116)$$

produces the required symmetric eigenvalue problem. In fact the matrix $T^{-1}D^{-\frac{1}{2}}AD^{-\frac{1}{2}}T$ has the remarkably simple form

$$T^{-1}D^{-\frac{1}{2}}AD^{-\frac{1}{2}}T = \begin{bmatrix} \alpha_1 d_1^{-1} & \left(\dfrac{\beta_2\gamma_2}{d_1 d_2}\right)^{\frac{1}{2}} & \\ \left(\dfrac{\beta_2\gamma_2}{d_1 d_2}\right)^{\frac{1}{2}} & \alpha_2 d_2^{-1} & \left(\dfrac{\beta_3\gamma_3}{d_2 d_3}\right)^{\frac{1}{2}} \\ & \left(\dfrac{\beta_3\gamma_3}{d_2 d_3}\right)^{\frac{1}{2}} & \alpha_3 d_3^{-1} \end{bmatrix}. \qquad (6.117)$$

For a real matrix, necessary for real eigenvalues, and for a real vector $Y$ we clearly need the diagonal elements of $D$ to be positive (or all to have the same sign) and to be non-zero, and we must also have $\beta_2\gamma_2 \geqslant 0$, $\beta_3\gamma_3 \geqslant 0$. In the finite-difference equations these elements depend on $h$, but if for all sufficiently small $h$ these various inequalities are satisfied then not only will the finite-difference matrix have real eigensolutions but the differential problem will also have this quality. More than symmetry, of course, is needed for the one-to-one correspondence of problem (6.51) between the order of the eigenvalue and the number of zeros in the eigenfunction.

*Section 6.4* It is possible to solve simultaneous first-order equations of eigenvalue type, and even to treat a second- or fourth-order eigenvalue problem in terms of simultaneous first-order equations. In the latter case, however, the eigenvalue does not appear in many of the finite-difference equations, and the matrix of the algebraic eigenvalue problem has the form $A - \lambda B$, where $B$ is singular. This case can be treated by computer library programs, but it is more difficult and lacks any symmetries contained in the matrices stemming from direct methods for equations of second or fourth order.

*Section 6.5* The qualification 'for sufficiently small $h$' with respect to equation

(6.61) is somewhat analogous to the partial stability criterion in step-by-step methods. It may happen that for some $h_0$ the finite-difference matrix is singular even though the differential problem is perfectly well-behaved, and in this event we can expect our results to apply only for $h < h_0$. Consider, for example, the problem

$$y'' + 9.55y = 1, \quad y(0) = 0, \quad y(1) = 0, \qquad (6.118)$$

which has a unique solution very near to

$$y(x) \sim 0.105\{1 - \cos(3.09x)\} - 4.08 \sin(3.09x). \qquad (6.119)$$

But at $h = 0.2$ the matrix of the equations corresponding to (6.34) is

$$A = \begin{bmatrix} -1.618 & 1 & & \\ 1 & -1.618 & 1 & \\ & 1 & -1.618 & 1 \\ & & 1 & -1.618 \end{bmatrix}, \qquad (6.120)$$

and if we perform the elimination process without interchanges, which is reasonably accurate since the multipliers are not very large, we produce something very close to the upper triangular matrix

$$\begin{bmatrix} -1.618 & 1 & & \\ & -0.99995303 & 1 & \\ & & -0.61795303 & 1 \\ & & & +0.00024597 \end{bmatrix}, \qquad (6.121)$$

the near-vanishing of the final diagonal element showing that the matrix is very nearly singular.

A diagonally dominant matrix cannot be singular, and for example if $g(x) < 0$ and $f(x) \equiv 0$ in (6.30) the corresponding finite-difference matrix is non-singular for all $h$. If $g(x) < 0$, moreover, we can always choose an $h$ for which, for any $f(x)$, the matrix is diagonally dominant and therefore non-singular. If $g(x) > 0$ then we cannot have a diagonally dominant matrix for any $f(x)$, and we may therefore find a singular matrix with some $h$. Fortunately, an $h$ for which the local truncation error is satisfactorily small will usually be small enough to avoid this singularity.

There is another context in which the size of the interval must be smaller than some particular number in order to get useful results. This might happen for equations like

$$y'' + 100y' - y = 0, \qquad (6.122)$$

whose solution is a combination of two exponential terms, here approximately $e^{-100.01x}$ and $e^{0.01x}$. The simple central-difference equations for (6.122) give at interval $h$ the difference equation

$$(1 + 50h)Y_{r+1} - (2 + h^2)Y_r + (1 - 50h)Y_{r-1} = 0. \qquad (6.123)$$

For $h = 0.2$ the solution of (6.123) is a combination of two terms, approximately $1^r$ and $(-0.9)^r$. Whereas the true solution is a nice steady curve, this approximation reveals some rather nasty oscillations. For a sufficiently small $h$ the matrix comprising (6.123) is diagonally dominant, and the solution reveals no oscillation. For $h = 0.01$, for example, the two relevant terms are $1^r$ and $(\frac{1}{3})^r$.

In some contexts it is unsatisfactory to have to use such a small interval, and an appropriate solution with the correct qualitative behaviour is obtained by approximating to the $y'$ term by the forward-difference formula

$$hY'_r = Y_{r+1} - Y_r. \tag{6.124}$$

In our example (6.123) becomes

$$(1 + 100h)Y_{r+1} - (2 + 100h + h^2)Y_r + Y_{r-1} = 0, \tag{6.125}$$

and with $h = 0.2$ the solution is a combination of approximate terms $1^r$ and $(\frac{1}{21})^r$, with conveniently smooth behaviour. The local truncation error of (6.125), of course, is somewhat larger with a term $O(h^3)$ than that of (6.123), but in some applications this can be tolerated. The procedure is commonly called 'upwind differencing'.

*Section 6.6(i)* We mention in Chapter 9 that the equations (6.61) for the method of the deferred approach to the limit are not necessarily valid for singular differential equations even if the solution which satisfies the boundary conditions is non-singular. We do not know whether or not this is true for eigenvalue problems, but the deferred approach table should give us the relevant information and provide a solution with reasonable accuracy.

Consider, for example, the problem

$$y'' + \lambda(1 + x)^{-1}y = 0, \quad y(-1) = 0 = y(1), \tag{6.126}$$

which has a singularity at the first boundary point. The exact finite-difference equations are

$$\left. \begin{array}{c} y_{r+1} - \{2 - h^2\lambda(1 + x_r)^{-1}\}y_r + y_{r-1} = c(y_r) \\ c(y_r) = (\frac{1}{12}\delta^4 - \frac{1}{90}\delta^6 + \frac{1}{560}\delta^8 - \cdots)y_r \end{array} \right\}. \tag{6.127}$$

For the deferred approach we neglect the $c(y_r)$ and obtain approximations at successively smaller intervals. Corresponding to Table 6.4 we find the results shown in Table 6.11, with a similar normalization for the eigenvector, $Y$ being specified to be unity at $x = 0$.

The final values in Table 6.11 are correct to within a unit in the last figure. We know this because we have the mathematical solution that $\lambda$ is a root of the equation

$$J_1(2\sqrt{2\lambda}) = 0, \tag{6.128}$$

**Table 6.11**

| $h$ | $Y(x=-0.5)$ | | | $\bar{\lambda}$ | | |
|---|---|---|---|---|---|---|
| 0.5 | 0.939693 | | | 1.871644 | | |
| | | 0.935981 | | | 1.834914 | |
| 0.25 | 0.936909 | | 0.936014 | 1.844097 | | 1.835247 |
| | | 0.936012 | | | 1.835226 | |
| 0.125 | 0.936236 | | | 1.837444 | | |
| Error | $O(h^2)$ | $O(h^4)$ | $O(h^6)$ | $O(h^2)$ | $O(h^4)$ | $O(h^6)$ |

and the eigenfunction is a suitable normalization of

$$y(x) = (1+x)^{\frac{1}{2}} J_1\{2(1+x)^{\frac{1}{2}}\}, \tag{6.129}$$

where $J_1$ is the well-known Bessel function. In general, however, we should need the figures in the relevant version of Table 6.11 to demonstrate by their consistency that the relevant equations are valid. For $\bar{\lambda}$ in Table 6.11 another approximate value, for $h = 0.0625$, is really needed to verify the results. In this example, of course, equation (6.129) shows that there is no singularity in the solution, the Bessel function $J_1(x)$ having an expansion in odd powers of $x$.

*Section 6.6(ii)* There are various devices which the experienced numerical analyst might use which will not yet have been incorporated in computer library routines. Consider, for example, problem (6.62) in which the direct solution of the second-order equation requires the use of the somewhat unsatisfactory equation (6.76) involved in the derivative boundary condition. This can be avoided by judicious use of the Taylor series at $x = 0$. The differential equation and the first boundary condition in (6.62) give $y''(0)$ in terms of $y(0)$, and successive differentiation of the differential equation gives successive derivatives in terms of $y(0)$. Computing enough of these, we can use the Taylor series to give $y(-h)$ in terms of $y(0)$, and with $h = 0.1$ we find

$$y(-0.1) = 1.0165247 y(0) - 0.0171008, \tag{6.130}$$

which we use as our boundary condition at $x = 0$ without involving any extra difference correction. At each stage in the deferred correction process the simultaneous algebraic equations for solution, with the $f, g, k$ and $c(y_r)$ terms taken appropriately from (6.62) and (6.73), are (6.130) and (6.33) for $r = 0, 1, \ldots, n-1$, and the equations used step-by-step are (6.33) for $r = -1, -2, \ldots,$ and $n, n+1, \ldots$.

In Table 6.12 we give first the error in the first approximation $Y_r^{(1)}$, next the difference correction $c(Y_r^{(1)})$ which uses the third and fourth differences of $Y_r^{(1)}$, and then the resulting $Y_r^{(2)}$. This in fact agrees with $y_r$ in all the figures tabulated,

**Table 6.12**

| $x$ | $-10^6(Y_r - Y_r^{(1)})$ | $10^6 c(Y_r^{(1)})$ | $Y_r^{(2)}$ | $\delta^2$ | $\delta^4$ | $\delta^6$ | $\delta^8$ |
|---|---|---|---|---|---|---|---|
| $-0.5$ | 1091 | | | | | | |
| $-0.4$ | 1068 | 43 | 2.336987 | | | | |
| | | | | $-167597$ | | | |
| $-0.3$ | 1055 | 42 | 2.169390 | 66938 | | | |
| | | | | $-100659$ | $-19061$ | | |
| $-0.2$ | 1044 | 41 | 2.068731 | 47877 | 8017 | | |
| | | | | $-52782$ | $-11044$ | $-3595$ | |
| $-0.1$ | 1032 | 41 | 2.015949 | 36833 | 4422 | 1796 | |
| | | | | $-15949$ | $-6622$ | $-1799$ | $-953$ |
| $0.0$ | 1015 | 42 | 2.000000 | 30211 | 2623 | 843 | $+518$ |
| | | | | $+14262$ | $-3999$ | $-956$ | $-435$ |
| $0.1$ | 992 | 43 | 2.014262 | 26212 | 1667 | 408 | 258 |
| | | | | $+40474$ | $-2332$ | $-548$ | $-177$ |
| $0.2$ | 958 | 44 | 2.054736 | 23880 | 1119 | 231 | 70 |
| | | | | $+64354$ | $-1213$ | $-317$ | $-107$ |
| $0.3$ | 913 | 46 | 2.119090 | 22667 | 802 | 124 | 53 |
| | | | | $+87021$ | $-411$ | $-193$ | $-54$ |
| $0.4$ | 853 | 48 | 2.206111 | 22256 | 609 | 70 | 39 |
| | | | | $+109277$ | $+198$ | $-123$ | $-15$ |

| | | | | | | | | | | | |
|---|---|---|---|---|---|---|---|---|---|---|---|
| 0.5 | 775 | 50 | 2.315388 | + 131731 | 22454 | + 684 | 486 | − 68 | 55 | − 34 | − 19 |
| 0.6 | 677 | 53 | 2.447119 | + 154869 | 23138 | + 1102 | 418 | − 47 | 21 | + 10 | + 44 |
| 0.7 | 555 | 56 | 2.601988 | + 179109 | 24240 | + 1473 | 371 | − 16 | 31 | − 28 | − 38 |
| 0.8 | 404 | 60 | 2.781097 | + 204822 | 25713 | + 1828 | 355 | − 13 | 3 | + 19 | + 47 |
| 0.9 | 221 | 64 | 2.985919 | + 232363 | 27541 | + 2170 | 342 | + 9 | + 22 | − 24 | − 43 |
| 1.0 | 0 | 68 | 3.218282 | + 262074 | 29711 | + 2521 | 351 | + 7 | − 2 | + 17 | + 41 |
| 1.1 | − 265 | 73 | 3.480356 | + 294306 | 32232 | + 2879 | 358 | + 22 | + 15 | | |
| 1.2 | − 580 | 78 | 3.774662 | + 329417 | 35111 | + 3259 | 320 | | | | |
| 1.3 | − 953 | 84 | 4.104079 | + 367787 | 38370 | | | | | | |
| 1.4 | − 1392 | 90 | 4.471866 | | | | | | | | |
| 1.5 | − 1907 | | | | | | | | | | |

and $Y_r^{(3)}$, which effectively takes account of fifth and sixth differences and $Y_r^{(4)}$, which uses seventh and eighth differences, also give results identical with the tabulated figures. Indeed, with extra figures we find that $Y_r^{(2)}$ has a maximum error in $0 \leqslant x \leqslant 1$ of only one unit in the seventh decimal place. This is all somewhat surprising, since differences of orders five to eight, given in the table, are by no means negligible, and although we would expect better results with the use of (6.130) than with that of (6.76), the spectacular nature of the improvement would not be expected in general. A little investigation reveals that near $x = 0$ there is almost exact cancellation in the contribution from the terms in $\mu \delta^5$ and $\delta^6$, and a similar phenomenon in the contributions from $\mu \delta^7$ and $\delta^8$ to the difference correction $c(Y_r)$ defined in (6.72). This, we repeat, is a somewhat unusual occurrence!

If in the difference-correction method we decide that we will use differences no higher than a particular order, there is no difficulty in deciding how far we have to extend the various solutions, and their corrections to points outside the boundary, in order to get accurate difference-correction values at internal points and any boundary points where the condition involves a derivative so that the boundary value is one of the unknowns. Consider, for example, the problem of equations (6.62), and assume that we want to use only the difference corrections given explicitly in (6.73), so that we compute successive approximations $Y_r^{(1)}$, $Y_r^{(2)}$, $Y_r^{(3)}$ and $Y_r^{(4)}$. At the origin, whose value is the most difficult unknown in this scheme, to get $Y_0^{(4)}$ we need to calculate $c^{(3)}(Y_0^{(3)})$, which means that we have to be able to calculate up to $\delta^8$ of $Y_0^{(3)}$, and this requires the computation of $Y_r^{(3)}$ out as far as $r = -4$. To get $Y_{-4}^{(3)}$ we must compute $Y_r^{(2)}$ out to $r = -7$, since the relevant difference correction now involves up to sixth-order differences of $Y_r^{(2)}$; and finally to get $Y_{-7}^{(2)}$ we must compute $Y_r^{(1)}$ out to $r = -9$, since we need up to fourth differences of $Y_r^{(1)}$ at $r = -7$. Hence the remark at the end of the paragraph containing equation (6.76).

Our work with the deferred approach method virtually confirms that problem (6.126) has a non-singular solution, but the infinite coefficient at the boundary $x = -1$ makes it difficult to compute external values for use with a method of deferred correction, which would seem to be legitimate just because the solution is non-singular.

There are various ways of extending the solution below $x = -1$, but we insist on as much smoothness as possible, so that at $x_r = -1$, corresponding to $r = 0$ in our usual notation, we want an equation with the same difference correction as in (6.127) but with computable terms on the left of that equation. This is easily achieved with the use of the Taylor series. Note first that if we multiply the differential equation in (6.126) by $1 + x$, differentiate successively and put $x = -1$ at each stage, we obtain the formulae

$$y_0'' = -\lambda y_0', \quad y_0''' = -\frac{\lambda}{2}y_0'' = \frac{\lambda^2}{2}y_0', \quad y_0^{iv} = -\frac{\lambda^3}{6}y_0', \quad y_0^{v} = \frac{\lambda^4}{24}y_0', \ldots. \quad (6.131)$$

Then we can compute $y_1$ and $y_{-1}$ from the Taylor series

$$y_{\pm 1} = y_0 \pm hy_0' + \frac{h^2}{2!}y_0'' \pm \frac{h^3}{3!}y_0''' + \cdots, \tag{6.132}$$

which with the use of (6.131) and the boundary condition $y_0 = 0$ can be written as

$$y_{\pm 1} = hy_0'(\pm P - Q), \tag{6.133}$$

where

$$\left. \begin{aligned} P &= 1 + \tfrac{1}{12}h^2\lambda^2 + \tfrac{1}{2880}h^4\lambda^4 + \cdots \\ Q &= \tfrac{1}{2}h\lambda + \tfrac{1}{144}h^3\lambda^3 + \tfrac{1}{86400}h^5\lambda^5 + \cdots \end{aligned} \right\}. \tag{6.134}$$

Finally,

$$\delta^2 y_0 - c(y_0) = h^2 y_0'' = -h^2\lambda y_0' = -h\lambda(y_1 - y_{-1})/2P, \tag{6.135}$$

from (6.131) and (6.133), so that with $y_0 = 0$ the required formula is

$$y_1\left(1 + \frac{h\lambda}{2P}\right) + y_{-1}\left(1 - \frac{h\lambda}{2P}\right) = c(y_0) = (\tfrac{1}{12}\delta^4 - \tfrac{1}{90}\delta^6 + \cdots)y_0. \tag{6.136}$$

In relation to (6.127) and the relevant iterative method this is used in the form

$$Y_1^{(s+1)}\left(1 + \frac{h\lambda^{(s+1)}}{2P^{(s+1)}}\right) + Y_{-1}^{(s+1)}\left(1 - \frac{h\lambda^{(s+1)}}{2P^{(s+1)}}\right) = c(Y_0^{(s)}), \tag{6.137}$$

for the computation of $Y_{-1}^{(s+1)}$ after $\lambda^{(s+1)}$ and $Y_1^{(s+1)}$ have been computed. Further external values come from (6.127) with appropriate $r$.

The routine which gave Table 6.6 for problem (6.68) produces for our current problem the values shown in Table 6.13. This gives values of $Y_r^{(1)}$ at interval $h = \tfrac{1}{4}$, together with three successive corrections evaluated from successive $Y_r^{(2)}$, $Y_r^{(3)}$ and $Y_r^{(4)}$ to show more clearly the rate of convergence. We also give corresponding values of $\lambda^{(s)}$ and its corrections. Notice, as we would expect, that it is $h^2\lambda^{(s)}$ and the corresponding corrections to $h^2\lambda^{(s)}$ which are more

**Table 6.13**

| $x$ | $Y_r^{(1)}$ | | $\delta Y_r$ | | $c(Y_r^{(1)})$ | $c(Y_r^{(3)})$ | $\delta^6(Y^{(0)})$ | $\delta^6(Y^{(1)})$ | $\delta^6(Y^{(2)})$ | $\delta^6(Y^{(3)})$ |
|---|---|---|---|---|---|---|---|---|---|---|
| $-1.00$ | 0 | 0 | 0 | 0 | $-1004$ | $-988$ | 31 | $-172$ | $-104$ | $-125$ |
| $-0.75$ | 580316 | $-539$ | 7 | $-1$ | $-887$ | $-878$ | $-237$ | $-73$ | $-136$ | $-108$ |
| $-0.50$ | 893092 | $-422$ | 4 | 0 | $-789$ | $-781$ | $-75$ | $-125$ | $-94$ | $-114$ |
| $-0.25$ | 1000000 | 0 | 0 | 0 | $-698$ | $-691$ | $-100$ | $-95$ | $-102$ | $-95$ |
| 0.00 | 953232 | 468 | $-4$ | 0 | $-616$ | $-609$ | $-92$ | $-91$ | $-90$ | $-92$ |
| 0.25 | 796600 | 799 | $-7$ | 0 | $-541$ | $-536$ | $-85$ | $-84$ | $-84$ | $-84$ |
| 0.50 | 566516 | 872 | $-7$ | 0 | $-473$ | $-469$ | $-79$ | $-77$ | $-77$ | $-77$ |
| 0.75 | 292903 | 615 | $-5$ | 0 | $-412$ | $-408$ | $-72$ | $-71$ | $-71$ | $-72$ |
| 1.00 | 0 | 0 | 0 | 0 | $-357$ | $-354$ | $-66$ | $-65$ | $-65$ | $-64$ |
| $\lambda^{(s)}$ | 1.844097 | $-8938$ | 87 | 1 | | | | | | |

comparable with the $Y_r$ and its corrections, and to compute $\lambda$ correct to six decimal places we should here probably need $Y_r$ correct to at least seven places. We also give the difference corrections $c(Y_r^{(1)})$ and $c(Y_r^{(3)})$ to show the relations between these and the corrections to $Y_r^{(s)}$ itself and to $\lambda^{(s)}$.

The tabulated six-decimal $Y_r$ values are rounded from a computer's eight-figure results, and we also give in the table values of sixth differences, correctly rounded to six decimals, of the first three approximations. We note a slight 'bump' in these differences near the first boundary point and a 'smoothing-out' effect in successive iterations. The bump is due to a lack of exact smoothness in the particular use of (6.136) in respect to the general (6.127). The form of the difference correction is the same in both equations, but the error in $\lambda^{(s)}$ is another factor which has slightly different effects in the respective formulae. Such a small bump causes a slight decrease in the expected rate of convergence, but the computed $Y_r^{(4)}$ is very accurate indeed, the errors in successive internal values having the remarkable values $10^{-8}(32, 5, 2, 0, -1, -1, 0)$. Subject to our expectation $\lambda^{(4)}$ is less accurate, though the computed value is in error by only three units in the seventh decimal place.

If the differences of early approximations are even less smooth than those of Table 6.13 it can no longer be expected that taking successive segments of difference corrections will increase by powers of two in $h$ the order of the current accuracy. Here it may be necessary to repeat the computation at some stage, without bringing in a higher difference, with the hope of smoothing out the bump.

For problems with singular solutions, however, both our correcting devices will be valid only for a modified problem in which the singularity is effectively removed. For example, it can be shown that for the problem

$$y'' + \{x/(1-x^2)\}y' + \lambda y = 0, \quad y(-1) = y(1) = 0, \tag{6.138}$$

the eigenfunction has an expansion

$$\begin{aligned} y &= (1-x^2)^{\frac{3}{2}}\{a_0 + a_1(1-x^2) + a_2(1-x^2)^2 + \cdots\} \\ &= (1-x^2)^{\frac{3}{2}}z(x), \end{aligned} \tag{6.139}$$

with an infinite second derivative at both boundaries. Equations (6.61) and (6.71) can now be expected to have no validity, and it is advisable to remove the singularity in the solution by working in terms of $z$ in (6.139) which satisfies the differential equation

$$z'' - (1-x^2)^{-1}(5xz' + 3z) + \lambda z = 0. \tag{6.140}$$

A non-singular solution must clearly satisfy the boundary conditions

$$\pm 5z' + 3z = 0 \quad \text{at } x = \pm 1, \tag{6.141}$$

and at these points the differential equation (6.140) reduces to

$$\tfrac{7}{2}z'' \pm 4z' + \lambda z = 0, \tag{6.142}$$

that is to

$$\tfrac{7}{2}z'' - \tfrac{12}{5}z + \lambda z = 0 \qquad (6.143)$$

with the use of (6.141).

A solution with error $O(h^2)$ is therefore obtainable by using simple finite-difference equations for (6.140) at internal points and (6.141) and (6.143) at both boundary points, the external values being eliminated from the simple finite-difference equations for (6.141) and (6.143) to produce a standard eigenvalue problem. We can also be sure that the standard equations will hold for the method of the deferred approach to the limit, and for this problem this method is considerably easier to handle than the corresponding difference-correction method.

We end this section on a more optimistic note. For a few problems there are stable and economic methods which give better than $O(h^2)$ accuracy in the first approximation. One of these is a second-order problem lacking an explicit first derivative, in differential equation and boundary condition, such as

$$y'' = f(x, y), \qquad y(a) = \alpha, \quad y(b) = \beta. \qquad (6.144)$$

The finite-difference equations which give this better accuracy are

$$Y_{r-1} - 2Y_r + Y_{r+1} = \tfrac{1}{12}h^2\{f(x_{r-1}, Y_{r-1}) + 10f(x_r, Y_r) + f(x_{r+1}, Y_{r+1})\},$$
$$r = 1, 2, \ldots, n-1. \qquad (6.145)$$

The resulting algebraic equations are still a triple-diagonal set and we have introduced no spurious solutions, so that the method is essentially stable, but the main error $y_r - Y_r$ is now the very desirable $O(h^4)$. In fact the equation for the error is (6.61) with $w_2(x_r) = 0$. Then the extrapolated value

$$\bar{Y}_r = Y_r(\tfrac{1}{2}h) + \tfrac{1}{15}\{Y_r(\tfrac{1}{2}h) - Y_r(h)\} \qquad (6.146)$$

has the even more desirable error $O(h^6)$. For eigenvalue problems both the eigenfunction and the eigenvalue have error $O(h^4)$ in the first approximation.

Formula (6.145) is also useful for initial-value problems, but for such problems there are other valuable techniques and the method is clearly more useful for good accuracy with boundary-value problems of a kind in which the generally preferred deferred-correction process might be replaced by the easier deferred-approach method.

*Section 6.8* The continuation method lacks theoretical results which would easily give us the best method of continuation for any particular problem, and quite often the scientific nature of the problem will provide the best suggestion. A fairly simple method, however, is also worth trying which is not exactly a method of continuation but which is of a somewhat analogous nature. The idea is to use successively smaller intervals, starting with a very large interval so that

**Table 6.14**

| $x$ | $Y(h = \frac{1}{2})$ | $Y(h = \frac{1}{4})$ | $Y(h = \frac{1}{8})$ | $Y(correct)$ |
|---|---|---|---|---|
| 0.000 | 0.0000 | 0.0000 | 0.0000 | 0.0000 |
| 0.125 | | | − 0.9075 | − 0.9269 |
| 0.250 | | − 1.5568 | − 1.6600 | − 1.6926 |
| 0.375 | | | − 2.0920 | − 2.1313 |
| 0.500 | − 1.2134 | − 1.9395 | − 2.0712 | − 2.1160 |
| 0.625 | | | − 1.6046 | − 1.6514 |
| 0.750 | | − 0.7128 | − 0.8322 | − 0.8705 |
| 0.875 | | | + 0.0818 | + 0.0615 |
| 1.000 | 1.0000 | 1.0000 | 1.0000 | + 1.0000 |

the finite-difference equations are not too difficult to solve. With simple interpolation the solution for one interval can then serve as the starting approximation at the next smaller interval, and we hope that we have a convergent process.

Consider, for example, problem (6.93) with $t = 1$, for which one continuation method at the end of Section 6.8 failed to find a solution. With interval $h = \frac{1}{2}$ there is only one internal mesh point, and with simple finite-difference equations the mesh value at $x = \frac{1}{2}$ is easily seen to be the solution of the cubic equation

$$Y^3 + Y + 3 = 0. \tag{6.147}$$

One solution is very close to $- 1.2$, and the other two are complex. Starting with $Y = 0, - 1.2, 1.0$ at $x = 0, 0.5, 1$ respectively, we interpolate as accurately as possible with the formulae

$$Y(\tfrac{1}{4}) = \quad \tfrac{3}{8}Y(0) + \tfrac{3}{4}Y(\tfrac{1}{2}) - \tfrac{1}{8}Y(1) = - 1.025 \left.\right\}$$
$$Y(\tfrac{3}{4}) = - \tfrac{1}{8}Y(0) + \tfrac{3}{4}Y(\tfrac{1}{2}) + \tfrac{3}{8}Y(1) = - 0.525 \left.\right\}, \tag{6.148}$$

start with the rounded $Y$ values $0, - 1, - 1.2, - 0.5$ and $1.0$ at interval $h = \frac{1}{4}$, and converge to a not particularly different solution. Repetition at $h = \frac{1}{8}$ produces the results of Table 6.14. These are obviously converging to a smooth solution, and indeed the values at $h = \frac{1}{8}$ have a difference from the true solution of less than 5 units in the second decimal place, as shown in the final column, which gives the correct values. Two extrapolations for the deferred approach at $x = 0.5$ give the very good result $- 2.1106$.

---

### EXERCISES

1. Find the $L$ and $U$ matrices which, with standard floating-point four-decimal arithmetic, represent the triangular splitting of the matrix $A$ in equations (6.11). With the same arithmetic solve also the equations $Lc = b$, where $b$ is

the vector on the right of (6.11), and verify that the computed $U$ and $c$ agree exactly with those involved in the equations taken from (6.11), (6.12) and (6.13) in which back-substitution is to be performed (but notice that this agreement depends on the order in which the additions are performed in the $LU$ splitting, the $u_{33}$ element coming from the equation $-1398 + 1400 + u_{33} = 0.1001$, and $c_3$ from $2218 - 2222 + c_3 = -0.2896$).

Show also that the *exact* product of the computed $L$ and $U$ differs from the matrix $A$ in (6.11) by about 0.63, 0.62, 0.87 and 0.11 in the respective positions $(2, 2)$, $(2, 3)$, $(3, 2)$ and $(3, 3)$.

2. Repeat Exercise 1 for the computation of $L$ and $U$ using the more accurate arithmetic described in the first part of Section 6.9. (Note that in the computation of $l_{rs}$ a division is performed, and in this computation the numerator is *not* rounded before the division.)

   Show that for this $LU$ the difference from $A$ has the smaller approximate components 0.37, 0.38, 0.12 and 0.0005 in the respective positions mentioned in Exercise 1.

3. With the standard arithmetic find the $L$ matrix which with the $U$ matrix in (6.17) gives an approximate triangular splitting of a particular row permutation of the matrix in (6.11). What is this row permutation? Show also that the difference between the row permutation of $A$ and the exact product of the computed $L$ and $U$ is less than half a unit in the fourth significant figure in every element.

4. Start with the approximation $y_1 = y_2 = y_3 = y_4 = 0$, and solve equations (6.20), correct to four decimal places, with the iterative method of equations (6.109) and (6.110). Repeat this with the first two equations interchanged (destroying diagonal dominance) and verify that this iteration diverges.

5. In analogy with (6.26) produce a similar array for (6.28), and show that the number of fill-in non-zero numbers created by elimination without interchanges is six below the diagonal and six above the diagonal.

   Verify that there is no fill-in with the array (6.29) using elimination without interchanges.

6. Consider equations (6.27) with unseparated boundary conditions $y_0 = z_3 + z_4$, $y_1 = z_2 + z_3$, $z_0 = y_3 + y_4$, $z_1 = y_2 + y_3$. Write down the algebraic equations obtained by taking unknowns in order $y_0, y_1, y_2, y_3, y_4, z_0, z_1, z_2, z_3$ and $z_4$, with the first two rows coming from the first two boundary conditions, rows 3 and 4 from (6.27) with $r = 1$, rows 5 and 6 from (6.27) with $r = 2$, rows 7 and 8 from (6.27) with $r = 3$, and rows 9 and 10 coming from the last two boundary conditions. After the elimination of $y_0, y_1$ and $y_2$, without interchanges, show that the matrix for the remaining seven unknowns has only nine zero elements, and that at this stage an interchange is certainly necessary.

7. If in Exercise 6 interchanges are made so that, at each stage, the amount of fill-in is as small as possible, show that at the stage of Exercise 6 the matrix of order seven has 14 zero elements.

8. In the solution of a fourth-order equation, give two reasons why it is desirable to exclude fifth and higher differences in equations (6.46) from the first set of finite-difference equations, incorporating them only later in the deferred correction process.

9. If the sign of the term in $y'$ in equation (6.122) is changed, show that central differences produce oscillations which are not removed by the use of (6.124). Suggest an analogous procedure which will have the desired effect, and verify the conclusions by calculating the various terms which contribute to the solutions.

10. Show that the analytical solutions of the eigenvalue problem

$$y'' + \lambda y = 0, \quad y(1) = 2y(2), \quad 2y'(1) = y'(2)$$

are

$$\lambda_{\pm n} = k_{\pm n}^2, \quad k_{\pm n} = -\sin^{-1}(0.6) \pm 2n\pi$$
$$y_{\pm n} = \sin k_{\pm n}(x-1) + 2\cos k_{\pm n}(x-1)$$

for $n = 0, 1, 2, \ldots$. Hence show that $y_{-n}$ and $y_n$ both have $2n$ internal zeros in the range $(1, 2)$.

11. Write down the simple finite-difference equations for the problem of Exercise 10. At interval $h = \frac{1}{4}$ these will involve the seven values at the points $x = 0.75$ (0.25) 2.25. By eliminating the first and the last two of these unknowns show, for interval $h = \frac{1}{4}$, that the relevant algebraic eigenvalue problem is of form (6.53), with the unsymmetric matrix

$$A = \begin{bmatrix} -2 & 1.6 & 0 & 0.8 \\ 1 & -2 & 1 & 0 \\ 0 & 1 & -2 & 1 \\ 0.5 & 0 & 1 & -2 \end{bmatrix}.$$

Find the smallest eigenvalue of this matrix, and also of the corresponding matrices with $h = \frac{1}{8}$ and $\frac{1}{16}$. Perform two deferred-approach extrapolations and show that your final result agrees well with the correct value $\lambda_0$ given in Exercise 10.

What do you notice about the computed approximations to the corresponding eigenfunction, normalized so that $Y(1.5) = 1$? How do they compare with the correct eigenfunction?

12. Consider the problem

$$y''' + yy'' + 0.1(y'^2 - 1) = 0, \quad y(0) = y'(0) = 0, \quad y'(4) = 0.$$

By treating the corresponding first-order equations find first approximations to the solutions at intervals $h = \frac{1}{2}, \frac{1}{4}, \frac{1}{8}, \dots$, and improve them by the method of the deferred approach to the limit, stopping when six figures are expected to be accurate.

13. Solve the problem of Exercise 12 using the method of deferred correction with interval $h = 0.25$.

14. Consider the fourth-order equation and boundary conditions

$$y^{iv} - y = 0, \quad y(0) = 1, \quad y'(0) = 0, \quad y''(1) = y(1),$$
$$y'''(1) = y'(1),$$

and experiment with various methods of solution on the lines of the discussion surrounding equations (6.65)–(6.67).

15. Solve the problem of Exercise 14 with the deferred-correction method, on the lines of the discussion of equations (6.74), (6.75).

16. Prove the remarks following Table 6.12 about the cancellation of the contributions of: (a) the fifth and sixth differences, and (b) the seventh and eighth differences.

17. Perform some computations for a first approximation to the first eigen-solution of (6.138) at intervals $\frac{1}{2}, \frac{1}{4}, \frac{1}{8}, \dots$, and decide whether or not equation (6.61) is valid. (Note that this eigenfunction is symmetric about $x = 0$, and that this fact can be used roughly to halve the number of algebraic equations requiring solution.)

18. Repeat Exercise 17 with the method suggested using equations (6.140)–(6.143). Also compute a few values external to the range and check the smoothness of the differences near the ends of the range.

19. Show that the problem

$$y'' + 5y^2 + By = 12 + 5(-5x + 6x^2)^2 + B(-5x + 6x^2),$$
$$y(0) = 0, \quad y(1) = 1,$$

has at least one real solution for any value of $B$, given by $y = -5x + 6x^2$. With the linear first approximation $y^{(1)} = x$ show that the first step of Newton iteration gives a correction $z$ satisfying the differential system

$$z'' + (10x + B)z + 5x^2 + Bx - \{12 + 5(-5x + 6x^2)^2 + B(-5x + 6x^2)\} = 0,$$
$$z(0) = z(1) = 0.$$

Hence show that the iterative method may have convergence difficulties if $B$ is fairly close to the value 4.7. (*Hint*: recall the ideas in the later part of Section 2.6.)

20. Using the simplest finite-difference equations try the iterative method of Exercise 19 with $B = 4.676$, probably failing to achieve convergence after quite a number of steps. Try also with $B = 5$, showing that after about ten iterations the process converges to a solution quite different from $-5x + 6x^2$.

21. Obtain a solution of problem (6.93) by using continuation in the form
$$y'' = 1 - ty^3 - 9(2 - t)y.$$

# 7
# Expansion methods

## 7.1 INTRODUCTION

The finite-difference methods of previous chapters have been based essentially on the idea of approximating functions by polynomials, and in fact such methods clearly give exact solutions when these are polynomials of appropriate degree. For more general functions, local truncation errors depend on the accuracy with which the solution can be approximated by a polynomial. It is sometimes more convenient to make quite explicit this relation with polynomials, and we may then be able to calculate the coefficients $b_r$ of an approximating expansion

$$y(x) = \sum_{r=0}^{\infty} b_r x^r, \tag{7.1}$$

rather than working with a representation of the approximate solution at a set of discrete points. For some simple classes of problems these methods are particularly useful for both initial-value and boundary-value problems.

It turns out, however, that (7.1) is not the best form to take for the approximating polynomial, and it is usually replaced by a formula like

$$y(x) = \sum_{r=0}^{\infty} a_r \phi_r(x), \tag{7.2}$$

where each $\phi_r(x)$ is a fixed polynomial of degree $r$. If (7.1) and (7.2) are both truncated after the term $r = N$ both series are equivalent to polynomials of degree $N$, but they are different polynomials with a different error, and suitable choices of $\phi_r(x)$ produce results with errors considerably smaller than the results from (7.1).

The most usual choices of $\phi_r(x)$ are Chebyshev polynomials and spline functions. We shall give illustrations of the use of the former for both initial-value and boundary-value problems, and of the latter for boundary-value problems only.

## 7.2 PROPERTIES AND COMPUTATIONAL IMPORTANCE OF CHEBYSHEV POLYNOMIALS

Elementary trigonometric identities show that $\cos r\theta$ can be expressed as a polynomial in $\cos \theta$. It is therefore convenient to define the Chebyshev polynomial of degree $r$ as

$$T_r(x) = \cos r\theta, \quad x = \cos \theta, \tag{7.3}$$

relevant only to the range $-1 \leqslant x \leqslant 1$ since $|\cos \theta| \leqslant 1$ for all real $\theta$. For approximation purposes the two most important properties of $T_r(x)$ are that $|T_r(x)| \leqslant 1$ and that $T_r(x)$ attains its maximum values of $\pm 1$ alternately at $r + 1$ points in the interval $-1 \leqslant x \leqslant 1$. It is easy to check these facts for the first few Chebyshev polynomials given by

$$T_0(x) = 1, \quad T_1(x) = x, \quad T_2(x) = 2x^2 - 1, \quad T_3(x) = 4x^3 - 3x, \tag{7.4}$$

successive members of the sequence being obtained from the second-order recurrence relation

$$T_{r+1}(x) = 2xT_r(x) - T_{r-1}(x). \tag{7.5}$$

The coefficient of $x^r$ in $T_r(x)$ is $2^{r-1}$, and the properties mentioned lead to the fact that of all polynomials of degree $r$ with leading coefficient $2^{r-1}$ the Chebyshev polynomial has the smallest maximum size of unity in $|x| \leqslant 1$. If we measure the 'magnitude' of a polynomial by its maximum size in this range then $T_r(x)$ is the 'smallest polynomial' of degree $r$, and this might be compared with the simplest corresponding polynomial $2^{r-1}x^r$ whose magnitude is the much larger $2^{r-1}$.

One can also show that for a general function $f(x)$ in $-1 \leqslant x \leqslant 1$ the best polynomial approximation $p_n(x)$ of degree $n$, 'best' here meaning with error of smallest maximum size, is such that the error $e_n(x) = f(x) - p_n(x)$ has $n + 2$ equal and alternating maximum and minimum values in this range, the 'equi-ripple' property of $T_{n+1}(x)$. All these facts indicate the value of the expansion (7.2) as a series of Chebyshev polynomials rather than a series of powers of $x$ which of course is just the Taylor series. For if we write

$$\left. \begin{array}{l} f(x) = b_0 + b_1 x + b_2 x^2 + \cdots \\ f(x) = a_0 + a_1 T_1(x) + a_2 T_2(x) + \cdots \end{array} \right\}, \tag{7.6}$$

and in each case truncate the series after the term of degree $n$, then the respective leading terms are $b_{n+1}x^{n+1}$ and $a_{n+1}T_{n+1}(x)$, and the latter is significantly smaller than the former and has the required behaviour.

For example, the relevant expansions of $(x + 2)^{-1}$ in $-1 \leqslant x \leqslant 1$ are approximately

$$\left. \begin{array}{l} (x + 2)^{-1} = 0.5 - 0.25x + 0.125x^2 - 0.0625x^3 + 0.03125x^4 - \cdots \\ (x + 2)^{-1} = 0.577 - 0.309T_1(x) + 0.083T_2(x) - 0.022T_3(x) + 0.006T_4(x) - \cdots \end{array} \right\} \tag{7.7}$$

and the far better convergence of the second is quite obvious. Truncation of each of (7.7) to provide a polynomial approximation of degree $n$ gives maximum errors decreasing with the Taylor series from 0.125 to 0.004 as $n$ increases from 2 to 7, and with the Chebyshev series the corresponding figures are 0.030 and 0.000042, much smaller and increasingly so as $n$ increases. Moreover, even for $n = 2$ the truncated Chebyshev series is the approximate quadratic polynomial

$$p_2(x) = 0.494 - 0.309x + 0.166x^2, \tag{7.8}$$

and its error has the approximate maximum values $+0.030$, $-0.025$, $+0.020$, $-0.018$ at the approximate points $x = -1$, $-0.6$, $+0.4$, $+1$, so that $p_2(x)$ is quite close to the best quadratic approximation.

## 7.3 CHEBYSHEV SOLUTION OF ORDINARY DIFFERENTIAL EQUATIONS

For certain classes of problems, whether of initial-value type or boundary-value type and with conditions either separated or unseparated or even involving more than two points in the range $-1 \leqslant x \leqslant 1$, it is not difficult to find solutions of the form

$$y(x) = \tfrac{1}{2}a_0 + a_1 T_1(x) + a_2 T_2(x) + \cdots, \tag{7.9}$$

the factor $\frac{1}{2}$ in the first coefficient having some technical convenience.

### (i) Special linear problems

The easiest such case is a linear problem with linear associated conditions and whose coefficients and other terms are themselves polynomials. The main characteristic of such problems is that there exist simple formulae which enable us to express both sides of the differential equation as a sum of Chebyshev polynomials, the coefficients on the left being linear combinations of the $a_r$. The important formulae are

$$x T_r(x) = \tfrac{1}{2}\{T_{r-1}(x) + T_{r+1}(x)\} \tag{7.10}$$

and

$$\left.\begin{aligned} T_r'(x) &= 2r\{T_{r-1}(x) + T_{r-3}(x) + \cdots + T_1(x)\} \quad &\text{for even } r \\ &= 2r\{T_{r-1}(x) + T_{r-3}(x) + \cdots + T_2(x) + \tfrac{1}{2}T_0(x)\} \quad &\text{for odd } r \end{aligned}\right\}. \tag{7.11}$$

Then, for example, for the differential equation

$$y' + (2x - 1)y = 1 - x + 2x^2, \tag{7.12}$$

substitution of (7.9) into (7.12) and use of (7.10) and (7.11) give

$$a_1 T_0 + 4a_2 T_1 + 6a_3(T_2 + \tfrac{1}{2}T_0) + 8a_4(T_3 + T_1) + \cdots$$
$$+ a_0 T_1 + a_1(T_0 + T_2) + a_2(T_1 + T_3) + a_3(T_2 + T_4) + a_4(T_3 + T_5) + \cdots$$
$$- (\tfrac{1}{2}a_0 T_0 + a_1 T_1 + a_2 T_2 + a_3 T_3 + a_4 T_4 + \cdots) = T_0 - T_1 + (T_0 + T_2). \tag{7.13}$$

Equating corresponding terms on the two sides of (7.13) then provides a system of linear equations for the computation of the $a_r$, and another equation is obtained from the initial condition. If we take this to be

$$y(-1) = 0 \tag{7.14}$$

we can then record the infinite set of equations

$$\left.\begin{array}{l}
\tfrac{1}{2}a_0 - a_1 + a_2 - a_3 + a_4 - \cdots = \quad 0 \\
-\tfrac{1}{2}a_0 + 2a_1 \quad\;\; + 3a_3 \qquad\quad + \cdots = \quad 2 \\
a_0 - a_1 + 5a_2 \qquad\;\; + 8a_4 + \cdots = -1 \\
a_1 - a_2 + 7a_3 \qquad\;\; + \cdots = \quad 1 \\
a_2 - a_3 + 9a_4 + \cdots = \quad 0 \\
a_3 - a_4 + \cdots = \quad 0 \\
a_4 + \cdots = \quad 0 \\
\vdots
\end{array}\right\} , \tag{7.15}$$

of which the first comes from (7.14) and the rest from (7.13) for the coefficients of $T_0, T_1, T_2, T_3, T_4, T_5, \ldots$.

We cannot solve the infinite set, but an approximation comes from treating a finite segment of these equations. For example, taking the first five equations and truncating them after the terms given explicitly in (7.15), we find the approximation

$$y_4(x) = 6.1644 T_0(x) + 5.3973 T_1(x) - 1.7397 T_2(x) - $$
$$- 0.8767 T_3(x) + 0.0959 T_4(x). \tag{7.16}$$

Better results are obtained, of course, with the retention of more coefficients and the solution of more equations in (7.15). For example, the sixth degree approximation is given by

$$y_6(x) = 5.7013 T_0(x) + 5.0214 T_1(x) - 1.5943 T_2(x) - 0.9327 T_3(x) + $$
$$+ 0.0715 T_4(x) + 0.0913 T_5(x) + 0.0015 T_6(x). \tag{7.17}$$

The true solution is

$$y = x + e^{2+x-x^2}, \tag{7.18}$$

and the maximum errors of (7.16) and (7.17) in the range $-1 \leqslant x \leqslant 1$ are respectively 0.75 and 0.043. In practice, of course, one determines the accuracy obtained by the consistency of successive approximations and the rate of decrease of the coefficients in the various series.

It is interesting to note that each polynomial approximation is the exact solution of a slightly different problem. To find (7.16), for example, we have effectively taken all of $a_5, a_6, a_7, \ldots$, to be zero, in which case all the equations of the infinite set (7.15) are satisfied except the last two given explicitly in (7.15), and these are also satisfied if we add on the right of the differential equation in

(7.12) the terms

$$(a_3 - a_4)T_4(x) + a_4 T_5(x) \sim -0.97T_4(x) + 0.10T_5(x). \tag{7.19}$$

The perturbations decrease rapidly as the degree of the approximation increases. For the approximation (7.17) they are about $0.0898T_6(x) + 0.0015T_7(x)$.

These approximations, of course, tend to the true solution only in the range $-1 \leqslant x \leqslant 1$, but we can in theory cover any finite range $a \leqslant x \leqslant b$ with the substitution

$$\xi = \frac{2x - a - b}{b - a}, \tag{7.20}$$

so that $\xi$ has the required range $-1 \leqslant \xi \leqslant 1$. The differential equation can easily be written with $\xi$ as independent variable, and there is no further change of technique. For example, for the same problem we can cover the range $-1 \leqslant x \leqslant 0$ via the substitution $\xi = 2x + 1$, expecting of course to get quite a bit better accuracy with the same degree of approximation. The fourth- and sixth-order approximations are

$$\left.\begin{array}{l} y_4(x) = 3.3613T_0(\xi) + 3.7677T_1(\xi) + 0.3516T_2(\xi) - 0.0744T_3(\xi) - 0.0197T_4(\xi) \\ y_6(x) = 3.3627T_0(\xi) + 3.7690T_1(\xi) + 0.3517T_2(\xi) - 0.0742T_3(\xi) - \\ \qquad - 0.0202T_4(\xi) - 0.0002T_5(\xi) + 0.0004T_6(\xi) \end{array}\right\}$$
$$\tag{7.21}$$

with maximum respective errors of 0.0027 and 0.00014 in $-1 \leqslant x \leqslant 0$.

The treatment of the range $0 \leqslant x \leqslant 1$ via equation (7.20), with $\xi = 2x - 1$, can be performed without using this variable but by using the expansion

$$y(x) = \tfrac{1}{2}a_0 T_0^*(x) + a_1 T_1^*(x) + \cdots + a_n T_n^*(x), \tag{7.22}$$

where $T_r^*(x)$ is the Chebyshev polynomial for the special range $0 \leqslant x \leqslant 1$, related to $T_r(x)$ through the equation

$$T_r^*(x) = T_r(2x - 1). \tag{7.23}$$

All the required properties of $T_r^*(x)$ follow without difficulty from those of $T_r(x)$. For example, (7.10) becomes

$$xT_r^*(x) = \tfrac{1}{4}\{T_{r-1}^*(x) + 2T_r^*(x) + T_{r+1}^*(x)\}, \tag{7.24}$$

and in (7.11) the only change is that the coefficient $2r$ is replaced by $4r$.

With the differential equation in (7.12) and the initial condition $y(0) = e^2$, which gives the same solution (7.18), we find the respective fourth-order and sixth-order approximations (the coefficients of $T_3^*(x)$ and $T_5^*(x)$ being zero),

$$\left.\begin{array}{l} y_4(x) = 8.9060T_0^*(x) + 0.5T_1^*(x) - 1.0488T_2^*(x) + 0.0318T_4^*(x) \\ y_6(x) = 8.9056T_0^*(x) + 0.5T_1^*(x) - 1.0487T_2^*(x) + 0.0324T_4^*(x) - 0.0007T_6^*(x) \end{array}\right\}$$
$$\tag{7.25}$$

with maximum respective errors of approximately 0.0018 and 0.000029.

There is no significant change in technique for boundary-value problems of the same class, though of course there is at least one more equation from the boundary conditions. Consider, for example, the boundary-value problem with differential equation

$$y'' + xy' + xy = 1 + x + x^2, \quad -1 \leqslant x \leqslant 1, \tag{7.26}$$

and boundary conditions

$$y(0) = 1, \quad y'(0) + 2y(1) - y(-1) = -1, \tag{7.27}$$

the second of which involves three points. To express (7.27) in terms of the coefficients $a_r$ we need merely observe that

$$T_0(0) = 1, \quad T_1(0) = 0, \quad T_2(0) = -1, \quad T_3(0) = 0, \quad T_4(0) = +1, \dots, \tag{7.28}$$

and that

$$T_0(\pm 1) = 1, \quad T_1(\pm 1) = \pm 1, \quad T_2(\pm 1) = 1, \quad T_3(\pm 1) = \pm 1, \quad T_4(\pm 1) = 1, \dots, \tag{7.29}$$

obtaining the equations

$$\left. \begin{array}{l} \tfrac{1}{2}a_0 \qquad - a_2 + a_4 + \cdots = 1 \\ \tfrac{1}{2}a_0 + 4a_1 + a_2 + a_4 + \cdots = -1 \end{array} \right\}. \tag{7.30}$$

Then by differentiating (7.11) and using (7.11) again we can express $T_r''$ in terms of $T_s(x)$, and the differential equation can be expressed in the form

$$(\tfrac{1}{2}a_1 + 6a_2 + 36a_4 + \cdots)T_0 + (\tfrac{1}{2}a_0 + a_1 + \tfrac{1}{2}a_2 + 30a_3 + 0a_4 + \cdots)T_1 +$$
$$+ (\tfrac{1}{2}a_1 + 2a_2 + \tfrac{1}{2}a_3 + 56a_4 + \cdots)T_2 + (\tfrac{1}{2}a_2 + 3a_3 + \tfrac{1}{2}a_4 + \cdots)T_3 +$$
$$+ (\tfrac{1}{2}a_3 + 4a_4 + \cdots)T_4 + (\tfrac{1}{2}a_4 + \cdots)T_5 = \tfrac{3}{2}T_0 + T_1 + \tfrac{1}{2}T_2. \tag{7.31}$$

So, as before, we find approximations to $a_0, a_1, \dots, a_4$ by solving the linear equations

$$\left. \begin{array}{l} \tfrac{1}{2}a_0 \qquad - a_2 \qquad + \quad a_4 = 1 \\ \tfrac{1}{2}a_0 + 4a_1 + a_2 \qquad + \quad a_4 = -1 \\ \qquad \tfrac{1}{2}a_1 + 6a_2 \qquad + 36a_4 = \tfrac{3}{2} \\ \tfrac{1}{2}a_0 + a_1 + \tfrac{1}{2}a_2 + 30a_3 \qquad = 1 \\ \qquad \tfrac{1}{2}a_1 + 2a_2 + \tfrac{1}{2}a_3 + 56a_4 = \tfrac{1}{2} \end{array} \right\}, \tag{7.32}$$

which give the approximate solution

$$y_4(x) = 1.2699 T_0(x) - 0.6373 T_1(x) + 0.2747 T_2(x) + \\ + 0.0077 T_3(x) + 0.0047 T_4(x), \tag{7.33}$$

with maximum error of about 0.011.

Again the neglected equations other than (7.32) will show that (7.33) is the exact solution of (7.26) and (7.27) with the addition of terms on the right-hand side of (7.26) given by

$$(\tfrac{1}{2}a_2 + 3a_3 + \tfrac{1}{2}a_4)T_3 + (\tfrac{1}{2}a_3 + 4a_4)T_4 + \tfrac{1}{2}a_4 T_5$$
$$\sim 0.163 T_3(x) + 0.023 T_4(x) + 0.002 T_5(x). \tag{7.34}$$

Again this perturbation decreases rapidly as the order of the approximation increases. So too does the error of the approximating solution, that of degree eight having a maximum error of only 0.000006.

## (ii) More general linear problems. Collocation

For linear differential problems of more general form we can still find linear equations for the coefficients $a_r$ in (7.9) by asking that the differential equation be satisfied exactly at a finite number of points in the range $-1 \leqslant x \leqslant 1$. This is usually called a 'collocation' method. If we truncate (7.9) after the term $a_n T_n(x)$ we have $n + 1$ unknowns to find, with some equations coming from the associated conditions and others from the satisfaction of the differential equation. For the first-order equation (7.12), for example, we have five coefficients for the quartic approximation and one initial condition, so that we satisfy the differential equation at four points.

A result like (7.19) gives the strong suggestion that suitable points are here the zeros of $T_4(x)$, and in general of $T_n(x)$. The equations for solution for the coefficients $a_r$ for a quartic approximation are therefore obtained from (7.13) in the form

$$a_0(T_1 - \tfrac{1}{2}T_0) + a_1(2T_0 + T_2 - T_1) + a_2(5T_1 + T_3 - T_2) + a_3(7T_2 + 3T_0 + T_4 - T_3)$$
$$+ a_4(9T_3 + 8T_1 + T_5 - T_4) = 2T_0 - T_1 + T_2, \tag{7.35}$$

there being four such equations for the respective values of $x = \pm\sqrt{1/2 \pm 1/2\sqrt{2}}$, the zeros of $T_4(x)$, and the one initial condition in the first of (7.15). Their solution gives the approximation

$$y_4(x) = 6.2222 T_0(x) + 5.4444 T_1(x) - 1.7778 T_2(x) -$$
$$- 0.8889 T_3(x) + 0.1111 T_4(x), \tag{7.36}$$

which is quite similar to that of (7.16) obtained with the previous method. The corresponding sextic approximation is

$$y_6(x) = 5.7019 T_0(x) + 5.0219 T_1(x) - 1.5945 T_2(x) - 0.9327 T_3(x) +$$
$$+ 0.0713 T_4(x) + 0.0913 T_5(x) + 0.0017 T_6(x), \tag{7.37}$$

which as expected is quite a bit closer to its counterpart in (7.17). The maximum absolute errors are respectively 0.83 and 0.045 for the approximations (7.36) and (7.37).

For the second-order problem (7.26) and (7.27) there are two equations from the boundary conditions, and for a quartic approximation like (7.33) we need to satisfy the differential equation at only three points, equation (7.34) suggesting that here these should be the zeros of $T_3(x)$ and in general of $T_{n-1}(x)$. Computation gives the result

$$y_4(x) = 1.2659\,T_0(x) - 0.6356\,T_1(x) + 0.2712\,T_2(x) + 0.0079\,T_3(x) + 0.0053\,T_4(x),$$
$$(7.38)$$

which is quite close to the approximation (7.33) of the previous method. The maximum absolute error of this approximation is about 0.016.

For approximations obtained by the methods of Section 7.3(i) the representation (7.9) in terms of Chebyshev polynomials is somewhat more useful than that of (7.1) in terms of powers of $x$. However, these methods are valuable only for a small class of problems. For the collocation method, available for more general problems, the representation (7.1) is perhaps somewhat easier to use. For example, for problem (7.12), the coefficients in the approximate representation $y = \sum_{r=0}^{4} b_r x^r$ are obtained from an obvious set of algebraic equations. The first of these, for the satisfaction of the initial condition, is just

$$b_0 - b_1 + b_2 - b_3 + b_4 = 0, \qquad (7.39)$$

and the others, obtained by satisfying the differential equation at four points $x_j, j = 1, 2, 3, 4$, come from

$$(b_1 + 2b_2 x_j + 3b_3 x_j^2 + 4b_4 x_j^3) + (2x_j - 1)(b_0 + b_1 x_j + b_2 x_j^2 + b_3 x_j^3 + b_4 x_j^4)$$
$$= 1 - x_j + 2x_j^2 \qquad (7.40)$$

written in the form

$$(2x_j - 1)b_0 + \{1 + x_j(2x_j - 1)\}b_1 + \{2x_j + x_j^2(2x_j - 1)\}b_2 + \{3x_j^2 + x_j^3(2x_j - 1)\}b_3$$
$$+ \{4x_j^3 + x_j^4(2x_j - 1)\}b_4 = 1 - x_j + 2x_j^2. \qquad (7.41)$$

If the $x_j$ are the zeros of $T_4(x)$ we produce a result essentially identical with (7.36), and in fact the relations between $a_r$ and $b_r$ are just

$$b_0 = \tfrac{1}{2}a_0 - a_2 + a_4, \quad b_1 = a_1 - 3a_3, \quad b_2 = 2a_2 - 8a_4, \quad b_3 = 4a_3, \quad b_4 = 8a_4.$$
$$(7.42)$$

The word 'essentially' here means 'with exact arithmetic'. Normally the $a_r$ are somewhat smaller than the $b_r$, so that the rounding errors in the $b_r$ representation may be slightly larger than those in the $a_r$ representation. This is unlikely to be significant, however, unless polynomials of high degree and involved, whereas in practice it is rather unusual to ask for polynomials of degree higher than five. If more accuracy is required, this is achieved more easily by splitting the range into several sub-ranges, as described in the next section, rather than increasing the degree of the polynomial representation.

The positions of the collocation points, of course, have some effect on the results, and the two most widely used choices are: (a) the zeros of the appropriate

Chebyshev polynomial, as in the example treated; and (b) the zeros of the appropriate Legendre polynomial, both adjusted by a change of variable, if necessary, to fit the required range. For the range $-1 \leqslant x \leqslant 1$ the Legendre polynomial is usually denoted by $P_r(x)$, where $P_0(x) = 1$, $P_1(x) = x$, and higher terms come from the recurrence relation

$$P_{r+1}(x) - \frac{2r+1}{r+1} x P_r(x) + \frac{r}{r+1} P_{r-1}(x) = 0. \tag{7.43}$$

We deduce that the zeros of $P_4(x)$ are $\pm \sqrt{(3 \pm 2\sqrt{1.2})/7}$, and the approximation from (7.39) and (7.41) with the relevant $x_j$ is

$$y_4(x) = 5.5512 T_0(x) + 4.9504 T_1(x) - 1.5596 T_2(x) - 0.8654 T_3(x) + 0.0934 T_4(x). \tag{7.44}$$

The corresponding sextic approximation is

$$y_6(x) = 5.7227 T_0(x) + 5.0357 T_1(x) - 1.6005 T_2(x) - 0.9366 T_3(x)$$
$$+ 0.0711 T_4(x) + 0.0963 T_5(x) + 0.0022 T_6(x). \tag{7.45}$$

These are respectively the Legendre counterparts of the Chebyshev approximations (7.36) and (7.37) and the maximum errors are respectively 0.39 and 0.011.

### (iii) Split-range collocation methods for boundary-value problems

For improved accuracy an alternative to increasing the degree of the polynomial approximation is the use of two or more sub-ranges, using polynomials of the same or nearly the same degree in the separate sub-ranges. This is not at all difficult in an initial-value problem, since the solution in any sub-range depends only on the specified or computed initial conditions in that sub-range, and the convenient length of each sub-range is reasonably easy to discover.

The problem is more difficult for differential systems of boundary-value type, in which the solution in each sub-range is dependent on the solution in all sub-ranges, and suitable lengths of sub-ranges are not so easily determined unless something quite definite about the behaviour of the solution is known in advance. However, it is obviously possible to do something rather similar to 'decreasing the interval' in finite-difference methods, and to rely on consistency for estimating the accuracy obtained.

The numerical technique is essentially quite obvious. Suppose we have a second-order equation, and select three sub-ranges in which we seek respective polynomials of degrees $m_1, m_2$ and $m_3$. In the respective sub-ranges we satisfy the differential equation at $m_1 - 1$, $m_2 - 1$ and $m_3 - 1$ points, the zeros of the relevant polynomials. This gives $m_1 + m_2 + m_3 - 3$ equations, and there are $m_1 + m_2 + m_3 + 3$ unknown coefficients. Of the missing six equations, two are given by the boundary conditions, and two more at each of the two junction points of the ranges, these coming from the need to equate both the values of the polynomials and their first derivatives at the junctions of regions 1 and 2 and 2 and 3 to give suitable continuity at these points.

These last equations are the only ones which connect the separate approximations in the three sub-regions, and the matrix has a sparse structure typified by

$$
A = \begin{bmatrix}
\times & \times & \times & \times & & & & & & & \\
\times & \times & \times & \times & & & & & & & \\
\times & \times & \times & \times & & & & & & & \\
\times & \times & \times & \times & \times & \times & \times & & & & \\
\times & \times & \times & \times & \times & \times & \times & & & & \\
& & & & \times & \times & \times & & & & \\
& & & & \times & \times & \times & \times & \times & \times & \times \\
& & & & \times & \times & \times & \times & \times & \times & \times \\
& & & & & & & \times & \times & \times & \times \\
& & & & & & & \times & \times & \times & \times \\
& & & & & & & \times & \times & \times & \times
\end{bmatrix}, \qquad (7.46)
$$

where in three sub-ranges we seek approximations of respective degrees 3, 2 and 3. The first and last rows in (7.46) refer to the boundary conditions, which we have here taken to be separated.

As a simple example we try again the second-order problem given by (7.26) and (7.27), using just the two sub-ranges $-1 \leqslant x \leqslant 0$ and $0 \leqslant x \leqslant 1$. Here we use the representation of type (7.1) rather than that of type (7.9), and consider cubic polynomial approximations

$$
y_3^{(I)}(x) = b_0^{(I)} + b_1^{(I)}x + b_2^{(I)}x^2 + b_3^{(I)}x^3, \qquad y_3^{(II)}(x) = b_0^{(II)} + b_1^{(II)}x + b_2^{(II)}x^2 + b_3^{(II)}x^3 \tag{7.47}
$$

in the separate regions. The matrix corresponding to (7.46) here has the form

$$
A = \begin{bmatrix}
\times & & & & & & & \\
\times & \times & \times & \times & & & & \\
\times & \times & \times & \times & & & & \\
\times & & & & \times & & & \\
& \times & & & & \times & & \\
& & & & \times & \times & \times & \times \\
& & & & \times & \times & \times & \times \\
\times & \times & \times & \times & \times & \times & \times & \times
\end{bmatrix}. \tag{7.48}
$$

The first equation satisfies the first boundary condition. The next two satisfy the differential equation at the points $-\frac{1}{2}(1 + 2^{-\frac{1}{2}})$, $-\frac{1}{2}(1 - 2^{-\frac{1}{2}})$, the zeros of the relevant Chebyshev polynomial in region I. The next two respectively satisfy the equalities of the functions and first derivatives at $x = 0$, the junction of the two sub-ranges. The next two satisfy the differential equation at the points

$\frac{1}{2}(1 - 2^{-\frac{1}{2}})$ and $\frac{1}{2}(1 + 2^{-\frac{1}{2}})$, the zeros of the relevant Chebyshev polynomial in region II. The last equation satisfies the second boundary condition, and this being unseparated it includes far more terms than the last of (7.46). The results are

$$y_3^{(\mathrm{I})}(x) = 1 - 0.6801x + 0.4306x^2 - 0.0655x^3, \qquad x \leqslant 0$$
$$y_3^{(\mathrm{II})}(x) = 1 - 0.6801x + 0.5142x^2 + 0.0941x^3, \qquad x \geqslant 0 \Big\}, \qquad (7.49)$$

with a maximum error about half that of (7.38), the quartic approximation obtained with Chebyshev collocation over the full range $-1 \leqslant x \leqslant 1$.

## 7.4 SPLINE SOLUTION OF BOUNDARY-VALUE PROBLEMS

A spline function is a generalization of the split-range approximation of the previous subsection. The relevant range $a \leqslant x \leqslant b$ is split into a number $N$ of sub-ranges $[a, x_1], [x_1, x_2], \ldots, [x_{N-1}, b]$, the points $x_1, x_2, \ldots, x_{N-1}$ being the *knots* in the relevant literature. In each sub-range the solution is approximated by a polynomial of the same degree $m$. The polynomials are of course different in successive sub-ranges, but as before we insist on some degree of continuity at the knots. The spline is the whole approximation comprising the separate polynomials, and if this is to have $s$ continuous derivatives everywhere we must satisfy the conditions typified by

$$p_j(x_j) = p_{j+1}(x_j), \quad p_j'(x_j) = p_{j+1}'(x_j), \ldots, \quad p_j^{(s)}(x_j) = p_{j+1}^{(s)}(x_j), \qquad (7.50)$$

where $p_j(x)$ is the polynomial expression in the interval $[x_{j-1}, x_j]$.

Such a spline is determined by the positions of the knots and by the two parameters $m$ and $s$, respectively the degree of the polynomial and the degree of the required continuity. Obviously $s < m$, since otherwise every polynomial is the same polynomial and we are doing nothing new. Each polynomial has $m + 1$ coefficients, and there are $N$ polynomials, so that there are $N(m + 1)$ unknown coefficients. Equations (7.50) give $(N - 1)(s + 1)$ relations between them, so that we need another $N(m + 1) - (N - 1)(s + 1)$ equations stemming from the boundary conditions and the satisfaction of the differential equation at the internal collocation points of each sub-range. The accuracy of the approximation depends, of course, on the number $N - 1$ of knots and the parameters $m$ and $s$. Detailed analysis has shown that the accuracy increases as $N$ increases and in general also as $m$ increases, but that there is usually no advantage in taking the degree of smoothness $s$ beyond $r - 1$, where $r$ is the order of the differential equation. For such a differential equation it is then usual to take $s = r - 1$, and with $r$ boundary conditions we still need $N(m - r + 1)$ collocation equations, which are obtained by satisfying the differential equation at $m - r + 1$ points in each sub-range.

The 'spline method' is then just a systematic way of applying the collocation procedure of the previous subsection. Much attention has been paid to the way in which each sub-range polynomial is expressed, for example as $\sum_{r=0}^{m} b_r x^r$ or

$\sum_{r=0}^{m} a_r T_r(x)$, or a number of other forms which appear in recent literature. As we have remarked, they will all give the same result with exact arithmetic, and with polynomials of low degree the rounding error problem does not appear to be very significant. There is, however, one representation of each polynomial, producing the *Hermite spline*, which has significant advantages both in the amount of work involved and in the form of the computed results.

This representation is very convenient for a cubic polynomial, which has four coefficients and can therefore be expressed in terms of the values of the polynomial and of its first derivative at the two ends of the sub-range. For example, for the sub-range $[x_{j-1}, x_j]$, with the simple change of variable

$$x = x_{j-1} + \theta(x_j - x_{j-1}), \tag{7.51}$$

the cubic polynomial can be written in the form

$$\begin{aligned} p_j(x) = (\theta - 1)^2(1 + 2\theta)p_{j-1} + \theta^2(3 - 2\theta)p_j \\ + \{\theta(\theta - 1)^2 p'_{j-1} + \theta^2(\theta - 1)p'_j\}(x_j - x_{j-1}), \end{aligned} \tag{7.52}$$

and its obvious advantage is that the continuity conditions on the function and its first derivative at the knots are already satisfied. The unknown quantities are the $2(N + 1)$ values of $p_j$ and $p'_j$ for $j = 0, 1, \ldots, N$. For a second-order equation they are obtained from the two boundary conditions and from collocation at two points in each sub-range, that is from $2(N + 1)$ equations. Other representations would have $4N$ unknowns, and $2(N - 1)$ extra equations would be needed from the continuity conditions. This is an obvious advantage of the Hermite cubic spline for second-order equations, and it can be extended with no great difficulty to differential equations of higher order and polynomials of higher degree. The second advantage is that we obtain immediately approximations to the values and first derivatives of the solution at a sequence of discrete points, rather on the lines of our finite-difference methods.

The accuracy of the results will, of course, depend on the choice of collocation points. An interesting result is that collocation with the zeros of the appropriate Legendre polynomial sometimes gives special accuracy at the knots. If the sub-ranges are of size $h$, the error of a cubic spline is everywhere of order $O(h^4)$, but if we use a quintic spline, with four collocation points in each sub-range, the general error is of order $O(h^6)$ while the error in the values at the knots is

**Table 7.1**

| Interval $h$ | Max. error on $[-1, 1]$ | Max. error at knots |
|---|---|---|
| 1.0 | 0.007 | 0.0023 |
| 0.5 | 0.00046 | 0.00015 |
| 0.25 | 0.00003 | 0.00001 |

**Table 7.2**

| Interval $h$ | Max. error on $[-1,1]$ | Max. error at knots |
|---|---|---|
| 1.0 | $0.2 \times 10^{-4}$ | $0.1 \times 10^{-5}$ |
| 0.5 | $0.7 \times 10^{-6}$ | $0.5 \times 10^{-8}$ |
| 0.25 | $0.1 \times 10^{-7}$ | $0.2 \times 10^{-10}$ |

of order $O(h^8)$. These results only refer to the behaviour of the error as the size of the sub-ranges tends to zero, but it is found in practice that the values of the solution at the knots are substantially more accurate than the intermediate values. This fact is often referred to in the literature as 'superconvergence', and its effect is to make current computer routines concentrate on Legendre rather than on Chebyshev polynomials. To demonstrate the method and these comments we apply it to the problem of equation (7.26) with the boundary conditions (7.27). With collocation at the zeros of the Legendre polynomial $P_2(x)$ we obtain the results of Table 7.1. With interval $h = 0.25$ we have eight sub-intervals, and seven internal knots.

Using splines of degree 5 instead of degree 3 we obtain the results of Table 7.2, which show clearly the improved accuracy at the knots.

The application of this type of method to non-linear problems is quite straightforward. Consider, for example, the problem previously discussed in Chapter 6, given by

$$y'' = 15y^3 - 30y + 1, \quad y(0) = 0, \quad y(1) = 1. \qquad (7.53)$$

If $u$ is an approximate solution which satisfies the boundary conditions we seek a correction $\eta$ so that $y = u + \eta$, and substitution in (7.53) and neglect of powers of $\eta$ higher than the first give for $\eta$ the differential system

$$\eta'' + (30 - 45u^2)\eta = -(u'' - 15u^3 + 30u - 1), \quad \eta(0) = 0, \quad \eta(1) = 0. \qquad (7.54)$$

This is a linear boundary-value problem for $\eta$ which can be solved by the spline collocation method.

If $u(x)$ is given initially as a spline, then $u(x) + \eta(x)$ is the new spline estimate for $y(x)$, and this we improve in the usual iterative process until there is no further correction to the required accuracy. Again, of course, we need a first approximation of sufficient accuracy to guarantee convergence. In Chapter 6 we mentioned and used a continuation method for this purpose. Here we use a slightly better first estimate of the solution, using a cubic spline with interval $h = 0.2$, which gives internal knots at $x = 0.2, 0.4, 0.6$ and $0.8$. For $u(x)$ we take the values $-1, -1, 0$ and $0.5$ at these respective knots, with zero derivatives at the knots and the two boundary points. The successful progress of the iteration is shown in Table 7.3.

**Table 7.3**

| $y(0)$ | 0.000 | 0.000 | 0.000 | 0.000 | 0.000 |
|---|---|---|---|---|---|
| $y'(0)$ | 0.000 | $-4.668$ | $-4.571$ | $-4.561$ | $-4.561$ |
| $y(0.2)$ | $-1.000$ | $-0.753$ | $-0.742$ | $-0.741$ | $-0.741$ |
| $y'(0.2)$ | 0.000 | $-2.168$ | $-2.300$ | $-2.293$ | $-2.293$ |
| $y(0.4)$ | $-1.000$ | $-0.817$ | $-0.856$ | $-0.853$ | $-0.853$ |
| $y'(0.4)$ | 0.000 | 1.493 | 1.149 | 1.154 | 1.154 |
| $y(0.6)$ | 0.000 | $-0.186$ | $-0.295$ | $-0.293$ | $-0.293$ |
| $y'(0.6)$ | 0.000 | 4.465 | 4.246 | 4.219 | 4.219 |
| $y(0.8)$ | 0.500 | 0.667 | 0.577 | 0.574 | 0.574 |
| $y'(0.8)$ | 0.000 | 3.298 | 3.618 | 3.617 | 3.617 |
| $y(1)$ | 1.000 | 1.000 | 1.000 | 1.000 | 1.000 |
| $y'(1)$ | 0.000 | 0.178 | 0.647 | 0.667 | 0.667 |

This iteration, of course, is just the Newton method applied directly to the given differential equation. In previous chapters we have applied the Newton method to the system of algebraic equations representing the differential equation, which depend on the method we are using the solve the given problem.

## 7.5 ADDITIONAL NOTES

*Section 7.3(i)* There is no essential change in technique for the solution of simultaneous first-order equations of appropriate form, though there are of course more linear algebraic equations to be solved to give all the coefficients for all the dependent variables. The corresponding matrix is of a significantly sparser kind if we work directly with polynomials rather than with Chebyshev series, remembering only that the induced perturbations must be Chebyshev polynomials and not just powers of $x$. We clarify all this with respect to the problems of (7.26) and (7.27) treated in the form

$$y' = z, \quad z' + xz + xy = 1 + x + x^2, \quad y(0) = 1, \quad z(0) + 2y(1) - y(-1) = -1,$$
(7.55)

and with cubic approximations

$$y = a_0 + a_1 x + a_2 x^2 + a_3 x^3, \quad z = b_0 + b_1 x + b_2 x^2 + b_3 x^3. \quad (7.56)$$

Substituting in the equations, we now equate corresponding powers of $x$, discovering that to get exact polynomial solutions we must for this problem apply the respective perturbations

$$\tau_1 T_3(x), \quad \tau_2 T_3(x) + \tau_3 T_4(x) \quad (7.57)$$

on the right of the two equations. And to equate coefficients of powers of $x$ we therefore write

$$y' = z + \tau_1(4x^3 - 3x),$$
$$z' + xz + xy = 1 + x + x^2 + \tau_2(4x^3 - 3x) + \tau_3(8x^4 - 8x^2 + 1). \tag{7.58}$$

The linear equations stemming successively from the two boundary conditions and from the coefficients of $1, x, x^2, x^3$ in the first equation and $1, x, x^2, x^3$ and $x^4$ in the second equation are

$$\left.\begin{array}{l} a_0 \qquad\qquad\qquad\qquad\qquad\qquad = 1 \\ a_0 + 3a_1 + a_2 + 3a_3 + b_0 \qquad\qquad = -1 \\ \quad a_1 \qquad\qquad - b_0 \qquad\qquad = 0 \\ \qquad 2a_2 \qquad\qquad - b_1 \qquad = -3\tau_1 \\ \qquad\qquad 3a_3 \qquad\qquad - b_2 \qquad = 0 \\ \qquad\qquad\qquad\qquad\qquad - b_3 = 4\tau_1 \\ \qquad\qquad\qquad b_1 \qquad\qquad = 1 + \tau_3 \\ a_0 \qquad\qquad\quad + b_0 \quad + 2b_2 \quad = 1 - 3\tau_2 \\ \quad a_1 \qquad\qquad + b_1 \quad + 3b_3 = 1 - 8\tau_3 \\ \qquad a_2 \qquad\qquad + b_2 \qquad = 4\tau_2 \\ \qquad\qquad a_3 \qquad\qquad + b_3 = 8\tau_3 \end{array}\right\}, \tag{7.59}$$

a total of eleven equations for the eight coefficients and the three $\tau$ terms. The cubic polynomials can, of course, then be expressed in Chebyshev form if we so desire, and we find

$$\left.\begin{array}{l} y_3(x) = 1.2844T_0(x) - 0.6422T_1(x) + 0.2844T_2(x) + 0.0072T_3(x) \\ z_3(x) = -0.6206T_0(x) + 1.1375T_1(x) + 0.0431T_2(x) + 0.0383T_3(x) \\ \tau_1 = -0.0383, \quad \tau_2 = 0.1638, \quad \tau_3 = 0.0277 \end{array}\right\}. \tag{7.60}$$

The derivative $z = y'$ is here, of course, likely to be as accurate as $y$ itself, and it is quite close to the differentiated form of (7.33) which gives

$$z_3 = -0.6142T_0(x) + 1.1364T_1(x) + 0.0438T_2(x) + 0.0376T_3(x). \tag{7.61}$$

The direct computation of the coefficients of the polynomial approximation $\sum_{r=0}^{n} b_r x^r$, with perturbing terms of type $\tau_s T_s(x)$ on the right-hand side of the differential equation, was used in the first exposition of this method, which therefore became known as the $\tau$ method. The relevant algebraic equations are easier to write down for the $b_r$ rather than for the Chebyshev coefficients, though the fact that the $\tau_s$ are now included as unknowns in these equations involves rather more computation. The two methods, of course, give identical results in the absence of computational rounding errors.

One should also note that simple eigenvalue problems for differential equations with polynomial coefficients are also easily soluble by these methods.

Consider, for example, a problem of the form given by

$$(x + 1)y'' + \lambda y = 0, \quad y(-1) = 0, \quad y(1) = 0. \tag{7.62}$$

With the approximation

$$y(x) = \tfrac{1}{2}a_0 T_0(x) + a_1 T_1(x) + a_2 T_2(x) + a_3 T_3(x), \tag{7.63}$$

the two boundary conditions and two of the equations of type (7.32) for the equating of the coefficients for $T_0$ and $T_1$ give the four equations

$$\left.\begin{array}{l}
\tfrac{1}{2}a_0 - a_1 + a_2 - a_3 = 0 \\
\tfrac{1}{2}a_0 + a_1 + a_2 + a_3 = 0 \\
\tfrac{1}{2}\bar{\lambda}a_0 \phantom{+ a_1} + 4a_2 + 12a_3 = 0 \\
\bar{\lambda}a_1 + 4a_2 + 24a_3 = 0
\end{array}\right\}, \tag{7.64}$$

from which we easily find for $\bar{\lambda}$ the quadratic equation

$$\bar{\lambda}^2 - 28\bar{\lambda} + 48 = 0. \tag{7.65}$$

Its smaller root is 1.8345, which is remarkably close to the true first eigenvalue of (7.62), which is about 1.8352. The eigenfunction approximation is

$$y_3(x) = 0.5T_0(x) - 0.0902T_1(x) - 0.5T_2(x) + 0.0902T_3(x), \tag{7.66}$$

and with this normalization it has the considerably larger error of about 0.05 compared with the true eigenfunction.

Two other points are worth noting. First, in the methods of this subsection the error in the coefficient of $T_r(x)$, compared with the correct coefficient in the Chebyshev series, is almost always largest for $r = n - 1$. Compared with (7.33), for example, with $n = 4$, the Chebyshev series to four decimals is

$$y(x) = 1.2686T_0(x) - 0.6348T_1(x) + 0.2744T_2(x) + $$
$$+ 0.0169T_3(x) + 0.0055T_4(x) + \cdots, \tag{7.67}$$

and in (7.33) the coefficient $a_3$ clearly has by far the largest error. This is caused by the rather large coefficients in the neglected terms and neglected equations in the full set of type (7.32). With an extra term the fourth equation in (7.32) is

$$\tfrac{1}{2}a_0 + a_1 + \tfrac{1}{2}a_2 + 30a_3 + 130a_5 = 1, \tag{7.68}$$

and the first neglected equation of (7.32) is dominated by the terms in

$$\tfrac{1}{2}a_2 + 3a_3 + \tfrac{1}{2}a_4 + 90a_5 = 0. \tag{7.69}$$

Then if $a_6, a_7, \ldots$, are really negligible, $a_5$ from (7.69) and (7.33) has the approximate value of $-0.0018$, and this term in (7.68) would imply an increase of about 0.0078 to the computed $a_3$, giving it a value considerably closer to that of (7.67). With first-order problems the same phenomenon applies but with less force.

Second, although these methods give exact polynomial solutions of problems perturbed slightly by terms dominated by a Chebyshev polynomial of some degree revealed by the computation, this unfortunately does not imply that the computed solution differs from the true solution by a multiple of a Chebyshev polynomial of the same or any degree. From what we said in Section 7.2 about the best polynomial approximation to a given function, we should in fact prefer to find a polynomial approximation $\bar{y}_n(x)$ for which

$$y(x) - \bar{y}_n(x) = kT_{n+1}(x). \tag{7.70}$$

One possibility of achieving something near this for a first-order equation is first to integrate the equation and then to apply essentially the method of this subsection to the integrated form. Two integrations are performed for a second-order equation, and the relevant Chebyshev integration formulae, analogous to (7.11) for the differentiation formulae, are given by

$$\left. \begin{aligned} \int T_0(x)\,dx = T_1(x), \quad \int T_1(x)\,dx = \tfrac{1}{4}\{T_0(x) + T_2(x)\} \\ \int T_r(x)\,dx = \frac{1}{2}\left\{\frac{T_{r+1}(x)}{r+1} - \frac{T_{r-1}(x)}{r-1}\right\} \quad r > 1. \end{aligned} \right\} \tag{7.71}$$

It is also sometimes advantageous in this context to perturb somewhat the associated conditions, and all this is discussed in the relevant literature. With these methods the results do not have the largest error in the coefficient $a_{n-1}$, the large coefficients in the differentiation formulae (7.11) being replaced by small coefficients in the integration formulae (7.71). .

*Section 7.4* We have mentioned in this section the application of spline collocation methods to the solution of boundary-value problems. The same approach can be used for initial-value problems, but the method is not used in practice as it does not usually seem to be competitive with other methods described in Chapters 3 and 4.

---

EXERCISES

1. By expressing $x^4 + 3x^3$ as a series of Chebyshev polynomials, find the cubic which is the best approximation to this quartic in $-1 \leqslant x \leqslant 1$, and show that its maximum error is $\tfrac{1}{8}$ in this range. At what points do the maximum errors occur?

2. In Exercise 1 the best approximation was obtained by omitting the last term of the finite Chebyshev series. In general this series will be infinite, but when it is converging rapidly the appropriate best polynomial approximation must

be quite close to the polynomial obtained by truncating the Chebyshev series at an appropriate stage. From equation (7.7), for example, show that the best cubic approximation to $(x+2)^{-1}$ in $-1 \leqslant x \leqslant 1$ is not far from

$$p_3(x) = 0.494 - 0.243x + 0.166x^2 - 0.088x^3,$$

and by direct calculation show that its maximum errors alternate in sign, and occur at the approximate points $x = -1, -0.75, -0.1, 0.65, 1$ with approximate values $+0.009, -0.006, +0.006, -0.005, +0.004$.

3. Perform the arithmetic which gives the first of (7.25) and show that in $0 \leqslant x \leqslant 1$ this approximation is the exact solution of (7.12) with the approximate perturbation $-0.0042T_4^*(x) + 0.0737T_3^*(x)$ on the right-hand side.

4. The function $y = (1 + \frac{2}{3}x)^{\frac{1}{2}}$ satisfies the differential equation

$$(3 + 2x)y' - y = 0, \quad \text{with } y(0) = 1.$$

Find quartic approximations to the solution of this problem in the range $0 \leqslant x \leqslant 1$ by: (a) the method of Section 7.3(i) and the suggestion of equation (7.20); and (b) the use of the approximation $\sum_0^n b_r x^r$ and the method analogous to the first part of Section 7.5. Show that the two solutions are identical, with (a) having the Chebyshev approximation

$$y_4(x) = 1.150124T_0^*(x) + 0.145204T_1^*(x) - 0.004602T_2^*(x)$$
$$+ 0.000295T_3^*(x) - 0.000023T_4^*(x)$$

and (b) the essentially identical approximation

$$y_4(x) = 1 + 0.333280x - 0.054685x^2 + 0.015357x^3 - 0.002953x^4.$$

5. Perform computations to give the table corresponding to Table 7.1 when the relevant Chebyshev zeros are used instead of the Legendre zeros for the collocation points.

6. Repeat Exercise 5 with the use of a quintic spline, the equation analogous to (7.52) now having the form

$$p_j(x) = (1 - \theta)^3(1 + 3\theta + 6\theta^2)p_{j-1} + \theta^3(10 - 15\theta + 6\theta^2)p_j$$
$$+ \{\theta(1 - \theta)^3(1 + 3\theta)p'_{j-1} + \theta^3(1 - \theta)(3\theta - 4)p'_j\}(x_j - x_{j-1})$$
$$+ \{\theta^2(1 - \theta^3)p''_{j-1} + \theta^3(1 - \theta)^2 p''_j\}(x_j - x_{j-1})^2/2.$$

Observe the lack of superconvergence with the use of Chebyshev collocation points.

7. Investigate the application of the split-range collocation method to a problem involving a pair of first-order equations such as

$$y' = f(x)y + g(x)z + k(x), \quad z' = p(x)y + q(x)z + r(x),$$

with $y(a) = A$, $z(b) = B$. Using three sub-regions, what is the structure of the matrix corresponding to (7.46)?

8. Use the methods of this chapter to solve problem (3.46) of Chapter 3. Find quartic approximations over the range $[0, 1]$ with $\lambda = 10$, $\lambda = 100$, $\lambda = -100$; compare the amount of computation involved, and the accuracy achieved, with the results of the methods of Chapter 3.

9. Solve problem (3.74) of Chapter 3 by a Chebyshev series method over the interval $[0, 2]$. Use the Newton method, beginning with the initial estimate $y^{(0)}(x) = 1$, and obtain a quartic polynomial approximation. Compare the accuracy obtained, and the labour involved, with the Taylor series method of Chapter 3.

# 8
# Algorithms

## 8.1 INTRODUCTION

Workers with access to computing machines will not usually have to write lengthy computer programs making use of the methods described in previous chapters for the solution of ordinary differential equations. Most computing services provide a reasonably large range of numerical software, the computer having permanently available in its store a number of routines for solving standard problems in numerical analysis, including the solution of ordinary differential equations. The user's tasks are to learn to understand: (a) the technical details of how to use these routines; (b) how to choose the routine most appropriate to his particular problem; and (c) how to interpret the numerical results which the machine has provided. The notes associated with the routines generally make the required information reasonably available, and it is hoped that the material of this book will also be helpful in these respects.

In this chapter we cannot discuss every routine which is used or exists, but we concentrate on a brief description of the main algorithms which appear in the more available and widely-used collections. Of the latter, the most well known are:

1. the collected algorithms published originally in the *Communications of the Association for Computing Machinery* (ACM), and later in *Transactions on Mathematical Software* (*TOMS*) now also published by ACM;
2. the NAG library, available from the Numerical Algorithms Group Ltd, Oxford;
3. the International Mathematical and Statistical Library (IMSL), Texas, USA.

These libraries, of course, have a good deal in common and our discussion will be quite general, usually without mentioning any particular routines by their code names. There are two main sections, one each for initial-value problems and boundary-value problems, with subsections mentioning the various possibilities for these separate problems.

## 8.2 ROUTINES FOR INITIAL-VALUE PROBLEMS

We start by considering routines for first-order problems of initial-value type, given by

$$y'_j = f_j(x, y_1, y_2, \ldots, y_n), \qquad j = 1, 2, \ldots, n, \tag{8.1}$$

with specified values of $y_j, j = 1, 2, \ldots, n$ at an initial point $x_0$. The first subsection discusses important and fairly general questions of how to obtain required accuracy, considerations which we have considered only briefly in the preceding chapters.

### (i) Error and accuracy considerations

With most initial-value techniques it is possible to make good and generally reliable estimates for the local truncation errors at each step. But these local errors do not necessarily give directly a good estimate of the global error, the difference between the true result and the calculated result after an arbitrary number of steps. In fact it is rather difficult to design an algorithm, economic in both space and time, which produces results with a specified global error. (This may be possible with some of the interval-analysis techniques mentioned in the next chapter, but these have not generally appeared as yet in compilations of numerical software and indeed they are currently very uneconomic in computer time.)

The reason for this is obvious, in that the error at any particular point is the result of an accumulation of the errors in all the steps made to reach that point. If the problem under discussion is even mildly inherently unstable it may be necessary to compute some previous values with increased accuracy in order to produce the required accuracy at a later point. But it is impossible to predict in advance whether or not this will be necessary, or how much greater the earlier accuracy should be, since the later behaviour of the solution is not yet known.

Most initial-value algorithms use a fairly simple approach which proves to work well in practice. For a single first-order equation an estimate of the local truncation error is made at each step, and the process adjusts the step size so that some measure of the local truncation error is roughly equal to a specified 'tolerance'. Some routines use the modulus of the local error, so that

$$|T_r| \sim \varepsilon, \tag{8.2}$$

and sometimes the 'error per unit step' is used, so that

$$|T_r| \frac{b - a}{h_r} \sim \varepsilon, \tag{8.3}$$

where $h_r$ is the current step size and $b - a$ is the length of the range under consideration, $a \leqslant x \leqslant b$. When a simple algorithm is used it is commonly found that the control of the 'error per unit step' gives a global error roughly of amount $\varepsilon$. But the algorithms based on linear multi-step methods, such as those of Adams type, adjust both the step size and the order of the method as the step-by-step process goes on, and the relation between $T_r$ and the global error may be much more complicated.

A useful method results, however, by thinking of $\varepsilon$ not as a formal bound on the errors but as a strategy which controls the choice of each step size. The algorithm is then designed so that the global error is reasonably proportional to $\varepsilon$, and two separate calculations with different values of $\varepsilon$ will provide a good estimate of the global error. Consider, for example, an algorithm of this kind used to solve the equation

$$y' = 1 + \lambda y^2, \quad y(0) = 0, \tag{8.4}$$

whose solution is

$$y = \lambda^{-\frac{1}{2}} \tan (x\lambda^{\frac{1}{2}}). \tag{8.5}$$

With $\lambda = 1.5$ we find for the errors at $x = 1$ the values

$$1.3 \times 10^{-4} \quad \text{for} \quad \varepsilon = 10^{-4}, \qquad 1.0 \times 10^{-5} \quad \text{for} \quad \varepsilon = 10^{-5}. \tag{8.6}$$

For this value of $\lambda$ the problem is slightly inherently unstable, and these results are quite satisfactory.

For larger values of $\lambda$ the inherent instability is more acute, and indeed the solution is infinite at $x = 1$ for $\lambda = (\pi/2)^2 \sim 2.46$. Some results are:

$$\left.\begin{array}{llll} \lambda = 2, & \text{error} = 9.3 \times 10^{-4}, & \varepsilon = 10^{-4}, & \text{and } 8.9 \times 10^{-5}, \quad \varepsilon = 10^{-5} \\ \lambda = 2.4, & \text{error} = 520 \times 10^{-4}, & \varepsilon = 10^{-4}, & \text{and } 610 \times 10^{-5}, \quad \varepsilon = 10^{-5} \end{array}\right\} \tag{8.7}$$

In the second of (8.7) the errors are very much larger than $\varepsilon$, but a reduction by a factor of 10 in $\varepsilon$ still reduces the error by a factor almost equal to this.

The actual value of this solution with $\lambda = 2.4$ is rather in excess of 30 at $x = 1$, which suggests that in this case a relative error test will be more useful than an absolute test. But in places where the solution is very small a relative test would suggest that a very small interval is needed, and it is therefore quite common for algorithms to provide a mixed test, for example

$$|T_r| \sim \varepsilon \max \{1, |y_r|\}, \tag{8.8}$$

which is the original absolute test (8.2) for $|y_r| \leqslant 1$ and a relative test for $|y_r| > 1$.

For a set of simultaneous first-order equations there are various possibilities for extending these ideas. For $n = 2$ in (8.1), for example, there are two components of local truncation error, and with obvious notation we might replace the absolute error test (8.2) by either

$$\max \{|T_r(y_1)|, |T_r(y_2)|\} \sim \varepsilon \tag{8.9}$$

or

$$\{T_r^2(y_1) + T_r^2(y_2)\}^{\frac{1}{2}} \sim \varepsilon. \tag{8.10}$$

It is unlikely that the difference between these will be significant, but the exact meaning of a relative error test may be more important. Two possibilities are

$$\max\left\{\left|\frac{T_r(y_1)}{y_1}\right|, \left|\frac{T_r(y_2)}{y_2}\right|\right\} \sim \varepsilon \tag{8.11}$$

and

$$\left\{\frac{T_r^2(y_1) + T_r^2(y_2)}{y_1^2 + y_2^2}\right\} \sim \varepsilon. \tag{8.12}$$

If the difference between these choices turns out to be important the user may have to examine the detailed specification of the algorithm to discover exactly what it does. It is worth noting that with (8.9) or (8.10) different effects may be obtained by writing $z_2 = ky_2$ where $k$ is a constant. A large value of $k$ will produce improved accuracy in $y_2$ compared with that of $y_1$.

Another important fact is that all error estimates rely for their theoretical justification on the interval size being sufficiently small, and a consequence of this is that it is difficult to obtain efficiently a reliable solution with low accuracy. In engineering problems an error of 10% might be quite acceptable, but for a step size which purports to give this accuracy the error estimate is likely to be quite unreliable. It is therefore necessary to perform the calculations to an accuracy perhaps much greater than what is actually required to ensure that the required accuracy has been achieved, and published algorithms commonly give a warning to this effect, suggesting for example that the tolerance parameter should not be set larger than $10^{-3}$, say.

Finally, it is obvious that in the very first step we have no estimate of the local truncation error and therefore cannot choose the step size by the methods suggested. In some routines the user is asked to provide an estimate of the first interval size, and there are some special cases when this is quite reasonably possible. For example, in some problems we may need to solve a system of equations containing a parameter, and a steady variation in this parameter will result in steady changes in the first interval size which can be recognized after one or two solutions.

Otherwise one might choose an arbitrary $h$, find the resulting local truncation error and adjust $h$ accordingly. The danger here is that the first choice may be so large that the estimate for the local truncation error is quite unrealistic and the choice of $h$ correspondingly misleading. Sophisticated routines recommend the initial choice of a very small $h$, possibly only a little larger than the machine rounding error. If the estimate of the truncation error, which should now be quite reliable, is then much smaller than the tolerance, we can multiply $h$ successively by a factor of, say, 10 until, in only a few steps, the interval becomes of reasonable size and the standard procedure takes over. Such an idea, crude though it may appear to be, will cope with virtually any situation, giving the 'robustness' which is an important ingredient of computer algorithms.

## (ii) Other considerations

Both the writer and the user of the algorithm must consider other points which we have not mentioned in earlier chapters. In most routines the user must specify the number $n$ of equations in (8.1), the given initial values, possibly the starting interval size which we have discussed, and must certainly provide a routine for computing the $n$ functions $f_j$ in (8.1) for given values of the arguments. Some methods of solution may also use the partial derivatives of the functions with respect to their arguments, and in this case subroutines for their computation must also be provided.

But it is also necessary to consider precisely what the user requires as the solution to his problem. For example, the client with problem (8.4) may in fact require values at the equidistant points $x = 0(0.1)\,1.0$. For $\varepsilon = 10^{-4}$ the routine which produced the results in (8.6) actually used, after some rejects, the mesh points $0.044, 0.133, 0.398, 0.587, 0.779, 0.933$ and $1$. Choosing instead the points at which the solution is required may fail, as in the first two steps here, where the interval $0.1$ is too large; or be uneconomic as in the later calculations. An alternative and much used method merely interpolates the values at the required points from those at mesh points, using for example a Hermite interpolating polynomial which needs the computed values of $y$ and $y'$ at a few successive neighbouring mesh points. The order of the error of the interpolating polynomial is deliberately made higher than that of the technique which produced the mesh values, so that the error of the latter calculation is affected only slightly by the interpolation.

Some comment is needed about the phrase 'rejection of mesh points', which currently is an essential feature of all algorithms. Having finished the step from $x_{r-1}$ to $x_r$ we also have an estimate of the local error $T_r$. We then choose $h_{r+1}$ so that $T_{r+1} = \varepsilon$, where $T_{r+1}$ is estimated by using the assumption $T_r \sim Ch_r^p$, where $C$ does not change much from step to step. This $h_{r+1}$ is used in the step from $x_r$ to $x_{r+1}$, and this produces a revised estimate of $T_{r+1}$. If this new estimate gives $T_{r+1} > \varepsilon$ the step must be rejected and we have to repeat it with a smaller interval $h_{r+1}$. The detailed design of the algorithm tries to arrange, to avoid waste of time, that rejection does not occur too often, and also that too frequent use of intervals which are smaller than necessary is also avoided. The former of these requirements is often met by choosing something like

$$T_{r+1} = 0.8\varepsilon, \tag{8.13}$$

which would give the formula

$$(h_{r+1})^p = 0.8\varepsilon \left( \frac{h_r^p}{T_r} \right). \tag{8.14}$$

Systems of simultaneous equations may present more complicated requirements, as for example the need for the values of only a few components of a large system. A more usual requirement is the need for a graphical display of the solutions, with the use of a device attached to the computer (analogous to the printer) which can be arranged to display results. For graphical purposes we merely need enough numerical values to enable some form of polynomial or spline interpolation to be prepared to produce intermediate values close enough together to simplify the curve-drawing process.

With the background information of this and the previous subsection we now proceed to consider some of the more usual and useful subroutines which will be found in the various program libraries.

### (iii) Runge–Kutta routines

Most computer libraries have a simple Runge–Kutta subroutine, usually of fourth order, for a system of first-order equations, there being no special version for a single equation. This essentially asks the user for the initial value of the independent variable and of all the dependent variables, together with the step length to be used. It provides the required results at the end of that step, and to obtain a solution over an extended range it is necessary only to call this subroutine the required number of times.

Some routines of this type use the Merson method of Chapter 3 for estimating the local truncation errors of the components of the solution at each mesh point. The basis of this method is the assumption that sufficiently locally we are dealing with a problem linear in both $x$ and $y$ such as $y' = ax + by + c$, and it is therefore possibly unreliable for highly non-linear problems. For the latter case a method due to England is often recommended. The relevant formulae are

$$
\left.
\begin{aligned}
&k_1 = hf(x, y), \quad k_2 = hf(x + \tfrac{1}{2}h, y + \tfrac{1}{2}k_1), \quad k_3 = hf(x + \tfrac{1}{2}h, y + \tfrac{1}{4}k_1 + \tfrac{1}{4}k_2) \\
&k_4 = hf(x + h, y - k_2 + 2k_3), \quad k_5 = hf(x + \tfrac{2}{3}h, y + \tfrac{7}{27}k_1 + \tfrac{10}{27}k_2 + \tfrac{1}{27}k_3) \\
&k_6 = f(x + \tfrac{1}{5}h, y + \tfrac{28}{625}k_1 - \tfrac{1}{5}k_2 + \tfrac{546}{625}k_3 + \tfrac{54}{625}k_4 - \tfrac{378}{625}k_5) \\
&Y_{r+1} = Y_r + \tfrac{1}{6}(k_1 + 4k_3 + k_4) \\
&T_{r+1} \sim \tfrac{1}{336}(-42k_1 - 224k_3 - 21k_4 + 162k_5 + 125k_6)
\end{aligned}
\right\}.
$$

$$(8.15)$$

When this method is applied to the highly non-linear problem (3.37) of Chapter 3 the error estimate $T_1$ at the end of the first step is $1.29 \times 10^{-6}$, while the actual error is $1.31 \times 10^{-6}$. This satisfactory estimate may be compared with that from Merson's method, noted at the end of Section 3.8. In this example the error estimate from (8.15) is much more satisfactory, and the error itself is also smaller.

Notice that the standard fourth-order Runge–Kutta method needs four evaluations of $f(x, y)$, the Merson device uses five, and the England formula uses six. Another published routine uses a Runge–Kutta method of fifth order, and with six evaluations of $f(x, y)$ produces a good estimate of the local error.

The program library will also usually contain two other more sophisticated routines. In the first the routine is provided with the initial conditions, a range beyond which no further results are required, and a value of the tolerance parameter $\varepsilon$, and it then proceeds without further interference, the computation being performed with step sizes calculated as described in Section 8.2(i). A second and very similar routine will also have facilities for calculating results at particular points specified in advance, and for suppressing if required the printing of specified components of the solution.

In Chapter 3 we mentioned the existence of implicit Runge–Kutta methods. Over the last 30 years explicit methods have been popular and convenient except for very stiff equations. For these equations suitable multi-step methods have been preferred to implicit Runge–Kutta methods, and though routines for the latter certainly exit, and are mentioned in Chapter 9, they have not yet been incorporated in the main program libraries.

### (iv) Multi-step methods

Any library will also have available a number of multi-step methods. There is, of course, nothing corresponding to the simple one-step Runge–Kutta routine described in the first paragraph of the preceding subsection, but there will be routines generally identical in ideas and facilities with the two Runge–Kutta routines of the penultimate paragraph of that subsection. Multi-step methods, of course, are considerably more complicated than Runge–Kutta methods, so that such routines are more difficult to write, though the instructions to the user considerably simplify his task.

The methods most commonly in evidence use combinations of predictor–corrector formulae of the Adams class typified by equations (4.34) and (4.35). They are again controlled by a prescribed tolerance parameter, but complications include changes not only of step size but also in the order of the formulae, the aim being to cover a specified range with maximum efficiency subject to the required accuracy. In this respect the start of the process needs and gets special consideration, with special techniques in the first few steps before enough mesh values are available for the standard procedure to be applied.

Though frequent step changes are inconvenient in multi-step methods, some such changes have to be made and the routines must achieve them as efficiently as possible. In Section 4.8 we gave some simple formulae for this purpose, but in modern computer routines other methods may be used which have advantages in other relevant contexts. Consider, for example, equation (4.34) truncated after the second difference term on the right-hand side. For using this formula the required information may be stored in various ways. They include the storage of

$$y_r, y_r', y_{r-1}', y_{r-2}', \tag{8.16}$$

or that of

$$y_r, y'_r, \nabla y'_r, \nabla^2 y'_r, \qquad (8.17)$$

or that of

$$y_r, hy'_r, t_2, t_3, \qquad (8.18)$$

where

$$t_2 = h(\tfrac{3}{4}y'_r - y'_{r-1} + \tfrac{1}{4}y'_{r-2}) = \tfrac{1}{2}h^2 y''_r + O(h^4), \qquad (8.19)$$

$$t_3 = h(\tfrac{1}{6}y'_r - \tfrac{1}{3}y'_{r-1} + \tfrac{1}{6}y'_{r-2}) = \tfrac{1}{6}h^3 y'''_r + O(h^4), \qquad (8.20)$$

the so-called Nordsieck approximations to the 'reduced derivatives' $(1/2!)h^2 y''$ and $(1/3!)h^3 y'''$ at the point $x_r$. The storage (8.18) then gives something very close to the first four terms for use in the Taylor series at interval $h$ at the point $x_r$.

Now if the interval is unchanged we next compute $y_{r+1}$ and then $y'_{r+1} = f(x_{r+1}, y_{r+1})$, and update the respective storage information by moving $y'_r$ to $y'_{r-1}$, etc., for storage (8.16); calculating $\nabla y'_{r+1} = y'_{r+1} - y'_r$, etc., for storage (8.17); and calculating $t_2$ and $t_3$ as linear combinations of $hy'_r, hy'_{r+1}$ and the current values of $t_2$ and $t_3$ for storage (8.18). The first of these is trivial and the second almost trivial, but the third involves some less trivial work.

Suppose now that at the current step the size of the interval is changed to $h^*$. For storage (8.16) we now have to interpolate the values of $y'$ at $x_r - h^*$ and $x_r - 2h^*$; for storage (8.17) we have to perform a minor modification of this; but for storage (8.18) all we have to do is to multiply the last three terms respectively by $(h^*/h)$, $(h^*/h)^2$ and $(h^*/h)^3$, a much more economic operation.

Notice also that both with storage (8.17) and also with storage (8.18) we can get a quick estimate of the local truncation error, which is very near to $-\tfrac{1}{24}h^4 y^{iv}$ for formula (4.34) truncated after the second difference. When updating the new values this may be estimated by

$$-\tfrac{1}{24}h(\text{change in } \nabla^2 y'_r) \qquad (8.21)$$

for storage (8.17), since $\nabla^2 y'_r$ is near to $h^2 y'''_r$; or by

$$-\tfrac{1}{4}(\text{change in } t_3) \qquad (8.22)$$

for storage (8.18). There is nothing so convenient for storage (8.16), though the method mentioned in the first part of Section 4.6 can be used and is still used in some routines.

Note finally that storage (8.18) is very convenient when values of the solution are required at specific points intermediate between the grid points. These values can be computed very easily from the Taylor series, using the approximate reduced derivatives which are easily available.

A method which uses a completely variable step, useful when the step size needs very frequent changes, was proposed by Krogh, but this is not always available in program libraries. The work of Nordsieck and of Krogh, mentioned in this section, have short descriptions and several references in reference [14] of Chapter 9.

### (v) Stiff equations

Systems of stiff equations arise in many scientific problems, particularly in the analysis of chemical reactions. The very rapidly decaying modes in the solutions insist that a very small interval will be needed to maintain stability in the Adams-type algorithms of the preceding subsection, and even more so with explicit Runge–Kutta methods. Most current algorithms are therefore based on the 'stiffly stable' formulae typified by equation (4.33). We do not associate with this a corresponding predictor formula, since the relevant iteration will not converge for the size of interval which we can and must use for economy when the decaying modes have ceased to make any significant contribution. The non-linear implicit equations are therefore solved by the Newton method, which requires the computation of the Jacobian matrix of the system.

Otherwise the routines have many of the features mentioned in connection with the Adams routines of the previous section, including special starting devices, facilities for changing the step size and the order, and the specification of tolerance values to control the local error. Some routines ask the user to provide a subroutine for the computation of the elements of the Jacobian matrix. Others perform this calculation mechanically and approximately by computing the derivative for two adjacent values of the relevant variable and using an approximation analogous to that of equation (2.92).

The major difficulty in solving stiff systems undoubtedly rests with the Newton process and particularly with the solution of the linear algebraic equations produced at each step of the iteration. In practice there can be a very large number of such equations, but they have a significant sparse structure so that special methods for sparse systems of linear equations are very useful. Curtis and others (reference [16] of Chapter 9) use such techniques in the routine FACSIMILE, which commonly copes with several hundred simultaneous first-order equations.

One important fact about stiff systems is that the degree or even the existence of stiffness may not be known in advance. If in any doubt it is quite proper to use a routine for stiff systems even if the current equations are not really or not very stiff. The use of Adams-type methods on stiff problems would require a prohibitively small step and be correspondingly rather or very uneconomic. If the problem is not stiff, both the Adams methods and the routines for stiff systems will give equally reliable results, with the latter taking just a bit longer.

## 8.3 ROUTINES FOR BOUNDARY-VALUE PROBLEMS

### (i) General considerations

With algorithms for initial-value problems we have seen that the information which the user must supply is largely the same for all the available routines. This is no longer true for routines for boundary-value problems, the initial information being provided in quite different ways for different routines. For example, any algorithm for a non-linear problem will require the user to supply initial estimates of the unknown quantities, which may be some unknown starting values for shooting methods or the unknown values of the complete solution for global methods. Non-linear problems, incidentally, are the rule rather than the exception, and few algorithms are designed especially for linear problems in standard software collections.

Algorithms for non-linear problems will include an iterative process for improving estimates of unknown quantities, and the iteration will converge only if the initial estimates are sufficiently accurate. In practical problems good estimates may not always be available and the situation is more difficult when the problem has more than one solution. We gave a simple example of the latter in the problem of equations (1.38) and (1.45) with solutions (1.46), and a more complicated example stems from the equation and conditions given by

$$y'' = ay^5 - \lambda y + (b\lambda - c), \quad y(0) = 1, \quad y'(1) = 0, \tag{8.23}$$

which actually produces new solutions from time to time as the parameter $\lambda$ increases in value relative to some constants $a, b$ and $c$. It could hardly be expected that a standard library routine could produce all the relevant solutions, or even the particular solution required, without some professional intervention. And in fact during the progress of the computation many of the more efficient standard routines will provide a good deal of information to help the user to determine better initial estimates to obtain any required solution.

### (ii) Shooting algorithms

Most computer libraries will have at least one shooting algorithm. Probably the simplest will allow for shooting from each boundary point with matching at an internal point prescribed by the user. A more sophisticated routine uses multiple shooting with the choice of sub-ranges determined automatically according to rules specified by the user, so that this routine is hardly more difficult to use than its simpler counterpart. All these routines require the problem to be expressed as a system of first-order equations, as do the similar routines for initial-value problems, and the user provides initial estimates of whatever unknown boundary values are needed to perform the shooting.

In fact the shooting is performed by one or other of the routines available for initial-value problems, which will frequently involve a Runge–Kutta process, and when the solution is finally obtained the routine will perform the usual interpolating process to produce the results at the required points.

The Newton method for improving initial estimates may contain some sophisticated features which help it to converge, and in almost all cases the partial derivatives required in this method are obtained approximately by the method typified by equation (2.92). This affects the rate of convergence, which is not now quite quadratic, but the retardation is not usually significant, whereas the accurate computation of the derivatives could be quite laborious.

We have also mentioned quite frequently that the Newton method, modified or not, may not produce the required convergence unless the initial estimates for the unknown starting values are sufficiently accurate. Some routines will therefore incorporate facilities for a continuation-type process as described in Chapter 6.

Some libraries have a general-purpose shooting routine, aimed at computing the values of any parameters which are needed for the satisfaction of extra boundary conditions. These include the computation of particular eigenvalues, using the methods mentioned in Chapter 5, and the computation of a boundary position at which more than one condition is satisfied, also treated in Chapter 5.

Many libraries include a routine for the solution of eigenvalue problems which we have not mentioned previously, but which is an interesting example of the prior use of a mathematical transformation. For the Sturm–Liouville problem

$$\frac{d}{dx}\left\{p(x)\frac{dy}{dx}\right\} + \{q(x) + \lambda r(x)\}y = 0, \tag{8.24}$$

with separated boundary conditions

$$a_1 y(-1) + a_2 y'(-1) = 0, \quad b_1 y(1) + b_2 y'(1) = 0, \tag{8.25}$$

the Prüfer transformation

$$y(x) = R(x)\sin\theta(x), \quad p(x)y'(x) = R(x)\cos\theta(x) \tag{8.26}$$

produces the pair of first-order equations

$$\left.\begin{aligned}p\theta' &= \cos^2\theta + (q + \lambda r)p\sin^2\theta \\ R'/R &= \sin\theta\cos\theta\left(\frac{1}{p} - q - \lambda r\right)\end{aligned}\right\}, \tag{8.27}$$

with easily calculated boundary values of at least $\theta(-1)$ and $\theta(1)$. Only the first of (8.27) needs to be solved to find an eigenvalue, and the usual technique is to shoot from both ends and match in the middle, the correct value of $\lambda$ providing the required agreement at the matching point. Full details are given in reference [46] in Chapter 9.

The accuracy obtained by the shooting algorithms depends largely on the degree of ill-conditioning of the boundary-value problem. If the corresponding initial-value problems are ill conditioned they may require the use of small intervals in step-by-step methods, but in general one will be able to obtain good and specified accuracy at this stage, perhaps by multiple shooting and the ε-tolerance methods already discussed. And in this event the boundary-value problem is likely to be well conditioned so that there is no further loss in accuracy. There is more difficulty if the boundary-value problem itself is ill conditioned, since to allow for this the corresponding initial-value solutions may need to be calculated with even extra accuracy. The sensitivity of the solution with respect to small changes in a particular parameter can, of course, always be obtained by operating the routine a second time with a relevant small change in the appropriate parameter, and as we remarked in Chapter 2 this technique should be practised more frequently.

### (iii) Finite-difference algorithms

The best available finite-difference algorithm, with deferred correction, is probably the algorithm called PASVA3, developed by Pereyra and others over several years (see reference [35] in Chapter 9). There are several versions, in all of which the routines deal with a set of first-order equations which occur directly or are derived from higher-order equations. Almost always the routines use the trapezoidal-rule type of finite-difference equations, and they solve these equations, almost always non-linear, by a Newton method with a few modifications designed to produce convergence or to accelerate it.

No differences are formed, the 'difference corrections' of successive orders being computed by Lagrangian formulae. Moreover, no values external to the range are computed, so that at points near the boundaries the 'difference corrections' will effectively be computed from some non central-difference-type formulae, with a corresponding lack of smoothness, as we mentioned in Chapter 6. But the programming is certainly more convenient, with no calculation and correction of initially unknown numbers of external values and with no need to find a suitable change in the nature of the differential equation as we pass through a boundary point which involves some special treatment, again as we illustrated in Chapter 6.

The user indicates the accuracy he requires, and usually specifies the initial constant interval and a first estimate of the solution. Computation proceeds as indicated in Chapter 6, first with no difference correction and then with subsequent subsets thereof, effectively stopping when successive approximations agree to the required number of figures. If such convergence does not take place quite quickly, the interval is reduced and the computation restarted. Convergence might be slower near the ends of the range, as for example near a not too distant singular point, or if affected less significantly by the use of sloping difference formulae, so that more points might be inserted here and the mesh no longer has a completely constant interval. The corresponding Lagrangian formulae depend on the positions of the mesh points, but they are not difficult to find and to use.

An example is provided by a particular PASVA3 routine applied to the problem of equation (6.62), solved in terms of two first-order equations. Starting with constant interval 0.1 the routine correctly decided that a requested three-decimal accuracy was obtained at this interval with one application of the difference correction. When nine-decimal accuracy was requested the routine decided that this was not being achieved at this interval with even four applications of the difference correction, and it finally produced the required accuracy, in both function and its first derivative, by using the mesh points 1.0, 1.016, 1.033, 1.066, 1.10 (0.05) 1.60 (0.10) 1.90, 1.95 and 2.00. The extra points near $x = 1$ are doubtless needed because this boundary is not too far from the singular point, and the extra mesh point at 1.95 is perhaps needed to deal with the slight lack of smoothness in the use of sloping differences at this point. Notice, incidentally, that any worries associated with ill-conditioning of the boundary-value problem are dealt with automatically by this method.

By the time of publication versions of PASVA3 may exist which use the technique of 'equilibration', in which after the computation of a first approximation the mesh points are adjusted so that the local truncation error is approximately equal in value at all of them. It is hoped that this will improve the rate of convergence of the technique.

We know of no routines in the major libraries which deal directly with higher-order equations or with eigenvalue problems with difference corrections. It is of course really very easy to set up the relevant equations, use library techniques for solving them, and correct the resulting approximations with the method of the deferred approach to the limit.

### (iv)  Other routines

There are, of course, some other routines of the kind we have mentioned in Chapters 5 and 6 which exist in private hands and which are frequently reported in numerical literature. These include direct solution of equations of higher order than the first, various forms of invariant embedding, and other topics mentioned in the references of Chapter 9. It is reasonable for users to write for information about these to the various authors, few of whom are allergic to questions about their own computer routines.

Included in this list are various expansion methods, including the use of Chebyshev series and spline approximation. A recent and very useful spline-collocation routine is the COLSYS routine (reference [58] of Chapter 9). This specifies an initial set of knots, and a Newton method is used to solve the collocation equations. Accuracy is determined by 'halving the interval', as in finite-difference methods, by inserting additional knots at the midpoints of the original intervals and repeating the computations. Further knots are introduced to obtain results of specified accuracy.

### (v) Final remarks

We have observed that significant work on the numerical solution of boundary-value problems has a much shorter history than that of initial-value problems. The construction of skilled and elegant numerical software is also a rather recent development, and for these reasons the major software libraries at present have rather a sparse display of routines for boundary-value problems. As we have said, there are currently many routines in private hands, and we should expect that within the next few years the software library will house a better selection and the numerical experts will have a better idea of the relative performance of the various algorithms on various problems.

Currently, however, the solution of boundary-value problems is considerably more hazardous than that of initial-value problems. Most initial-value problems can be solved by some known routine, and the main important considerations are the reliability of the error estimates and the amount of computational labour required to produce the required accuracy. But a non-linear boundary-value problem, even of the simplest kind, may or may not have a unique solution and any numerical method will require some sort of information about this solution. If there is a solution, the time required for its computation is usually dominated by the task of constructing a satisfactory first approximation, other refinements such as the deferred correction in finite-difference methods or the successive reduction of the sizes of the sub-ranges in spline collocation methods being really a minor part of the whole calculation. In practice it is quite common to have to try several different algorithms, and perhaps several different forms of continuation, before finding one which produces the required solution. But above all the client with a boundary-value problem should do all in his power to find useful first approximations *before* using the computer, by looking at the results of somewhat similar problems or by enlisting the services of competent mathematical analysts and applied mathematicians, or by whatever combination of such devices may be available to him.

# 9

# Further notes and bibliography

In this chapter we try to give a list of references to books and papers, the contents of which will reflect as far as possible the sentiments and mathematical simplicity of this book. Occasionally a topic or topics we consider important may have no such simple written publicity, in which case we may have to give a reference involving more elaborate mathematics.

In step with this bibliography we mention some usually more difficult problems or techniques which we have not discussed at all or in any detail in previous chapters, occasionally giving some solution suggestions and some references.

CHAPTER 1

From time to time we have mentioned the importance of relevant mathematics, and success with the solution of ordinary differential equations is made more likely by adding to a knowledge of numerical methods a corresponding knowledge of analytical methods, their aims and their limitations. The following books provide some relevant information.

[1] H.T.H. Piaggio (1971), *An Elementary Treatise on Differential Equations and Their Applications*, G. Bell and Sons. This book had its first edition as long ago as 1920, but despite its age it is still very useful indeed in our particular context.

[2] C.G. Lambe and C.J. Tranter (1961), *Differential Equations for Engineers and Scientists*, English Universities Press. This is a somewhat larger book than [1], with the addition of some relatively modern material and many examples.

There are a number of still larger mathematical books for non-mathematicians which include one or more useful chapters on ordinary differential equations, a good example of which is

[3] K.F. Riley (1983), *Mathematical Methods for the Physical Sciences*, Cambridge University Press.

The three books quoted all involve reasonably easy mathematics. A good deal more advanced in this respect is

[4] D.W. Jordan and F. Smith (1971), *Nonlinear Ordinary Differential Equations*, Oxford University Press.

Most books on ordinary differential equations include some relevant discussion of difference equations, but a more comprehensive account, with not very difficult mathematics, appears in

[5] L.M. Milne-Thomson (1933), *The Calculus of Finite Differences*, Macmillan.
We mentioned in equation (1.9) the more general implicit form

$$\phi(x, y, y') = 0 \tag{9.1}$$

of first-order equations. Mathematical and computational difficulties occur at points at which (9.1) is satisfied but where also

$$\frac{\partial}{\partial y'} \phi(x, y, y') = 0. \tag{9.2}$$

An early analytical discussion of the mathematical problem is given in reference [1], and a similar elementary numerical discussion occurs in

[6] L. Fox and D.F. Mayers (1981), On the numerical solution of implicit ordinary differential equations. *I.M.A.J. Num. Anal.*, **1**, 377–401.

A somewhat more detailed mathematical discussion of the problem, and a different numerical method, is given in

[7] A. Jepson and A. Spence (1984), On implicit ordinary differential equations. *I.M.A.J. Num. Anal.*, **4**, 253–274.

Problems may also involve several simultaneous implicit equations of type (9.1), in some of which the derivatives do not appear explicitly so that we have a system comprising both differential and algebraic equations. These problems have been considered by several writers, one recent relevant paper being

[8] C.W. Gear and L.R. Petzold (1982), Differential algebraic systems, pp. 75–89, *Lecture notes in mathematics* No. 973. Springer.
An algorithm DASSL by L.R. Petzold for the solution of such problems is scheduled to appear quite soon.

CHAPTER 2

The main ideas of this chapter appear in most books on numerical analysis, though the sensitivity of boundary-value problems has had little previous discussion. For recurrence relations with initial conditions, and indeed for other material relevant to this and also to a few other chapters of our current book, we must of course mention

[9] L. Fox and D.F. Mayers (1968), *Computing Methods for Scientists and Engineers*, Clarendon Press, Oxford. The spirit and level of this is very similar to that of the current book.

J.C.P. Miller was a very important early worker on the proper use of recurrence relations, and many of his writings appear in the prefaces to several volumes of mathematical tables produced by the British Association and by the Royal Society. What is referred to frequently in the literature as 'Miller's algorithm', and which is mentioned in Chapter 2 in the paragraph following that containing equation (2.47) and in the paragraph following Table 2.7, is described and analysed in some mathematical detail in

[10]  F.W.J. Olver (1964), Error analysis of Miller's recurrence algorithm. *Math. Comp.*, **18**, 65–74.

## CHAPTERS 3 AND 4

Most of the early work on numerical methods for ordinary differential equations concentrated on initial-value problems, and there is a significant amount of relevant literature. Two of the most useful books in our current context, with reasonably simple mathematics, are

[11]  J.D. Lambert (1973), *Computational Methods in Ordinary Differential Equations*, Wiley; and

[12]  C.W. Gear (1971), *Numerical Initial-value Problems in Ordinary Differential Equations*, Prentice-Hall.

The former of these concentrates largely on the theory, and the latter perhaps more on numerical practice. In fact reference [12] gives a complete fully annotated Fortran program for both stiff and non-stiff systems, which is the basis for many of the methods in current use.

A little more mathematically advanced is

[13]  L.F. Shampine and M.K. Gordon (1975), *Computer Solution of Ordinary Differential Equations: The Initial-value Problem*, Freeman.

Two other important books treat both initial-value problems and boundary-value problems:

[14]  G. Hall and J.M. Watt (Eds.) (1976), *Modern Numerical Methods for Ordinary Differential Equations* (proceedings of Liverpool-Manchester Summer School), Clarendon Press, Oxford; and

[15]  I. Gladwell and D. Sayers (Eds.) (1980), *Computational Techniques for Ordinary Differential Equations* (proceedings of an I.M.A. Symposium), Academic Press.

The first of these covers a large amount of material in relatively short space, in mathematics which is not too difficult but of the nature of a mathematician rather than a scientist. The second has less material, but brings some important matters up to date in some detail. Both cover theoretical and practical aspects, contain very many references, and could be perhaps the most important starts for further reading for anybody who has read and followed this current book.

Important new developments in the detailed programming for large sets of stiff systems, perhaps particularly in the numerical solution of the sparse sets of simultaneous algebraic equations involved in the Newton iteration, are contained in the subroutine called FACSIMILE, written by A.R. Curtis, and for which the most easily accessible article is probably

[16] A.R. Curtis (1980), The FACSIMILE numerical integrator for stiff initial-value problems. Chapter 3 of reference [15].

The Taylor series method has a long history and possibly quite a future. One of its difficulties for automatic computation is the need for successive differentiation of complicated expressions. Some years ago techniques were devised which made this quite automatic at the expense of solving a number of simultaneous first-order equations, sometimes a far greater number than that of the original problem. A computer routine was written for an early Cambridge University machine, and this and the method are described in

[17] D. Barton, J.M. Willers and R.V.M. Zahar (1972), Taylor series method for ordinary differential equations (in *Mathematical Software*, Ed. J.R. Rice), Academic Press.

An extension to the more difficult case of stiff systems is discussed in

[18] D. Barton (1980), On Taylor series and stiff systems. *TOMS*, **6**, 280–294.

For many years the original method for the Cambridge machine was not easily adaptable to other computers, but the paper

[19] G. Corliss and Y.F. Chang (1982) Solving ordinary differential equations using Taylor series. *TOMS*, **8**, 114–144 describes a fully portable Fortran program which uses the Taylor series method.

In Section 3.8(ii) we mentioned implicit Runge–Kutta methods and block methods, and promised references to relatively recent relevant literature. Block methods are discussed, for example, in

[20] J.R. Cash (1983), Block Runge–Kutta methods for the numerical integration of initial-value problems in ordinary differential equations. I The non-stiff case. *Math. Comp.*, **40**, 175–191.

For the non-stiff case we can reasonably use explicit formulae, and we can illustrate the technique in a simple way with a second-order method applied to the single equation $y' = f(x, y)$. Starting at the point $x_r$, we compute

$$\left. \begin{aligned} k_1 &= f(x_r, y_r) \\ y_{r+1}^{(1)} &= y_r + hk_1 \end{aligned} \right\}, \tag{9.3}$$

$$\left. \begin{aligned} k_2 &= f(x_r + h, y_r + hk_1) \\ y_{r+2}^{(1)} &= y_r + h(k_1 + k_2) \end{aligned} \right\}, \tag{9.4}$$

$$k_3 = f(x_r + 2h, y_r + 2hk_2), \tag{9.5}$$

$$\left. \begin{aligned} y_{r+1}^{(2)} &= y_r + \tfrac{1}{2}h(k_1 + k_2) \\ y_{r+2}^{(2)} &= y_r + h(\tfrac{1}{2}k_1 + k_2 + \tfrac{1}{2}k_3) \end{aligned} \right\}. \tag{9.6}$$

This block method of size two moves forward over two steps, computing both $y_{r+1}$ and $y_{r+2}$. The accepted values of these quantities are given in (9.6), and the differences $y_{r+1}^{(2)} - y_{r+1}^{(1)}$ and $y_{r+2}^{(2)} - y_{r+2}^{(1)}$ give estimates of the local truncation error. The corresponding treatment of simultaneous first-order equations uses formulae analogous to those of formula (3.94) in relation to (3.92) for those of a single equation. Generally speaking they are less expensive than the standard explicit Runge–Kutta methods in the necessary number of evaluations of the function $f(x, y)$ for an accuracy of particular order.

For stiff equations the block methods can be associated with implicit Runge–Kutta techniques, and a recent paper of this kind is

[21] J.R. Cash (1983), Block Runge–Kutta methods for the numerical integration of initial-value problems in order differential equations. II The stiff case. *Math. Comp.*, **40**, 193–206.

The general idea is contained in the simple formulae

$$\left.\begin{array}{l} k_1 = f(x_r + h, y_r + hk_1) \\ y_{r+1} = y_r + hk_1 \end{array}\right\}, \tag{9.7}$$

$$\left.\begin{array}{l} k_2 = f(x_r + 2h, y_r + hk_1 + hk_2) \\ y_{r+2} = y_r + h(k_1 + k_2) \end{array}\right\}, \tag{9.8}$$

analogous to those of the explicit method just described. Here the implicit nature of the method is contained in the first of (9.7) and (9.8) which are respectively non-linear in $k_1$ and $k_2$.

The major work on implicit Runge–Kutta methods, however, is that of Butcher and his associates, resulting in a detailed subroutine and description in

[22] J.C. Butcher, K. Burrage and F.H. Chipman (1979), STRIDE: Stable Runge–Kutta integration for differential equations. Report Series No. 150 (Computational Mathematics No. 20), Dept. of Mathematics, University of Auckland.

In a way similar to that discussed in Chapter 8 for linear multi-step methods, the routines are provided with quite elaborate strategies for choosing and modifying the step size and the order of the method as the step-by-step process continues. Reference [22] has a good list of relevant earlier papers, but there is still much current research in this field, and it is not yet known whether or not STRIDE will replace Gear's algorithm as the standard library program for stiff systems.

The deferred correction method is rarely used for initial-value problems, but an early paper which uses the trapezoidal rule method and its difference correction is

[23] L. Fox and E.T. Goodwin (1949), Some new methods for the numerical integration of ordinary differential equations. *Proc. Camb. Phil. Soc.*, **45**, 373–388.

We have not mentioned the direct solution of initial-value problems of second and higher order, that is without first transforming them to a system of first-order equations. But one can quite easily use for this purpose finite-difference equations essentially identical with those applied in Chapter 6 to boundary-value problems, the set of algebraic equations now being soluble in succession rather than simultaneously. The first approximation, with error $O(h^2)$, can be improved with the deferred approach to the limit or with the method of deferred correction.

In practice, however, these methods may be somewhat uneconomic compared with the higher-order Runge–Kutta or multi-step methods, at least for general problems. They are, on the other hand, very useful for some special problems. One of these, also mentioned in connection with boundary-value methods in Chapter 6, has a differential equation lacking an explicit first derivative, in the form

$$y'' = f(x, y). \tag{9.9}$$

The approximating three-term recurrence relation

$$Y_{r+1} - 2Y_r + Y_{r-1} = \tfrac{1}{12}h^2(f_{r+1} + 10f_r + f_{r-1}) \tag{9.10}$$

has local truncation error of order $O(h^6)$, notwithstanding its simple form which clearly introduces no spurious solution, so that stability is essentially guaranteed with a small enough interval. This therefore gives some possibility of obtaining a global error of order $O(h^4)$. Note, however, that the initial conditions will specify $y(a)$ and $y'(a)$, and the simplest central-difference approximation for the derivative is not quite accurate enough for our purpose. We could, however, use a fourth-order Runge–Kutta or multi-step method to produce the required $Y_1$ correct to $O(h^4)$, this first step being continued with the successive use of (9.10) to obtain an error $O(h^4)$ at all points.

Various other special problems of this kind appear periodically in the literature. Aptly enough, certain problems with equation (9.9) and specified initial value and first derivative are known to have a periodic solution. Stable and accurate methods for obtaining such a solution are described in

[24] R.K. Jain, N.S. Kambo and R. Goel (1984), A sixth-order P-stable symmetric multi-step method for periodic initial-value problems of second-order differential equations. *I.M.A.J. Num. Anal.*, **4**, 117–125.

This paper gives several other references for this problem.

Another problem of some frequent occurrence is one in which the coefficients or terms in the differential equation change at certain values of either the independent or dependent variable. For example, in one of the problems encountered at a meeting of the Oxford Study Groups with Industry, the position of a body acted on by a force $P(x)$, depending on position, and also a simple friction force $F(\dot{x})$, depending on the velocity, satisfies the equation

$$m\ddot{x} = P(x) + F(\dot{x}). \tag{9.11}$$

A simple model of friction will be a constant limiting force $K$ acting in the direction opposite to that of the motion. The equations for solution may then have the form

$$\dot{x} = u, \quad \dot{u} = \frac{1}{m}\{P(x) + F(u)\}, \tag{9.12}$$

where

$$\left.\begin{array}{l} F(u) = -K \text{ if } u > 0; \quad = +K \text{ if } u < 0 \\ \quad\quad = -P(x) \text{ if } u = 0 \text{ and } |P(x)| \leqslant K \\ \quad\quad = -K \text{ if } u = 0 \text{ and } P(x) > K \\ \quad\quad = +K \text{ if } u = 0 \text{ and } P(x) < -K \end{array}\right\}, \tag{9.13}$$

the last three of (9.13) perhaps varying depending on the details of the physical modes of friction and 'stiction' under consideration.

Equation (9.12) has singularities, since $\dot{u}$ has a jump discontinuity of magnitude $2K/m$ at the points where $u$ passes through zero. 'Standard' numerical methods will fail or be extremely uneconomic, and the only really satisfactory way to obtain reliable results from this sort of problem is to determine the position of the singularities. Suppose, for example, that the initial conditions give $u > 0$. Then we begin by solving (9.12) with $F(u) = -K$, using a step-by-step method and continuing until $u$ changes sign. Then a simple process of inverse interpolation will determine the point $t_1$ at which $u = 0$; standard versions of some library routines will do this automatically. The process is then restarted from $t_1$ with the new initial conditions and with $K$ replaced by $-K$, and continued until $u$ changes sign again.

Then there are problems in which the argument is not the same in every term of the differential equation. A problem which was also presented to the Oxford Study Groups with Industry, in connection with the collection of current in an overhead electric railway system, was formulated to produce the relevant equations

$$y'(x) = Ay(\lambda x) + By(x), \quad y(0) = \alpha, \tag{9.14}$$

where $\lambda$, $A$ and $B$ are constants. More generally, $y$ may be a vector and $A$ and $B$ matrices with constant coefficients. For $\lambda < 1$ there is a unique solution, and this is the important practical case. With $\lambda > 1$, which is mathematically very interesting but lacking practical importance in this context, there may be several solutions. Both cases have been discussed in several papers, the first of which considered both mathematical and numerical methods in

[25] L. Fox, D.F. Mayers, J.R. Ockendon and A.B. Taylor (1971), On a functional differential equation. *J. Inst. Maths. Applics.*, **8**, 271–307.

Of more frequent occurrence is the 'delay' differential equation typified by

$$y'(x) = f\{x, y(x), y(x - \lambda(x))\}, \quad \lambda(x) > 0, \tag{9.15}$$

with a suitable initial condition. A short account of problems and methods is given in reference [14], and a longer account with some mathematical rigour in

[26] C.W. Cryer (1972), Numerical methods for functional differential equations, pp. 97–102 of *Delay and Functional Differential Equations and Their Applications*. (Ed. K. Schmitt), Academic Press.

We observed in Chapter 8 that the computation of an accurate and useful bound for the global error of any of our methods is not particularly easy. For some problems a useful bound can be obtained by a method of 'interval analysis', in which each correct mesh value is guaranteed to lie within computed intervals whose widths depend both on computer rounding errors and also on the local truncation error of the particular method. The mathematics of interval analysis is by no means easy, and perhaps the least difficult account of it is in

[27] E. Hansen (Ed.) (1969), *Topics in Interval Analysis*, Clarendon Press, Oxford.

This contains a chapter on initial-value problems:

[28] F. Kruckeberg (1969), Ordinary differential equations (from reference [27]).

A more recent account, with perhaps more difficult mathematics but again with some comments on initial-value problems, is in

[29] K.L.E. Nickel (1980), *Interval Mathematics*. Academic Press.

We have so far said very little about singularities in initial-value problems, but they are very common and can arise in various ways. Here we give a few typical examples. As we mentioned in Chapter 1, the positions of any singularities in linear equations are known in advance, and they can usually be dealt with by obtaining series solutions in the neighbourhood of the singularities. For example, Bessel's equation of order zero, given by

$$y'' + \frac{1}{x}y' + y = 0, \tag{9.16}$$

can have a singularity only at $x = 0$, but with the initial conditions $y(0) = 1$, $y'(0) = 0$ the series solution about the origin is

$$y = 1 - \tfrac{1}{4}x^2 + \tfrac{1}{64}x^4 - \cdots \tag{9.17}$$

and there is no singularity in this solution, the function and all its derivatives being finite everywhere. There is an apparent numerical difficulty at the start, for example with the solution of the first-order system

$$y' = z, \quad z' + \frac{1}{x}z + y = 0, \quad y(0) = 1, \quad z(0) = 0, \tag{9.18}$$

since we need the limiting value of $x^{-1}z$ as $x \to 0$. This is obtained from the series (9.17), and there is no further numerical problem with any of our methods.

More complicated is the equation

$$y'' + \frac{1}{x}y' + \left(1 - \frac{1}{4x^2}\right)y = 0, \tag{9.19}$$

which has only one solution which is finite at $x = 0$, with the expansion

$$y = x^{\frac{1}{2}}(1 - \tfrac{1}{6}x^2 + \cdots). \tag{9.20}$$

All its derivatives are infinite at $x = 0$, and this clearly complicates significantly any numerical method starting at $x = 0$. We have two reasonable procedures. First, we could compute $y$ for small values of $x$ from the series (9.20) and then continue the solution by numerical methods. Second, we could make the change of dependent variable given by

$$\eta = x^{-\frac{1}{2}}y, \tag{9.21}$$

find that this satisfies the differential equation

$$\eta'' + \frac{2}{x}\eta' + \eta = 0, \tag{9.22}$$

and, as in the previous example, the solution has no singularity if $\eta(0) = 1$ and $\eta'(0) = 0$.

For non-linear problems, as we mentioned in Chapter 1, the singularities may be rather more elusive. In some cases it may be possible to obtain analytical solutions which determine the positions of singularities, but in most problems this is unlikely and we would hope that our numerical method will deal with this.

Consider, for example, the problem

$$y' = k^2 + y^2, \quad y(0) = 0. \tag{9.23}$$

Its solution is $y = k \tan kx$, which approaches infinity as $x \to \pi/2k$. As $x$ increases towards this number, $y(x)$ will get quite large, and to find the position of the singularity and a reasonable approximation to the true solution as we approach it and even go through it, some change of variables is obviously required. Here the simplest possibility is to choose

$$\eta = y^{-1}, \tag{9.24}$$

and to solve numerically the resulting system

$$\eta' = -(1 + k^2\eta^2), \quad \eta(x_c) = \{y(x_c)\}^{-1}, \tag{9.25}$$

where $x_c$ is a suitable point at which to make this change. The singularity of $y$ then corresponds to a zero of $\eta$, which can be found by a simple process of inverse interpolation on computed values of $\eta$ at mesh points.

Some other change of variable may of course be more useful. In fact a knowledge of the nature of the singularity is always very valuable, and this is one area in which good mathematics could be combined with numerical analysis with considerable benefits.

## CHAPTERS 5 AND 6

Shooting methods for boundary-value problems are usually mentioned in books on initial-value problems, and of course reference [14] has a chapter on shooting methods. This includes multiple shooting methods, but these also have a literature of their own. A fairly easy account is given in the early paper

[30] M.R. Osborne (1969), On shooting methods for boundary-value problems. *J. Math. Anal. Appl.*, **27,** 417–433,
and a more modern version, in more difficult mathematics, is that of
[31] P. Deuflhard (1980), Recent advances in multiple shooting techniques, Chapter 10 of reference [15].

The invariant embedding method is treated in some detail, with relatively easy mathematics in the practical parts, in
[32] G.H. Meyer (1973), *Initial-value Methods for Boundary-value Problems.* Academic Press, New York,
and it is mentioned briefly in reference [14].

Boundary-value methods of Chapter 6 were probably first treated at any length in
[33] L. Fox (1957), *The Numerical Solution of Two-point Boundary-value Problems in Ordinary Differential Equations.* Clarendon Press, Oxford.
This book uses very simple mathematics, and though it is now rather out of date it contains much useful advice about a good deal of the material of this current book.

There are very few modern books of the required simple mathematical standard, but reference [14] has a rather small section, and a more recent paper which includes this topic in addition to other comments on boundary-value problems, all in easy mathematics, is
[34] L. Fox (1980), Numerical methods for boundary-value problems. Chapter 9 of reference [15].

The idea of the difference correction was first introduced in reference [33]. The more modern theory and application is largely due to Pereyra.

References to his earlier work are given in [14] and [34], but his main paper which describes the best computer program is

[35] V. Pereyra (1978), PASVA3. An adaptive finite-difference Fortran program for first-order non-linear boundary-value problems. Chapter 4 of reference [62].

The word 'adaptive' here means the deliberate choice of unequal intervals, for example after early experimentation increasing the number of mesh points in regions of large local truncation error at the original intervals. This is also discussed in

[36] M. Lentini and V. Pereyra (1977), An adaptive finite-difference solver for nonlinear two-point boundary-value problems with mild boundary layers. *SIAM J. Num. Anal.*, **14**, 91–111.

He also treated higher-order equations directly in

[37] V. Pereyra (1973), High-order finite-difference solution of differential equations. Comp. Sci. Dept. Stanford University report STAN-CA-73-348.

A routine for improving still further the $O(h^4)$ accuracy of methods relevant to equation (9.9) for boundary-value problems has been suggested in

[38] J.W. Daniel and A.J. Martin (1977), Numerov's method with deferred corrections for two-point boundary-value problems. *SIAM J. Num. Anal.*, **14**, 1033–1050,

Numerov being the original discoverer of the famous equation (9.10).

There are a few special problems which may require special methods. One such problem is that of rapid oscillation about a very smooth curve, for example with the system

$$y'' + (314.16)^2 y = (314.16)^2 x, \quad y(0) = 10^{-5}, \quad y(1) = 1, \qquad (9.26)$$

which has the unique solution

$$y = x - 0.0136\ldots\sin(314.16)x + 0.00001\cos(314.16x). \qquad (9.27)$$

Most computer routines confidently give the solution $y = x$, and the small and rapid oscillations around this line are revealed only with a very small interval. This interval is needed if the correct solution is required everywhere, but in some contexts it may be enough to compute the 'envelopes' given by the broken lines in Fig. 9.1.

We known of no certain method of achieving this in all cases, but there is some discussion for the corresponding initial-value problem in

[39] C.W. Gear (1980), Initial-value problems: practical theoretical developments. Chapter 7 of reference [15].

There is a longer discussion, with useful examples and a long bibliography, in

[40] R.E. Scheid (1984), The accurate solution of highly oscillatory ordinary differential equations. *Math. Comp.*, **41**, 487–509.

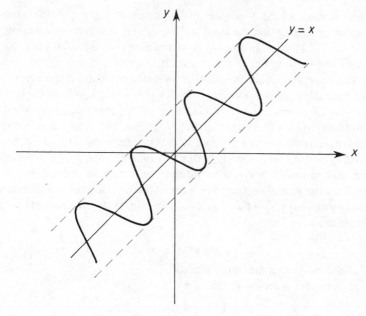

**Figure 9.1**

The mathematics here is not too difficult if the proofs of certain theorems are accepted without a reading.

Another important problem connected with boundary-layer theory and occurring in other scientific contexts is the so-called 'singular perturbation' problem with equations like

$$\varepsilon y'' + f(x, y, y') = 0, \quad y(a) = \alpha, \quad y(b) = \beta, \tag{9.28}$$

where $\varepsilon$ is small. The solution may have 'fierce' behaviour in the short boundary-layer regions and quite smooth behaviour in the central region. In some problems there are internal 'turning points' with 'spiky' behaviour. The following four references cover most of the relevant problems.

[41] W.L. Miranker (1981), *Numerical Methods for Stiff Equations and Singular Perturbation Problems*, Reidel, Dordrecht.

[42] B. Kreiss and H-O Kreiss (1981), Numerical methods for singular perturbation problems. *SIAM J. Num. Anal.*, **18**, 262–276.

[43] H-O Kreiss, N.K. Nichols and D.L. Brown (1983), Numerical methods for stiff two-point boundary-value problems. University of Wisconsin Math. Res. Center Summary Report 2599.

[44] E.P. Doolan, J.J.H. Miller and W.H.A. Schilders (1980), *Uniform Numerical Methods for Problems with Initial and Boundary Layers*. Boole Press, Dublin.

This last reference also includes a fair amount of relevant mathematics. It is also worth noting that the word 'stiff' in this context has some analogy with that for initial-value problems, here meaning that the solution varies over several different scales of magnitude in the relevant region.

An important variation in boundary-value problems is the case in which one boundary point is at infinity. Such a problem might have the condition $y(x) \to 0$ as $x \to \infty$, and it is tempting to approximate to this with the condition $y(X) = 0$ for some large $X$. But this can be a very poor approximation, needing a very large $X$ for accurate solution at medium-sized values of $x$. It is always better to attempt to discover mathematically the way in which $y(x) \to 0$ at infinity, and then to use an appropriate condition at $x$ involving perhaps a derivative or a difference. For example, if it is known that $y(x) \to Cx^{-1}$ as $x \to \infty$ for constant $C$, then the approximate condition

$$y_{r+1} = (x_r/x_{r+1})y_r \tag{9.29}$$

is useful at quite a medium value of $x_r$.

In a similar manner, for the simple problem

$$y'' = k^2 y, \quad y(0) = \alpha, \quad y(\infty) = 0 \tag{9.30}$$

it is clear that we need to eliminate from our method, and as early as possible, the term $e^{kx}$ in the complementary solution, and this is achieved exactly, for any $x$, with the condition

$$y' + ky = 0. \tag{9.31}$$

Useful ideas in this area, and in many others, are given in

[45] H.B. Keller (1976), Numerical solution of two-point boundary-value problems. Regional Conference Series in Applied Mathematics 24, SIAM, Philadelphia,

though the mathematics is in places somewhat advanced.

Similar situations arise in eigenvalue problems, and in this connection another useful paper is

[46] P.B. Bailey, M.K. Gordon and L.F. Shampine (1978), Automatic solution of the Sturm–Liouville problem. *TOMS*, **4**, 193–208,

which also analyses the Prüfer transformation mentioned in Chapter 8.

Another interesting paper involving eigenvalue problems, with an infinite range and some singularities, is

[47] J.K. Lund and B.V. Riley (1984), A sinc-collocation method for the computation of the eigenvalues of the radial Schrödinger equation. *I.M.A.J. Num. Anal.*, **4**, 83–98.

Mathematical analysis, of course, is useful in many more contexts in boundary-value problems. One of these is the treatment of singularities, and even in simple problems this is considerably more involved than that for similar types of initial-value problems, for example with equations like (9.16) and (9.19). It may happen, for example, that with specified boundary values the first derivative is then infinite at one of the boundaries. Treatment as two simultaneous first-order equations then has obvious problems. Moreover, even though the direct treatment of the second-order equations may produce reasonably satisfactory results at most points for the function values, the successive computation of the first and higher derivatives would be very inaccurate near the dangerous boundary. Again, even if the first infinite derivative is $y''$ at some boundary we can hardly expect the correcting devices of the deferred approach to the limit or the deferred correction to have any significant success.

In fact the deferred approach method can have strange results. For example, the function $y = \frac{1}{2}x^2 \log x$ is the solution of both

$$xy'' - y' - x = 0, \quad y(0) = 0, \quad y(1) = 0 \tag{9.32}$$

and

$$x^2 y'' - 2y - \frac{3}{2}x^2 = 0, \quad y(0) = 0, \quad y(1) = 0, \tag{9.33}$$

but the error in the approximate finite-difference solutions for (9.32) and (9.33) start with the respective forms

$$y_r - Y_r = Ah^2 + Bh^2 \log h, \quad y_r - Y_r = Ah^2 + Bh^4. \tag{9.34}$$

Moreover, the solution of the singular equation

$$xy'' + y' = x^2 \tag{9.35}$$

has no singularity with $y(0) = 0$, $y(1) = \frac{1}{9}$, being the simple $y = \frac{1}{9}x^3$, but the finite-difference error starts with multiples of $h^2$ and $h^2/\log h$. All this is analysed in

[48] D.F. Mayers (1964), The deferred approach to the limit in ordinary differential equations. *Comp. J.*, **7**, 54–57.

For non-linear equations it is sometimes possible to detect the nature of a boundary singularity, but an internal singularity is much more complicated. Consider, for example, the problem

$$y'' + 3yy' + y^3 + y = 0, \quad y(0) = -2, \quad y(\pi/2) = 0, \tag{9.36}$$

which according to the analytical solution of equation (1.33) here has the solution given by

$$y = \cos x/(\sin x - \frac{1}{2}), \tag{9.37}$$

with a singularity at $x = \pi/6$. This could hardly be known in advance without finding the analytical solution, which here is a rather surprising bonus. With an infinity in the solution, at a point generally unknown, one could hardly expect that finite-difference boundary-value methods could produce much useful information, and even the shooting method would have to be used in a sophisticated manner, perhaps with several changes of the dependent variable to avoid the infinities. We know of no useful references for problems of this kind.

Mathematics can frequently also be used in a powerful way to obtain good first approximations to the solution of non-linear problems with iterative techniques, a good start, as we have pointed out, being one of the most important and elusive ingredients for iterative methods. In another Oxford Industry Study Groups problem concerned with the design of mooring and towing systems, the problem involves four simultaneous first-order equations with independent variable $s$ and dependent variable $x, z, \phi$ and $T$, the tension in the cable, with reference to Fig. 9.2. The boundary conditions specify $x$ and $z$ at $s = 0$ and $s = 1$.

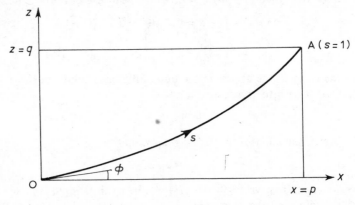

**Figure 9.2**

For some time all iterative methods failed, subsequently traced to a poor first approximation to the angle $\phi$. All these approximations assumed $\phi$ to be positive, but a later idea that the cable OA probably had the shape of a catenary suggested a negative $\phi$, and with the further thought that the tension in a catenary is proportional to $\varepsilon$, the difference between the length of the cable and the straight line OA, one obtained approximate formulae for $x, z, \phi$ and $T$, in terms of a small $\varepsilon$, which not only gave a good start to the iteration but which were really quite accurate for quite substantial $\varepsilon$. Some details of this problem are given in reference [34].

This is in a sense a type of continuation method. Other continuation methods, and a convergence analysis, are discussed in Chapter 18 of reference [14], which also includes a few more relevant references. The methods may soon become a significant feature of non-linear boundary-value routines.

Special equations of type (9.9) may also appear in boundary-value problems, and for example specified boundary values and the use of equation (9.10) in a simultaneous sense will guarantee a solution with $O(h^4)$ error if there is no singularity in equation and solution. This $O(h^4)$ error provides the reason for solving such problems directly rather than as simultaneous first-order equations, and a number of workers have extended the idea to provide highly accurate approximations even to more general problems like

$$y'' = f(x, y, y'), \quad g_1\{y(a), y'(a)\} = 0, \quad g_2\{y(b), y'(b)\} = 0, \quad (9.38)$$

a typical paper being that of

[49] M.M. Chawla (1978), A fourth-order tridiagonal finite-difference method for general two-point boundary-value problems with nonlinear boundary conditions. *J.I.M.A.*, **22**, 89–97. The ideas of some relevant papers can be applied also to give highly accurate first approximations for certain eigenvalue problems.

As with initial-value problems, so here the differential equation may change its form at a fixed value of either the independent or dependent variable. With the former, a typical case is given by

$$y'' = f(x, y, y') \text{ in } a \leqslant x \leqslant c, \qquad y'' = g(x, y, y') \text{ in } c \leqslant x \leqslant b, \quad (9.39)$$

with boundary conditions of usual type at $x = a$ and $x = b$ and of a derivative 'jump discontinuity' type at $x = c$ such as

$$y(1) = y(2), \quad y'(1) - y'(2) = K, \quad (9.40)$$

where $K$ is a constant and (1) and (2) denote the respective regions $a \leqslant x \leqslant c$, $c \leqslant x \leqslant b$.

With reference to Fig. 9.3, the simple finite-difference equations at $x = c$ for equations (9.39) are

$$\left. \begin{array}{l} Y_1(1) - 2Y_0(1) + Y_{-1}(1) = h^2 f\{c, Y_0(1), \tfrac{1}{2}h^{-1}(Y_1(1) - Y_{-1}(1))\} \\ Y_1(2) - 2Y_0(2) + Y_{-1}(2) = h^2 g\{c, Y_0(2), \tfrac{1}{2}h^{-1}(Y_1(2) - Y_{-1}(2))\} \end{array} \right\}, \quad (9.41)$$

and the conditions (9.40) at $x = c$ give

$$Y_0(1) = Y_0(2), \quad \{Y_1(1) - Y_{-1}(1)\} - \{Y_1(2) - Y_{-1}(2)\} = 2hK. \quad (9.42)$$

We would clearly like to eliminate $Y_1(1)$, $Y_{-1}(2)$ and one of $Y_0(1)$ and $Y_0(2)$ from these four equations to give a single equation containing $Y_{-1}(1)$, $Y_1(2)$ and one of $Y_0(1)$, $Y_0(2)$, and this is clearly possible in theory

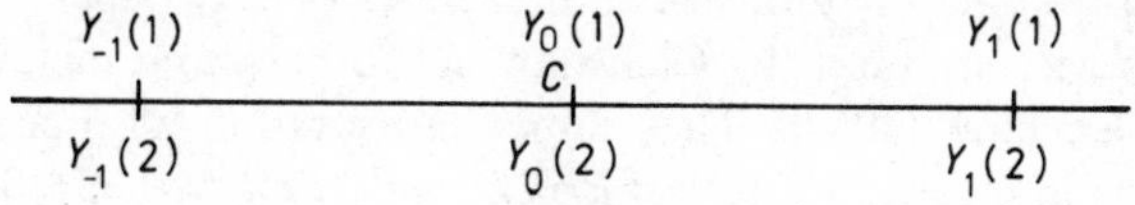

**Figure 9.3**

even if in practice it is preferable to keep all four equations in the solving process. When difference corrections are included we must, of course, extend the 'fictitious' values $Y_r(1)$ further to the right in region 2 and of $Y_r(2)$ further to the left in region 1, not attempting to difference the mesh values... $Y_{-2}(1)$, $Y_{-1}(1)$, $Y_0(1) = Y_0(2)$, $Y_1(2)$, $Y_2(2)$,... through the point $c$ at which the discontinuity occurs.

The corresponding treatment of the relevant first-order equations gives no extra problem, and we need not give all the equations involved.

With a change in form at a particular value of $y$, something like

$$y'' = f(x, y, y'), \quad y < 0; \quad y'' = g(x, y, y'), \quad y > 0, \tag{9.43}$$

the finite-difference equations form a set of linear or non-linear algebraic equations subject to linear constraints, and these may be solved by a method of linear or non-linear programming. If the positions of changes of sign in $y$ are known approximately from physical considerations, then first approximations to the solution can be found by solving the relevant equations, and a little further iteration will normally work quite adequately and produce an accurate solution.

Finally, interval analysis can be used to give guaranteed bounds for the solution of some boundary-value problems by various methods, an early account being that of

[50] E. Hansen (1969), On solving two-point boundary-value problems using interval arithmetic. From reference [27].

A later paper, reasonably easy to read, is that of

[51] L. Fox and M.R. Valenca (1980), Some experiments with interval methods for two-point boundary-value problems in ordinary differential equations. *B.I.T.*, **20**, 67–82.

CHAPTER 7

The first simple account of the $\tau$ method for ordinary differential equations probably appears in the book

[52] C. Lanczos (1957), *Applied Analysis*. Prentice-Hall, New York; Pitman, London,
  and a later account of this and other Chebyshev methods is given in
[53] L. Fox and I.B. Parker (1968), *Chebyshev Polynomials in Numerical Analysis*. Oxford Mathematical Handbooks (Eds. J. Crank and C.C. Ritchie), Oxford University Press,
  again written in easy terms. A similar series of easy papers on Chebyshev methods was produced by Clenshaw and others, and includes
[54] C.W. Clenshaw (1957), The numerical solution of linear differential equations in Chebyshev series. *Proc. Camb. Phil. Soc.*, **53**, 134–149,
  and

[55] C.W. Clenshaw and H.J. Norton (1963), The solution of nonlinear ordinary differential equations in Chebyshev series. *Computer J.*, **6**, 88–92.

Another series of papers by Ortiz and others on Chebyshev methods includes

[56] E.L. Ortiz, W.F.C. Purser and F.J. Rodriguez (1972), Automation of the $\tau$ method. Royal Irish Academy Conference on Numerical Analysis, Dublin.

A Chebyshev collocation method is described in

[57] K. Wright (1964), Chebyshev collocation methods for ordinary differential equations. *Computer J.*, **6**, 358–363,

and a spline-collocation method in

[58] V. Ascher, J. Christiansen and R.D. Russell (1978), COLSYS – A collocation code for boundary-value problems. Chapter 12 of reference [62].

CHAPTER 8 AND GENERAL

Descriptions of computer library routines, sometimes with useful annotations about the methods, are given in some of the major libraries of computer software. New routines appear regularly in the journal *Transactions on Mathematical Software* (*TOMS*), published by the Association for Computing Machinery.

Useful accounts of the performance of standard routines in a selected list of problems of varying difficulty appear for initial-value problems in

[59] W.H. Enright and T.E. Hull (1976), Test results on initial-value methods for non-stiff ordinary differential equations. *SIAM J. Num. Anal.*, **13**, 944–961,

and in

[60] W.H. Enright, T.E. Hull and B. Lindberg (1975), Comparing numerical methods for stiff systems of ordinary differential equations, *B.I.T.*, **15**, 10–48.

A more recent discussion of testing routines is given in

[61] T.E. Hull (1980), Comparison of algorithms for initial-value problems. Chapter 6 of reference [15].

For boundary-value problems a number of codes are discussed and analysed in

[62] B. Childs, M. Scott, T.W. Daniel, E. Denman and P. Nelson (1978), Codes for boundary-value problems in ordinary differential equations. Lecture Notes in Computer Science 76, Springer-Verlag, Berlin,

and a more recent survey is that of

[63] I. Gladwell (1980), A survey of sub-routines for solving boundary-value problems in ordinary differential equations. Chapter 11 of reference [15].

Finally, journals on numerical mathematics and computational methods, in which are likely to appear general new techniques, improved error analysis or new applications to particular problems, are included in the following short list, in rough order of increasing mathematical difficulty.

*Transactions on Mathematical Software*
*Journal of Computational Physics*
*B.I.T.*
*Mathematics of Computation*
*SIAM Reviews*
*SIAM J. Num. Anal.*
*I.M.A.J. Num. Anal.*
*Numerische Mathematik.*

The publishers of these and other relevant journals are given in *Mathematical Reviews*, published by the American Mathematical Society, and this journal also gives short reviews of virtually all published papers in mathematics, including numerical mathematics. Other useful publications are the *SIGNUM Newsletter*, a quarterly publication of the ACM Special Interest Group in Numerical Mathematics, and the *IMANA Newsletter*, a frequent publication of the corresponding IMA Special Interest Group.

# 10

# Answers to selected exercises

NOTE. In the numerical problems, computing machines of different word lengths could give quite dissimilar results in problems which exhibit any sort of instability, either inherent or induced. This is particularly true of Exercises 12, 13 and 14 in Chapter 5.

## CHAPTER 1

[5] The general solution is

$$y = A + Bx + Cx^2 + \tfrac{1}{6}x^3 + \tfrac{1}{24}x^4.$$

The particular solution with $y(0) = y'(0) = y''(0) = 1$ is

$$y = 1 + x + \tfrac{1}{2}x^2 + \tfrac{1}{6}x^3 + \tfrac{1}{24}x^4.$$

[6] The general solution is

$$y = A \sin x - B \cos x, \qquad z = A \cos x + B \sin x.$$

The particular solution with $y(0) = 0$, $y(\pi/2) = 1$ is

$$y = \sin x, \quad z = \cos x.$$

With the conditions $y(0) = 0$, $y(\pi) = 1$ there is no solution.

[7] The analytical solution (1.29) gives $A = -16$ for $y(0) = \tfrac{1}{4}$. There is therefore a singularity at $x = 2$ where $x^4 + A = 0$.

The condition $y(0) = 0$ gives $A = \infty$, with solution $y(x) = 0$ for all $x$. This is also implied by the Taylor series at $x = 0$, with all derivatives zero.

The condition $y(0) = -1$ gives $A = 4$, with no singularity since $x^4 + 4 \neq 0$ for any real $x$.

[8] The general solution is

$$y_r = A + Br + Cr^2.$$

[9] The general solution of the recurrence relation is

$$y_r = A\{\tfrac{1}{2}(1 + \sqrt{5})\}^r + B\{\tfrac{1}{2}(1 - \sqrt{5})\}^r.$$

If $y_0 = 0$ then $A \neq 0$, so that for large $r$ the first term in the general solution ultimately dominates, and $y_{r+1}/y_r \to \frac{1}{2}(1 + \sqrt{5})$ as $r \to \infty$.

[10] If in Exercise 9 we have $y_0 = 1$, $y_1 = \frac{1}{2}(1 - \sqrt{5})$, then $A = 0$ and $B = 1$, and the ratio $y_{r+1}/y_r = \frac{1}{2}(1 - \sqrt{5})$ for all $r$. (But see Exercise 13 of Chapter 2.)

## CHAPTER 2

[2] With the data of equation (2.11) we find, with neglect of $\delta g$, the expression

$$\delta x \sim 0.625 \delta V + 21.65 \delta \alpha,$$

so that the absolute uncertainty in $x$ is considerably greater than that in $y$. But the absolute value of $x$ when $y = H$ again is 6.25 feet, so that the relative uncertainties with the given $\delta V$ and $\delta \alpha$ are approximately 2% for velocity change and 0.91% for angle change, with a combined effect probably slightly smaller than that for $y$.

[4] With the notation of Section 2.4(i) the absolute uncertainty in $y_r$ is here

$$\delta y_r = \frac{1}{2} \cdot \frac{3}{2} \cdot \frac{5}{2} \cdot \ldots \cdot \frac{2r - 1}{2} \delta \alpha.$$

In the first recurrence $\delta \alpha = 0.0008$, and in the second $\delta \alpha = -0.0002$. The required result is therefore the first integer value of $r$ for which

$$\left( \frac{1}{2} \cdot \frac{3}{2} \cdot \frac{5}{2} \cdot \ldots \cdot \frac{2r - 1}{2} \right)(0.0008 + 0.0002) \text{ exceeds unity.}$$

Simple calculation gives $r = 7$.

[9] With our seven-decimal computer we find with forward recurrence the values

| $r$ | 0 | 1 | 2 | 5 | 10 | 20 |
|---|---|---|---|---|---|---|
| $y_r$ | 0 | $\bar{e}$ | 0 | 2.71828 | $-0.000013$ | $-187809.0$ |

and with backward recurrence the values

| $r$ | 20 | 19 | 18 | 10 | 5 | 0 |
|---|---|---|---|---|---|---|
| $y_r$ | 0 | $-\bar{e}$ | 0 | $-1.14003$ | $-6421.4$ | $34271.9$ |

where $\bar{e}$ is the computer-stored value of $e$. The true solution never exceeds $e \sim 2.718\ldots$ in absolute value.

[11] If $k = \pi - \varepsilon$ the solution is very close to

$$y = \frac{1}{\pi^2} + \frac{2}{\pi^2 \varepsilon} \sin kx - \frac{1}{\pi^2} \cos kx,$$

and if $k = \pi + \eta$ the solution is very close to

$$y = \frac{1}{\pi^2} - \frac{2}{\pi^2 \eta} \sin kx - \frac{1}{\pi^2} \cos kx.$$

The coefficients of the middle terms are large and opposite in sign, so that the two solutions are quite different in form.

[12] For inherent instability we need consider only the homogeneous forms of the equations and boundary conditions, for which the solution is

$$y = \cos kx + \left(\frac{\cos k\pi + \sin k\pi}{\cos k\pi - \sin k\pi}\right)\sin kx.$$

The coefficient of $\sin kx$ is very large if the denominator is very small, and the first possibility of this is when $k$ is near to $\frac{1}{4}$. The maximum difference between the two numerical solutions of the non-homogeneous problem, with $k = 0.249$ and $k = 0.251$, is about 49 000 at $x = \pi$, about twice the largest absolute value of each solution.

[13] The solution of the recurrence with $y(0) = 1$, $y(1) = \frac{1}{2}(1 - \sqrt{5}) + \alpha$, is

$$y_r = \frac{\alpha}{\sqrt{5}}\{\tfrac{1}{2}(1 + \sqrt{5})\}^r + \left(1 - \frac{\alpha}{\sqrt{5}}\right)\{\tfrac{1}{2}(1 - \sqrt{5})\}^r.$$

Computer storage would produce a non-zero $\alpha$, and for large $r$ the first term would dominate, so that the ultimate ratio $y_{r+1}/y_r$ would still be $\frac{1}{2}(1 + \sqrt{5})$. Inaccurate computer arithmetic would accentuate this.

## CHAPTER 3

[1] The last line of the required table corresponding to that of Table 3.1 has the approximate values

1.0 3.21828 0.46208 0.24881 $- 0.65692$ $- 0.29608$ $- 0.01479$ $- 0.00366$.

[2] The last lines of the required tables corresponding to those of Tables 3.3 and 3.4 have the respective approximate values

1.0 0.50000 0.05559 7.71214 $-0.05853$ $-0.01989$ 0.00016 * 0.00034
1.0 0.50000 0.02702 0.00067 very large $-0.02331$ $-0.00007$ 0.00000

The symbol * indicates that the TR method breaks down in the very first step for $h = 0.1$ and $\lambda = 100$ for this particular problem. For the FE method with $\lambda = - 10$, $h = 0.1$ the error in $0 \leqslant x \leqslant 1$ has a maximum absolute value of 0.014 at $x = 0.2$, but at $x = 1.5$, 2.0 and 2.5 the errors have respective absolute values of approximately 0.0002, 0.0008 and 0.3158.

[3] With method (a) successive estimates of the required value are 0.9, 0.91998 and 0.92012, while for method (b) they are 0.9, 0.91512, 0.91886, 0.91980, 0.92004 and 0.92010.

[6] With $h = 0.18$, 0.20 and 0.22 the respective errors in the computed values of

$Y_r$ after five steps have the absolute values 0.02591, 0.15421 and 1.03358.

[7] The 'deferred approach' tables for $Y(0.8)$ and $Z(0.8)$, correctly rounded to nine decimal places, are respectively as follows.

| $h$ | $Y(0.8)$ | | | | |
|---|---|---|---|---|---|
| 0.4 | 1.645489413 | | | | |
| | | 1.638331116 | | | |
| 0.2 | 1.640120690 | | 1.638877844 | | |
| | | 1.638843673 | | 1.638842139 | |
| 0.1 | 1.639162927 | | 1.638842697 | | 1.638842624 |
| | | 1.638842758 | | 1.638842622 | |
| 0.05 | 1.638922800 | | 1.638842623 | | |
| | | 1.638842632 | | | |
| 0.025 | 1.638862674 | | | | |
| Error $O(h^2)$ | $O(h^4)$ | $O(h^6)$ | $O(h^8)$ | $O(h^{10})$ | |

| $h$ | $Z(0.8)$ | | | | |
|---|---|---|---|---|---|
| 0.4 | 1.621969000 | | | | |
| | | 1.611348269 | | | |
| 0.2 | 1.614003452 | | 1.611005445 | | |
| | | 1.611026871 | | 1.611018775 | |
| 0.1 | 1.611771016 | | 1.611018566 | | 1.611018580 |
| | | 1.611019085 | | 1.611018580 | |
| 0.05 | 1.611207068 | | 1.611018580 | | |
| | | 1.611018612 | | | |
| 0.025 | 1.611065726 | | | | |
| Error $O(h^2)$ | $O(h^4)$ | $O(h^6)$ | $O(h^8)$ | $O(h^{10})$ | |

The correct values are respectively 1.638842623 and 1.611018580 to nine decimal places, so that we have obtained these results within hardly more than a rounding error. In practice we rely on consistency, and the most reliable consistency paths are the 'forward difference lines'. For $Y(0.8)$ the best such line is the 0.2 forward difference line, the last two entries here essentially guaranteeing seven decimal places of accuracy. The corresponding line for $Z(0.8)$ gives effectively eight accurate decimals. To guarantee utterly and completely the nine-decimal accuracy of the final numbers in both cases, we really need the results for $h = 0.0125$.

[9] The global error would be expected to be of the form

$$y(X) - Y(X, h) = h^2 z_2(x) + h^3 z_3(x) + \cdots.$$

The corresponding 'deferred approach to the limit' table is approximately as follows.

| $h$ | $Y_r(0.4)$ | | |
|-----|-----------|---|---|
| 0.4 | 2.240000 | | |
| | | 2.205501 | |
| 0.2 | 2.214126 | | 2.206018 |
| | | 2.205953 | |
| 0.1 | 2.207996 | | |
| Error | $O(h^2)$ | $O(h^3)$ | $O(h^4)$ |

The table suggested for the FE method is approximately as follows.

| $h$ | $Y_r(0.4)$ | | |
|-----|-----------|---|---|
| 0.2 | 2.094444 | | |
| | | 2.201256 | |
| 0.1 | 2.147850 | | 2.205984 |
| | | 2.204802 | |
| 0.05 | 2.176326 | | |
| Error | $O(h)$ | $O(h^2)$ | $O(h^3)$ |

The correct value to five decimal places is 2.20611.

[10] The values obtained by this method, with just two extrapolations at each point, are:

$$\left. \begin{array}{ccccccc} x & 0.0 & 0.2 & 0.4 & 0.6 & 0.8 & 1.0 \\ y & 2 & 2.054736 & 2.206110 & 2.447119 & 2.781096 & 3.218282 \end{array} \right\}.$$

The last number was obtained to this accuracy in Table 3.7, but of course this problem and its solution are very smooth indeed in this range.

[13] For two simultaneous equations the formulae corresponding to (3.101) and (3.102) are

$$\left. \begin{aligned} k_1 &= hf(x, y, z) \\ k_2 &= hf(x + \tfrac{1}{3}h, y + \tfrac{1}{3}k_1, z + \tfrac{1}{3}m_1) \\ k_3 &= hf(x + \tfrac{1}{3}h, y + \tfrac{1}{6}k_1 + \tfrac{1}{6}k_2, z + \tfrac{1}{6}m_1 + \tfrac{1}{6}m_2) \\ k_4 &= hf(x + \tfrac{1}{2}h, y + \tfrac{1}{8}k_1 + \tfrac{3}{8}k_3, z + \tfrac{1}{8}m_1 + \tfrac{3}{8}m_3) \\ k_5 &= hf(x + h, y + \tfrac{1}{2}k_1 - \tfrac{3}{2}k_3 + 2k_4, z + \tfrac{1}{2}m_1 - \tfrac{3}{2}m_3 + 2m_4) \end{aligned} \right\},$$

where $m_r$ is computed from $g(x, y, z)$ with the same arguments as those for $k_r$, and

$$\left. \begin{aligned} y^{(1)} &= y + \tfrac{1}{2}k_1 - \tfrac{3}{2}k_3 + 2k_4, & z^{(1)} &= z + \tfrac{1}{2}m_1 - \tfrac{3}{2}m_3 + 2m_4 \\ y^{(2)} &= y + \tfrac{1}{6}k_1 + \tfrac{2}{3}k_4 + \tfrac{1}{6}k_5, & z^{(2)} &= z + \tfrac{1}{6}m_1 + \tfrac{2}{3}m_4 + \tfrac{1}{6}m_5 \end{aligned} \right\}.$$

Then (3.103) become

$$y(x+h) - y^{(2)} \sim \tfrac{1}{5}(y^{(2)} - y^{(1)}) = \tfrac{1}{30}(-2k_1 + 9k_3 - 8k_4 + k_5) \left.\right\}$$
$$z(x+h) - z^{(2)} \sim \tfrac{1}{5}(z^{(2)} - z^{(1)}) = \tfrac{1}{30}(-2m_1 + 9m_3 - 8m_4 + m_5) \left.\right\}.$$

For the first step with $h = 0.1$ in Exercise 6 the computed answers are $y(0.1) = 1.919014$, $z(0.1) = 1.111060$, with estimated errors $-0.000270$ and $0.000105$ respectively, compared with the actual errors of $0.000127$ and $-0.000051$. The estimates have a tolerably satisfactory order of magnitude but with the wrong signs, the non-linearity of both $f(x, y, z)$ and $g(x, y, z)$, in $x$ if not in $y$, being of rather a large order.

[14] The error at $x = 1$ is here $-0.00001$, smaller than that of the stable BE method and comparable with that of the stable TR method in Table 3.4.

   The partial stability restriction on $h$ for equation (3.92) is that of the corresponding fourth-order Taylor series method, approximately $h|\lambda| < 2.78$, so that here $h < 0.0278$ approximately.

[15] For the FE method on the left and the TR method on the right we obtain the following solutions for various values of $h$.

| $h$ | $Z(2)$ | $T(2)$ | $Y(2)$ | $Z(2)$ | $T(2)$ | $Y(2)$ |
|---|---|---|---|---|---|---|
| 0.1 | 2.787 | 3.488 | 1.252 | 3.028 | 3.851 | 1.272 |
| 0.05 | 2.896 | 3.652 | 1.261 | 3.023 | 3.842 | 1.271 |
| 0.025 | 2.956 | 3.743 | 1.266 | 3.021 | 3.840 | 1.271 |

The deferred approach to the limit in the first two tables give values differing by no more than 0.001 in every case.

## CHAPTER 4

[4] The equation corresponding to (4.26) is here

$$(1 - \tfrac{3}{8}h\lambda)p^3 - (1 + \tfrac{19}{24}h\lambda)p^2 + \tfrac{5}{24}h\lambda p - \tfrac{1}{24}h\lambda = 0.$$

With $h\lambda = -6$ the largest negative root is about $-1.455$, so that this $|h\lambda|$ must be too large for stability. The other two roots, incidentally, are complex with quite small modulus. Further calculation shows that the limit for partial stability is exactly $h < 3/|\lambda|$.

[5] With $h = 0.39$ the errors in the computed $Y$ and $Z$ values after 50 steps are respectively 0.011 and 0.021; with $h = 0.41$ these figures are 0.015 and 0.025, and with $h = 0.6$ they are 11.0 and 4.2 approximately.

   Here the onset of instability is rather gradual with $h$ only a little greater than the limit. In the corresponding Exercise 6 of Chapter 3 the growth of error is much more striking.

[6] The formula is convergent, being consistent and zero-stable. The equation

governing partial stability is

$$f(p) = p^4 + \tfrac{8}{3}\mu p^3 - \tfrac{4}{3}\mu p^2 + \tfrac{8}{3}\mu p - 1 = 0, \quad \mu = -h\lambda > 0.$$

For various values of $p$ we have for the sign of $f(p)$ the table

$$\left.\begin{array}{ccccccc} p & -\infty & -1 & 0 & 1 & \infty \\ f(p) & + & - & - & + & + \end{array}\right\}.$$

These signs are true for all $h > 0$, so that for all such $h$ there is a zero for $f(p)$ in the interval $-\infty < p < -1$.

The same argument gives a similar result for the Simpson rule method, and equation (4.37) also confirms this.

[9] The relevant formula is here

$$Y_{r+1} = (1 + \tfrac{13}{12}h\lambda + \tfrac{15}{24}h^2\lambda^2)Y_r - (\tfrac{1}{12}h\lambda + \tfrac{5}{24}h^2\lambda^2)Y_{r-1},$$

and Table 4.3 looks like:

$$\left.\begin{array}{ccccccc} h\lambda & 1 & 0 & -1 & -2 & -3 & -4 \\ p_1 & 2.60 & 1 & C & C & 2.79 & 6.18 \\ p_2 & 0.11 & 0 & 0.35 & 0.82 & 0.58 & 0.49 \end{array}\right\},$$

where $C$ denotes that the roots are complex with modulus given below. The limit for partial stability is clearly between $h\lambda = -2$ and $h\lambda = -3$, and more detailed computation gives $h < 2.4/|\lambda|$, the coefficient being exact.

[10] The results obtained at the first set of points, first directly and then with two halvings of the intervals, are

| $x$ | 0.2 | 0.4 | 0.5 | 0.6 |
|---|---|---|---|---|
| $Y$ | 0.834722 | 0.715955 | 0.668207 | 0.626406 |
| $Y$ | 0.833472 | 0.714457 | 0.666824 | 0.625144 |
| $Y$ | 0.833349 | 0.714305 | 0.666685 | 0.625016 |

with global errors $C_1h^3 + C_2h^4 + \cdots$. The two obtainable values with error $O(h^4)$ at the respective mesh points are

$$\begin{array}{cccc} 0.833293 & 0.714243 & 0.666626 & 0.624964 \\ 0.833331 & 0.714283 & 0.666665 & 0.624998 \end{array}$$

and the single obtainable values with errors $O(h^5)$ are

$$0.833334 \quad 0.714286 \quad 0.666668 \quad 0.62500$$

all correct within a small rounding error in all six decimals.

## CHAPTER 5

[4] With interval $h = 0.1$ successive estimates of $y''(0)$ are the specified 0.3 for $n = 0$, and approximately 0.319487 and 0.319499 for $n = 1$ and 2 respectively. The correct value is very close to 0.319500.

[5] Corresponding to the results of Exercise 4, successive estimates by this method are 0.3, 0.319590 and 0.319500. Successive pairs of values of $t$ for the computation of $y_t$ were taken as (0.4, 0.3) and (0.3, 0.319590) to minimize the amount of computation involved. The rates of convergence in Exercises 4 and 5 are clearly very similar in this problem.

[6] Successive values of $y''(0)$ obtained by this method, with $h = 0.1$, are 0.3, 0.319613 and 0.319500.

[7] With $h = 0.1$ successive estimates for $p$ and $q$ are respectively

$$\left.\begin{array}{ccccc} p & -1 & -0.944 & -0.959 & -0.9590 \\ q & -0.5 & -0.332 & -0.341 & -0.3411 \end{array}\right\}.$$

[8] With $h = 0.25$ the computed results are $\bar{\lambda} = -0.9187$, and

| $x$ | 0.0 | 0.25 | 0.50 | 0.75 | 1.00 | 1.50 | 2.00 | 3.00 | 4.00 |
|---|---|---|---|---|---|---|---|---|---|
| $Y$ | 0.000 | 0.328 | 0.580 | 0.766 | 0.893 | 1.000 | 0.954 | 0.567 | 0.000 |

The maximum $Y$ value is at $x = 1.5$.

[9] One would expect somewhat better results since the numbers will not grow so rapidly. With $N = 6$, 12 and 18 the respective maximum errors, with $\bar{y}_1^{(p)} = \frac{1}{6}$, are 0.000001, 0.007812 and 4099, using our computer with a word length of effectively 15 decimals.

[11] For this problem the matrix in (5.77) is approximately

$$\begin{bmatrix} 1.000001 & 1.000004 \\ 0.999999 & 0.999997 \end{bmatrix},$$

which is obviously nearly singular.

In Exercise 10 the complementary functions have rapidly increasing values, and the corresponding initial-value problem is very ill conditioned. In Exercise 11 the complementary functions do not produce large numbers. The corresponding initial-value problem is well conditioned, whereas the boundary-value problem is obviously ill conditioned.

[12] The correct solution to the problem is $y = 1 + x^2$, $z = 3 + x$. The suggested method, with $h = \frac{1}{2}$, gives the quite unsatisfactory results

$$y(12) \sim 3.5 \times 10^{13}, \quad z(12) = 0.$$

[13] Matching at the half-way point $x = 6$ gives, with the suggested Runge–Kutta method at various intervals, the results

| $h$ | $Y(6)$ | $Y(12)$ |
|---|---|---|
| 1 | 57.063 | 145.01 |
| $\frac{1}{2}$ | $-8.000$ | 145.00 |
| $\frac{1}{4}$ | 56.000 | 145.00 |

There is clearly some improvement, but more sub-ranges are obviously needed.

[14] Consider the differential equations

$$y' = ly + mz + n, \qquad z' = qy + rz + s,$$

where the coefficients are functions of $x$, and the most general unseparated boundary conditions

$$\left.\begin{array}{l} \alpha_1 y(a) + \beta_1 z(a) + \gamma_1 y(b) + \delta_1 z(b) = \varepsilon_1 \\ \alpha_2 y(a) + \beta_2 z(a) + \gamma_2 y(b) + \delta_2 z(b) = \varepsilon_2 \end{array}\right\}.$$

In sub-region $i$, which starts, say, at $x = x_i$ and finishes at $x = x_{i+1}$, the general solutions of the differential equations are

$$\left.\begin{array}{l} y_i = y_i^{(p)} + \lambda_i y_i^{(1)} + \mu_i y_i^{(2)} \\ z_i = z_i^{(p)} + \lambda_i z_i^{(1)} + \mu_i z_i^{(2)} \end{array}\right\},$$

where $\lambda_i$ and $\mu_i$ are unknown constants, $y_i^{(p)}$ and $z_i^{(p)}$ satisfy the differential equations with $y_i^{(p)}(x_i) = z_i^{(p)}(x_i) = 0$, $y_i^{(1)}$ and $z_i^{(1)}$ satisfy the homogeneous differential equations with $y_i^{(1)}(x_i) = 1$, $z_i^{(1)}(x_i) = 0$ and $y_i^{(2)}$ and $z_i^{(2)}$ satisfy the homogeneous equations with $y_i^{(2)}(x_i) = 0$, $z_i^{(2)}(x_i) = 1$.

Then with $N$ sub-regions we have $2N$ unknowns, and these are obtained from the two linear equations given by the boundary conditions and the $2(N-1)$ equations obtained by equating respectively the two $y$ and two $z$ values at each interface.

With these sub-ranges, the very good results corresponding to those of Exercise 13 are here

| $h$ | $Y(4)$ | $Y(8)$ | $Y(12)$ |
|-----|--------|--------|---------|
| $\frac{1}{2}$ | 15.384 | 42.741 | 1003.09 |
| $\frac{1}{4}$ | 17.000 | 65.000 | 145.06 |
| $\frac{1}{8}$ | 17.000 | 65.000 | 145.000 |

[15] Use the simultaneous first-order equations and boundary conditions

$$y' = z, \quad z' = y + 1; \quad y(0) = 0, \quad y(b) = e^b, \quad z(b) = -1.$$

Equations (5.108) and (5.109) give

$$w = -1 + \operatorname{sech} x, \quad t = -\tanh x,$$

and (5.105) gives $y = w - tz$, so that at $x = b$ we have

$$e^b = -1 + \operatorname{sech} b - \tanh b, \quad \text{or } e^{2b} + 2e^b - 1 = 0.$$

Then $e^b = \pm\sqrt{2} - 1$, and the plus sign gives the only real value of $b$. Finally, equation (5.111) with $z(b) = -1$ gives $z = \sinh x$, and (5.105) gives $y = -1 + \cosh x$.

The table of $x$ and the computed $t(x)$ and $w(x)$ and the quantity $e^x - t(x) - w(x)$ is as follows.

| $x$ | $t(x)$ | $w(x)$ | $e^x - t(x) - w(x)$ |
|---|---|---|---|
| $-0.1$ | $-0.004979$ | $0.099668$ | $0.810149$ |
| $-0.2$ | $-0.019672$ | $0.197375$ | $0.641028$ |
| $-0.3$ | $-0.043372$ | $0.291312$ | $0.492878$ |
| $-0.4$ | $-0.074993$ | $0.379949$ | $0.365364$ |
| $-0.5$ | $-0.113181$ | $0.462117$ | $0.257595$ |
| $-0.6$ | $-0.156450$ | $0.537049$ | $0.168212$ |
| $-0.7$ | $-0.203295$ | $0.604367$ | $0.095513$ |
| $-0.8$ | $-0.252301$ | $0.664036$ | $0.037594$ |
| $-0.9$ | $-0.302206$ | $0.716297$ | $-0.007521$ |
| $-1.0$ | $-0.351946$ | $0.761593$ | $-0.041767$ |
| $-1.1$ | $-0.400666$ | $0.800497$ | $-0.066960$ |

The required quantity is the value of $x$ for which $e^x - t(x) - w(x) = 0$. This can be obtained by inverse interpolation, for example by using equation (3.110) in the iterative form

$$0 = f_r + p^{(n+1)}\delta f_{r+\frac{1}{2}} + \tfrac{1}{4}p^{(n)}(p^{(n)} - 1)(\delta^2 f_r + \delta^2 f_{r+1}) + \cdots,$$

where $p^{(n)}$ is taken to be zero for $n = 0$. Here, of course, $r$ represents the point $x = -0.8$, and three iterations give $p = 0.83329$ and the required $x = 0.883329$, very close to the correct value.


## CHAPTER 6

[4] Successive iterations with (6.20) give the values

| | | | | | |
|---|---|---|---|---|---|
| $y_1$ | 0 | 0.1337 | 0.1614 | 0.1659 | 0.1666 |
| $y_2$ | 0 | 0.2806 | 0.3258 | 0.3331 | 0.3333 |
| $y_3$ | 0 | 0.4288 | 0.4986 | 0.5000 | 0.5000 |
| $y_4$ | 0 | 0.6596 | 0.6665 | 0.6667 | 0.6667 |

With the first two rows interchanged the values are

| | | | |
|---|---|---|---|
| $y_1$ | 0 | $-2.7000$ | $-289.3$ |
| $y_2$ | 0 | $-28.6200$ | $-2923.6$ |
| $y_3$ | 0 | $-2.4327$ | $-289.0$ |
| $y_4$ | 0 | $0.3763$ | $-28.0$ |

which are obviously not converging.

[6] After the elimination of $y_0, y_1$ and $y_2$ the matrix looks like

$$\begin{bmatrix} 0 & 0 & \times & \times & \times & \times & \times \\ \times & 0 & \times & \times & \times & \times & \times \\ \times & 0 & \times & \times & \times & \times & \times \\ \times & \times & \times & \times & \times & \times & \times \\ \times & \times & \times & \times & \times & \times & \times \\ \times & \times & \times & 0 & 0 & 0 & 0 \\ \times & 0 & \times & \times & \times & \times & \times \end{bmatrix}$$

where $\times$ is in general a non-zero number.

[7] After elimination with pivotal rows 1, 2 and 10 the matrix of order seven looks like

$$\begin{bmatrix} \times & 0 & \times & \times & \times & \times & \times \\ \times & 0 & \times & \times & \times & \times & \times \\ \times & 0 & 0 & \times & \times & \times & 0 \\ \times & 0 & 0 & \times & \times & \times & 0 \\ \times & \times & 0 & \times & \times & \times & \times \\ \times & \times & 0 & \times & \times & \times & \times \\ \times & \times & \times & 0 & 0 & 0 & 0 \end{bmatrix}.$$

[8] The finite-difference equations will contain more non-zero elements, and will also involve 'spurious solutions' which might affect adversely the stability of the method.

[9] For the equation $y'' - 100y' - y = 0$, central differences with $h = 0.2$ give $p_1 = 1$, $p_2 = -1.2$ and large oscillations. Forward difference for $y'$ gives $p_1 = 1$, $p_2 \sim -\frac{1}{19}$, with a smaller oscillation. The replacement of $y'$ with a backward difference in the form $hy'_r \sim y_r - y_{r-1}$ gives two positive roots. With boundary conditions $y(0) = 0$, $y(1) = 1$ the results of the three methods are as follows.

| $x$ | Exact $y$ | Central differences | Forward differences | Backward differences |
|------|-----------|----------------------|----------------------|-----------------------|
| 0.0 | 0.000 | 0.000 | 0.000 | 0.000 |
| 0.2 | 0.000 | 0.594 | 1.061 | 0.000 |
| 0.4 | 0.000 | − 0.135 | 1.003 | 0.000 |
| 0.6 | 0.000 | 0.756 | 1.004 | 0.002 |
| 0.8 | 0.000 | − 0.336 | 1.002 | 0.048 |
| 1.0 | 1.000 | 1.000 | 1.000 | 1.000 |

Here the backward-difference method is clearly preferable.

[11] The eigenvalues and the 'deferred approach' table are

| $h$ | $\lambda$ | | |
|---|---|---|---|
| $\frac{1}{4}$ | 0.413201 | | |
| | | 0.414093 | |
| $\frac{1}{8}$ | 0.413870 | | 0.414094 |
| | | 0.414094 | |
| $\frac{1}{16}$ | 0.414038 | | |
| Error | $O(h^2)$ | $O(h^4)$ | $O(h^6)$ |

The correct eigenvalue is very close to 0.414094. At each interval the computed eigenfunction agrees exactly with the correct eigenfunction, with formulae using the second central difference for $h^2 y''$ and the first (mean) central difference for $hy'$.

[12] At $x = 4$ some values and 'deferred approach' tables are

| $h$ | $Y''$ | | | $Y$ | | |
|---|---|---|---|---|---|---|
| $\frac{1}{2}$ | 0.292817 | | | $-0.585537$ | | |
| | | 0.291410 | | | $-0.586816$ | |
| $\frac{1}{4}$ | 0.291762 | | 0.291401 | $-0.586496$ | | $-0.586778$ |
| | | 0.291402 | | | $-0.586780$ | |
| $\frac{1}{8}$ | 0.291492 | | | $-0.586709$ | | |
| Error | $O(h^2)$ | $O(h^4)$ | $O(h^6)$ | $O(h^2)$ | $O(h^4)$ | $O(h^6)$ |

Consistency suggests we have at least five-decimal accuracy at this stage.

[13] Successive values of $Y(4)$, where the error in the first approximation is largest, are $-0.586496$, $-0.586838$, $-0.586776$, $-0.586778$, the final value being correct to six decimals.

[15] The successive approximations at $x = 1$, where the error in the first approximation is largest, are 1.547030, 1.543016, 1.543082, 1.543081, the final value being correct to all the figures quoted.

[17] If equations (6.61) and (6.71) are valid, the deferred approach tables for $Y(0.5)$ (with normalization such that $Y(0) = 1$) and for $\lambda$ look like

| $h$ | $Y$ | | | | $\lambda$ | | | |
|---|---|---|---|---|---|---|---|---|
| $\frac{1}{2}$ | 0.645497 | | | | 2.83602 | | | |
| | | 0.615462 | | | | 3.43523 | | |
| $\frac{1}{4}$ | 0.622971 | | 0.608355 | | 3.28543 | | 3.52841 | |
| | | 0.608799 | | 0.606343 | | 3.52259 | | 3.54915 |
| $\frac{1}{8}$ | 0.612342 | | 0.606374 | | 3.46330 | | 3.54883 | |
| | | 0.606526 | | | | 3.54719 | | |
| $\frac{1}{16}$ | 0.607980 | | | | 3.52622 | | | |
| Error | $O(h^2)$ | $O(h^4)$ | $O(h^6)$ | $O(h^8)$ | $O(h^2)$ | $O(h^4)$ | $O(h^6)$ | $O(h^8)$ |

The lack of convergence indicates that neither (6.61) nor (6.71) is valid for this problem. The correct figure for the eigenvalue is about 3.55928.

[18] The corresponding deferred approach table for the eigenvalue, and the difference table for the function $Z(x)$ at interval $h = \frac{1}{8}$, are given respectively in the following tables.

| $h$ | $\lambda$ | | |
|---|---|---|---|
| $\frac{1}{2}$ | 3.57204 | | |
| | | 3.55912 | |
| $\frac{1}{4}$ | 3.56235 | | 3.55928 |
| | | 3.55927 | |
| $\frac{1}{8}$ | 3.56004 | | 3.55928 |
| | | 3.55928 | |
| $\frac{1}{16}$ | 3.55947 | | |

| Error | $O(h^2)$ | $O(h^4)$ | $O(h^6)$ |
|---|---|---|---|

| $x$ | $10^6 Z$ | | | |
|---|---|---|---|---|
| 0.625 | 894939 | | −6665 | 119 |
| | | −43464 | | 842 |
| 0.750 | 851475 | | −5823 | 103 |
| | | −49287 | | 945 |
| 0.875 | 802188 | | −4878 | 60 |
| | | −54165 | | 1005 |
| 1.000 | 748023 | | −3873 | 65 |
| | | −58038 | | 1070 |
| 1.125 | 689985 | | −2803 | 7 |
| | | −60841 | | 1077 |
| 1.250 | 629144 | | −1726 | −1 |
| | | −62567 | | 1076 |
| 1.375 | 566577 | | −650 | |
| | | −63217 | | |
| 1.500 | 503360 | | | |

Equation (6.71) for the deferred approach for the eigenvalue is clearly valid in this problem, and calculations for the eigenfunction reveal that (6.61) is also valid.

The differences of $Z$ reveal no more than a very small bump in the last figure, with no bump at all in the values rounded to five decimals. The function $Z$ is symmetric about $x = 0$, so that the same small bump appears also at $x = -1$.

[20] With $B = 4$, and interval $h = 0.1$, successive 'corrections' at $x = 0.5$ are approximately $-16.3, 8.1, 4.0, 1.9, 0.7, 0.1, 0.006$ and $0.00001$, the process converging to a nice smooth solution.

With $B = 4.676$ the corresponding 'corrections' start with $7364, -3682, -1841, -921$, and reduce by a factor very close to 2 for the first nine iterations. Successive corrections are then $-28.8, -24.8, 26.1, -16.2, -8.7, 0.7, -3.9, 6.2, \ldots$, with no sign of convergence.

With $B = 5$ successive 'corrections' are approximately $29.1, -13.6, -8.8$,

$-2.4, -2.3, 0.1, -0.6, -0.07, -0.003, -0.000002$, and there is convergence to a quite different solution from that with $B = 4$.

[21] Continuation in steps of $\frac{1}{4}$ in $t$ leads to the following solutions. The calculations are made with interval 0.1 in $h$, but the results are quoted at twice this interval.

| $x$ | $Y(t = 0)$ | $Y(t = \frac{1}{4})$ | $Y(t = \frac{1}{2})$ | $Y(t = \frac{3}{4})$ | $Y(t = 1)$ |
|-----|-----------|----------------------|----------------------|----------------------|-----------|
| 0.0 | 0.0000 | 0.0000 | 0.0000 | 0.0000 | 0.0000 |
| 0.2 | $-0.7480$ | $-0.8180$ | $-0.9726$ | $-1.1983$ | $-1.3980$ |
| 0.4 | $-0.9435$ | $-1.0969$ | $-1.3795$ | $-1.7768$ | $-2.1408$ |
| 0.6 | $-0.4520$ | $-0.6624$ | $-0.9827$ | $-1.3922$ | $-1.7449$ |
| 0.8 | 0.3884 | 0.2150 | $-0.0145$ | $-0.2792$ | $-0.4870$ |
| 1.0 | 1.0000 | 1.0000 | 1.0000 | 1.0000 | 1.0000 |

This is obviously a respectably smooth solution.

## CHAPTER 7

[1] $x^4 + 3x^3 = \frac{3}{8}T_0(x) + \frac{9}{4}T_1(x) + \frac{1}{2}T_2(x) + \frac{3}{4}T_3(x) + \frac{1}{8}T_4(x)$.

Omission of the last term gives the best cubic approximation, with maximum error of size $\frac{1}{8}$ at points where $|T_4(x)| = 1$, that is at $x = -1$, $-1/\sqrt{2}, 0, 1/\sqrt{2}, 1$.

[5] Corresponding to Table 7.1 we find, with collocation points at the relevant Chebyshev zeros, the following table.

| $h$ | Max. error on $[-1, 1]$ | Max. error at knots |
|-----|-------------------------|---------------------|
| 1.0 | 0.01 | 0.01 |
| 0.5 | 0.0031 | 0.0031 |
| 0.25 | 0.00081 | 0.00081 |

[6] The required table is as follows.

| $h$ | Max. error on $[-1, 1]$ | Max. error at knots |
|-----|-------------------------|---------------------|
| 1.0 | 0.000091 | 0.000068 |
| 0.5 | 0.0000037 | 0.0000037 |
| 0.25 | 0.00000021 | 0.00000021 |

These, of course, are more accurate than those of Exercise 5, but there is no superconvergence at knots.

[7] Suppose that we seek polynomials of degrees $m_i$ for $y_i$ and $z_i$ in sub-region $i$, where we take $i = 1, 2$ and 3. Then we satisfy both differential equations in

region $i$ at $m_i$ points, giving $2(m_1 + m_2 + m_3)$ equations for the coefficients. There are $2(m_1 + m_2 + m_3) + 6$ unknown coefficients. The extra six equations needed for their calculation come from the satisfaction of the two boundary conditions and the respective equalities of $y_i, y_{i+1}$ and $z_i, z_{i+1}$ at $i = 1, 2$.

If $m_1 = 3$, $m_2 = 2$, $m_3 = 3$, and successive columns refer to the coefficients of $y^{(I)}, z^{(I)}, y^{(II)}, z^{(II)}$ and $y^{(III)}, z^{(III)}$, the first few rows of the matrix are

$$
\begin{bmatrix}
\times & \times & \times & \times & & & & \\
\times & \times & \times & \times & \times & \times & \times & \times \\
\times & \times & \times & \times & \times & \times & \times & \times \\
\times & \times & \times & \times & \times & \times & \times & \times \\
\times & \times & \times & \times & \times & \times & \times & \times \\
\times & \times & \times & \times & \times & \times & \times & \times \\
\times & \times & \times & \times & \times & \times & \times & \times \\
\times & \times & \times & \times & & & \times & \times & \times & \times \\
& & \times & \times & \times & \times & & & & \times & \times & \times & \times \\
\end{bmatrix},
$$

comprising the initial boundary condition, the satisfaction of the two differential equations in region I, and the equalities of $y^{(I)}, y^{(II)}$ and $z^{(I)}, z^{(II)}$ at the junction of regions I and II. The full matrix has 22 rows and columns.

[8] The $\tau$ method of Section 7.3(i) gives, for $\lambda = 10$, 100 and $-100$, the approximate quartic solutions

$$y_4(x) = 0.7100 T_0^*(x) - 0.2374 T_1^*(x) + 0.0454 T_2^*(x) -$$
$$- 0.0049 T_3^*(x) + 0.0023 T_4^*(x),$$
$$y_4(x) = 0.7074 T_0^*(x) - 0.2421 T_1^*(x) + 0.0422 T_2^*(x) -$$
$$- 0.0066 T_3^*(x) + 0.0017 T_4^*(x),$$
$$y_4(x) = 0.7073 T_0^*(x) - 0.2423 T_1^*(x) + 0.0420 T_2^*(x) -$$
$$- 0.0068 T_3^*(x) + 0.0016 T_4^*(x).$$

The collocation method of Section 7.3(ii) gives corresponding results

$$y_4(x) = 0.7059 T_0^*(x) - 0.2494 T_1^*(x) + 0.0401 T_2^*(x) -$$
$$- 0.0071 T_3^*(x) - 0.0025 T_4^*(x),$$
$$y_4(x) = 0.7073 T_0^*(x) - 0.2415 T_1^*(x) + 0.0410 T_2^*(x) -$$
$$- 0.0034 T_3^*(x) + 0.0069 T_4^*(x),$$
$$y_4(x) = 0.7069 T_0^*(x) - 0.2429 T_1^*(x) + 0.0401 T_2^*(x) -$$
$$- 0.0046 T_3^*(x) + 0.0056 T_4^*(x).$$

The correct solution has coefficients

$$0.7071 \qquad -0.2426 \qquad 0.0416 \qquad -0.0071 \qquad 0.0012$$

with the addition of $-0.0002 T_5^*(x)$ to this accuracy.

Note that by each method the different values of $\lambda$ have no great effect. The labour involved depends on the library facilities available. If computer programs are available to solve general problems of this type, with polynomial coefficients, the series expansion method is probably more efficient, particularly if high accuracy is required. But if a new computer program has to be written, the use of the TR method with deferred approach will be rather easier.

[9] The maximum changes in the coefficients in successive iterates are 0.1460, 0.0057 and 0.0000, giving the quartic Chebyshev approximation

$$y_4(x) = 1.1129\,T_0(\xi) + 0.1403\,T_1(\xi) + 0.0221\,T_2(\xi)$$
$$- 0.0048\,T_3(\xi) + 0.0005\,T_4(\xi)$$

where, in accord with equation (7.20), $\xi = 2x - 3$. These results agree, to this precision, with the coefficients of the infinite Chebyshev expansion.

This method requires the solution of a system of non-linear algebraic equations, and for high accuracy it may require more labour than an efficient step-by-step method, especially the Taylor series method when there is no significant problem of partial instability. The current method, however, is unaffected by this instability.

# Index